INORGANIC
AND
METAL-CONTAINING
POLYMERIC MATERIALS

INORGANIC AND METAL-CONTAINING POLYMERIC MATERIALS

Edited by

John E. Sheats
Rider College
Lawrenceville, New Jersey

Charles E. Carraher, Jr.
Florida Atlantic University
Boca Raton, Florida

Charles U. Pittman, Jr.
Mississippi State University
Mississippi State, Mississippi

Martel Zeldin
Indiana University – Purdue University at Indianapolis
Indianapolis, Indiana

Brian Currell
Thames Polytechnic
London, England

PLENUM PRESS • NEW YORK AND LONDON

Library of Congress Cataloging-in-Publication Data

American Chemical Society International Symposium on Inorganic and
  Metal-containing Polymeric Materials (1989 : Miami Beach, Fla.)
   Inorganic and metal-containing polymeric materials / edited by
John E. Sheats ... [et al.].
       p.   cm.
   "Proceedings of an American Chemical Society International
Symposium on Inorganic and Metal-containing Polymeric Materials,
held September 11-14, 1989, in Miami Beach, Florida"--T.p. verso.
   "An expanded version of the papers presented at the Symposium on
Inorganic and Organometallic Polymers at the National Meeting of the
American Chemical Society in Miami Beach in September 1989"--Pref.
   ISBN 0-306-43819-4
   1. Inorganic polymers--Congresses. 2. Organometallic polymers-
-Congresses.   I. Sheats, John E. II. American Chemical Society.
III. American Chemical Society. Meeting (1989 : Miami Beach, Fla.)
IV. Symposium on Inorganic and Organometallic Polymers (1989 : Miami
Beach, Fla.) V. Title.
QD196.A44 1989
547.7--dc20   91-8183
                                                                   CIP

Proceedings of an American Chemical Society International Symposium
on Inorganic and Metal-Containing Polymeric Materials,
held September 11–14, 1989, in Miami Beach, Florida

ISBN 0-306-43819-4

© 1990 Plenum Press, New York
A Division of Plenum Publishing Corporation
233 Spring Street, New York, N.Y. 10013

Printed in the United States of America

PREFACE

Research on metal-containing polymers began in the early 1960's
when several workers found that vinyl ferrocene and other vinylic
transition metal π -complexes would undergo polymerization under the
same conditions as conventional organic monomers to form high polymers
which incorporated a potentially reactive metal as an integral part of
the polymer structures. Some of these materials could act as semiconductors and possessed one or two dimensional conductivity. Thus
applications in electronics could be visualized immediately. Other
workers found that reactions used to make simple metal chelates could
be used to prepare polymers if the ligands were designed properly. As
interest in homogeneous catalysts developed in the late 60's and early
70's, several investigators began binding homogeneous catalysts onto
polymers, where the advantage of homogeneous catalysis - known reaction
mechanisms and the advantage of heterogeneous catalysis - simplicity
and ease of recovery of catalysts could both be obtained. Indeed the
polymer matrix itself often enhanced the selectivity of the catalyst.
      The first symposium on Organometallic Polymers, held at the
National Meeting of the American Chemical Society in September 1977,
attracted a large number of scientists interested in this field, both
established investigators and newcomers. Subsequent symposia in 1977,
1979, 1983, and 1987 have seen the field mature. Hundreds of papers
and patents have been published. Applications of these materials
as semiconductors and one-dimensional conductors, as radiation shields
or as photo-resists, as catalysts, as controlled release agents for
drugs and biocides and a wide variety of applications have been
studied (see Chapter 1).
      This book is an expanded version of the papers presented at the
Symposium on Inorganic and Organometallic Polymers at the National
Meeting of the American Chemical Society in Miami Beach in September
1989. It is designed to give the reader an introduction to the wide
varieties of structures that can be created and the many novel
applications that can arise when traditional polymer science, involving
primarily compounds of C, H, N, O, S, Si and the halogens is combined
with the rest of the periodic table. Three review chapters give an
overview of the entire field in the United States, Japan, and Western
Europe (Chapter 1), the Soviet Union (Chapter 2), and the People's
Republic of China (Chapter 3). Topics such as conductive polymers
(Chapters 4,5,6), coordination polymers (Chapter 10), condensation
polymers (Chapters 17-21), polyphosphazenes (Chapter 16), carborane
siloxanes (Chapter 14), polymer bound catalysts (Chapters 3 and 6),
and biopolymers (Chapters 19-20) have been covered in previous
monographs cited in Chapter 1. Two new fields, pre-ceramic polymers
and polysilanes have emerged since publication of our previous
monograph in 1985. Polysilanes (Chapters 13 and 15) had been long
sought after but were considered not to be stable. The first example
of this class of compounds was reported in 1984. At the fall 1990 ACS

meeting, there were 143 papers on these materials. Pre-ceramic materials (Chapters 11-12) have emerged as a means of taking soluble, tractable polymeric materials which can be spun into fibers, woven into fabrics, cast into films, or fabricated into various shapes and pyrolyzing them to form inorganic materials such as metal oxides, carbides, nitrides, sulfides, or selenides which retain their original shape or else become highly porous, permitting their possible applications as catalysts. The discovery of "high temperature" superconductors in 1986 has given a further impetus to this field.

In summary, the possibilities for research and for new useful materials is as diverse as properties of the elements these materials contain. It is exciting to anticipate how this field will continue to develop in the coming years.

John E. Sheats
Charles E. Carraher, Jr.
Charles U. Pittman, Jr.
Martel Zeldin
Brian Currell

CONTENTS

*Asterisks on opening chapter pages identify contact authors

INORGANIC AND METAL-CONTAINING POLYMERS. AN OVERVIEW

Charles U. Pittman, Jr.*

University/Industry Chemical Research Center, Department of Chemistry
Mississippi State University, Mississippi State, MS 39762

Charles E. Carraher, Jr., College of Sciences
Florida Atlantic University, Boca Raton, FL 33431

John E. Sheats, Department of Chemistry, Rider College, Lawrenceville, NJ 08648-3099

Mark D. Timken, Department of Chemistry, Widener University, Chester, PA 19013-5792

Martin Zeldin, Department of Chemistry, Indiana University-Purdue
University of Indianapolis, Indianapolis, IN 46205-2810

INTRODUCTION

Research on organometallic and inorganic polymers has increased enormously since about 1970. This chapter will give the reader an introduction to the variety of structures and applications which have been studied without any attempt to be comprehensive. Good recent reviews of organometallic[1-8] and inorganic polymers[8-19] are available and the reader is referred to them for a more complete introduction. The purpose of this chapter is to stimulate the reader to think about what might result if organic and inorganic chemistry are combined in polymer science. What interesting materials might result. To this end we will first look at some organometallic polymers and then examine some inorganic polymers. The division between these classes is somewhat arbitrary so the authors' prejudices will show.

Organic polymers generally are made from fewer than 10 elements (C, H, N, O, S, P halides) but with organometallic polymers well over 40 additional elements can be included. Polymers might include those with main group metals, such as silicon or germanium, or they could contain transition metals or rare earth elements the variety of structural variations seems endless. The metals may be pendant, as in poly(η^5-vinylcyclopentadienyltricarbonylmanganese),[20,21] **1**, formed by the addition polymerization of its vinyl precursor. Metals may be present in the main chain as is found in titanium polyether, **2**, a product of interfacial polycondensation (see Scheme 1).

A large number of polycondensations have been performed as indicated in Scheme 1 (next page) to produce polymers with such metals as Zr, Hf, Ge, Sn, Pb and Bi in the main chain.[22-26]

Similarly, a wide range of η^5-vinylcyclopentadienyl monomers containing various transition metals have been both homo- and copolymerized.[20,21,27-42] The aromatic cyclopentadienyl ring behaves as an extremely strong electron donating substituent (Alfrey-Price values of e = -1.9 to -2.1 for such monomers versus -0.80 for styrene, -1.37 for p-N,N-dimethylaminostyrene and -1.96 for 1,1'-dianisylethylene).[43] Copolymers with styrene, acrylnitrile, methyl methacrylate, N-vinylpyrrolidone and other monomers have been made for such (η^5-vinylcyclopentadienylcarbonylmetal) systems and for metal containing acrylates, methacrylates and related systems.

CONDENSATION POLYMERIZATION

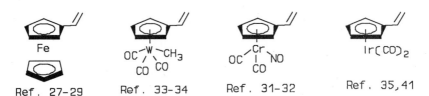

CARRAHER, 1974, 1977

Carraher 1973
1983

OTHER METAL HALIDES HAVE BEEN POLYMERIZED

R_2ZrX_2	R_2PbX_2		DIOLS	DITHIOHYDRAZIDES
R_2HfX_2	RBX_2	with	DITHIOLS	HYDRAZIDES
R_2GeX_2	R_3BiX_2		DIAMINES	UREA
R_2SnX_2			HYDRAZINES	
			DITHIOAMIDES	
			DIOXIMES	

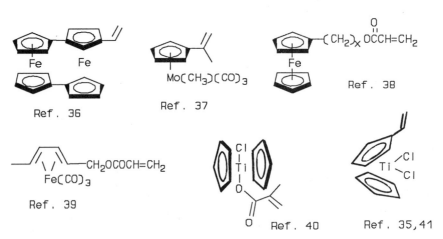

Ref. 27-29

Ref. 33-34

Ref. 31-32

Ref. 35,41

Ref. 36

Ref. 37

Ref. 38

Ref. 39

Ref. 40

Ref. 35,41

Scheme 1

The η^5-vinylcyclopentadienyl metal monomers are polymerized by radical initiation[27-42] and they are inert to anionic initiation.[43,44] Cationic initiators have given low molecular weights or no polymers[41,44-45] except for 1,1′-divinylferrocene[46] which undergoes cyclolinear polymerization and molecular weights up to 35,000 have been obtained using $BF_3.OEt_2$. Radical initiation has led to unusual homopolymerization kinetics in all systems studied carefully so far. Due to the propensity of metals to oxidize, peroxide initiators have not usually been suitable but azo initiators are satisfactory. Unlike vinylcyclopentadienyl monomers, ferrocenylmethyl methacrylate (and acrylate) are readily initiated with anionic systems[43,47] and the molecular weight may be varied regularly using $LiAlH_4$.[47] $LiAlH_4$/TMEDA produced living polymers of these ferrocene acrylic monomers from which block copolymers were achieved with acrylonitrile and MMA.[47]

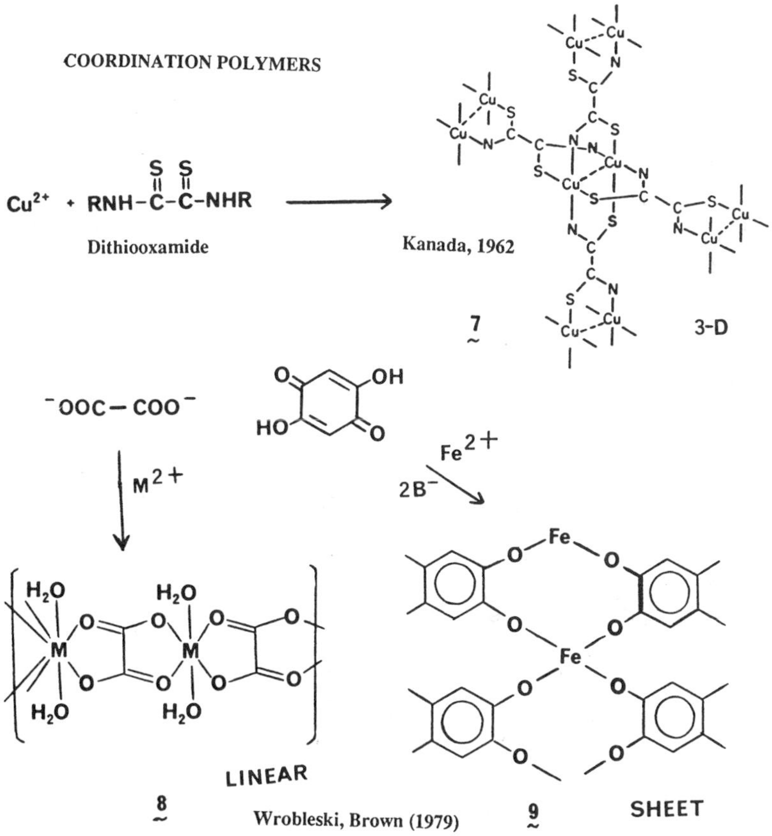

Scheme 2

Organometallic polymers come in a huge variety of structures. Coordination polymers illustrate this phenomena. For example, iron complexes of the dianion of oxalic acid are linear chain structures (see Scheme 2). However, if 2,5-hydroxyquinone is used as the chelating agent, a 2-dimensional planar sheet network is formed.[48] Coordination polymers may also form in three-dimensional networks as represented by the catana-μ-(N,N′-di-substituted dithioxamido)copper complex which results from the reaction of Cu(II) with dithio-oxamides.[49] Such materials are insoluble. The polymeric metal phosphinates shown in Scheme 3 can have single, double or triple-bridged structures. They have been made using Al, Be, Co, Cr, Ni, Ti and Sn as the metal.[14] They form films with thermal stabilities as high as 450°C and the chromium(III) species have been used as thickening agents for silicone greases for high pressure uses.

3

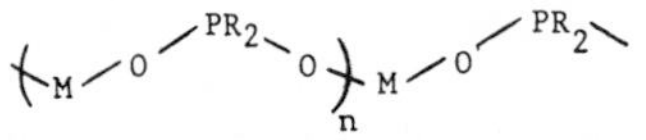

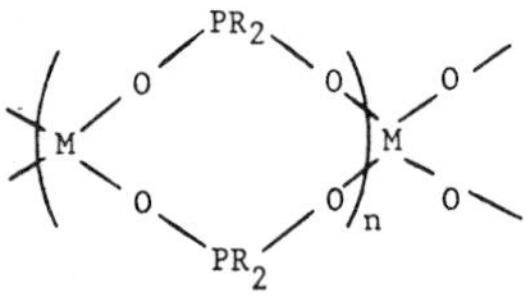

$$\left(M-O-PR_2-O\right)_n M-O-PR_2$$

$$\left(\begin{array}{c} O-PR_2-O \\ M \quad M \\ O-PR_2-O \end{array}\right)_n$$

$$\left[\begin{array}{c} PR_2 \\ O \quad O \\ M-O-PR_2-O-M \\ O \quad O \\ PR_2 \end{array}\right]_n$$

M = Al, Be, Co, Cr,
Ni, Ti and Sn

Thermally stable films (to 450⁰)
Cr(III) Polymer - Thickening agent
for silicone greases for high
pressure uses.

POLY(METAL PHOSPHINATES)
BLOCK (1970)

Scheme 3

Other wonderful shapes can also be found. The metal phthalocyanine structure with its bridging pyrazine units, shown in Scheme 4, resembles a shiskabob.[50] The metal and pyrazine chain forms the "skewer" which "pierces" the phthalocyanine groups. A braided struc-ture is also represented in Scheme 4. 5-Phenyltetrazolate reacts with metals from all three transition metal series to give the loops shown. The Ni(II) and Fe(II) adducts give extremely viscous aqueous solutions from which both flexible sheets and threads have been made.[51]

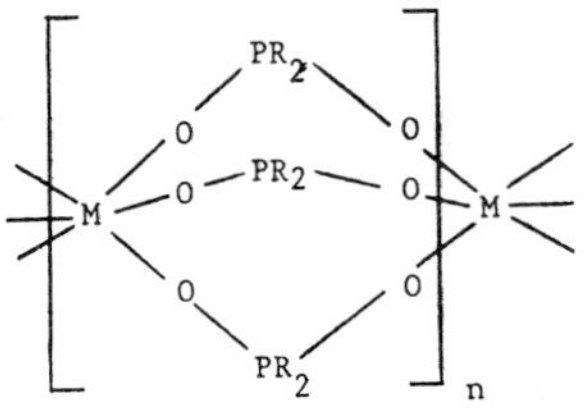

DIEL, MARKS, 1984

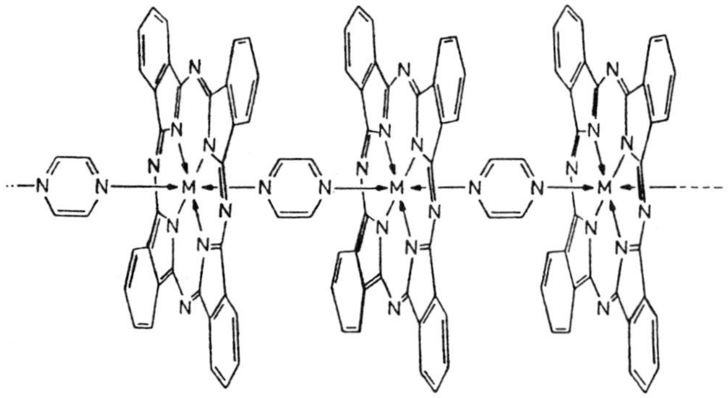

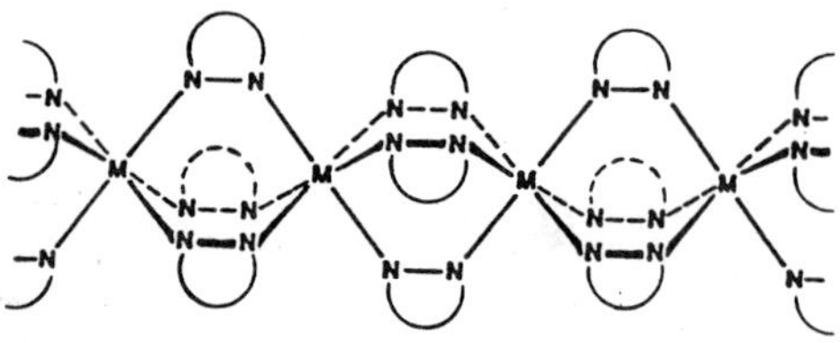

$$\bigcap_{N-N} = \begin{array}{c} R \\ N \ominus N \\ N-N \end{array}$$

M = Fe²⁺, Ni²⁺

THREADS AND FLEXIBLE SHEETS
FROM VISCOUS SOLUTIONS

RICHARDS ET AL. 1985

Scheme 4

4

Rigid rods are another shape into which coordination polymers and organometallic polymers with metals in the main chain have been made. One-to-one complexes of 2,5-dihydroxy-p-benzoquinone with Cu(II), Ni(II) and Cd(II) form polymers (see Scheme 5). The Cu(II) polymers were rigid rods 1000-2000Å long.[5-53] Transition metal-polyene polymers have been prepared with rod-like structures,[54-60] as illustrated in Scheme 5, by the reaction of cuprous iodide with *trans*-bis(tri-n-butylphosphine)platinum. The linear structures result from all-trans configurations at Pt or Pd atoms. This class of polymers, further illustrated in Scheme 6, form nematic liquid crystals. Some of them align their main chain in a direction perpendicular to an applied magnetic field while others align parallel to the field.[55,56] The nature of the transition metal within the chain determines its magnetic properties. The rigid rod structure is indicated by large Mark-Houwink exponents, the independence of $|\eta|$ values from solvent and the agreement of sedimentation-equilibrium experiments with parameters derived from an ellipsoid revolution model for rigid rods.

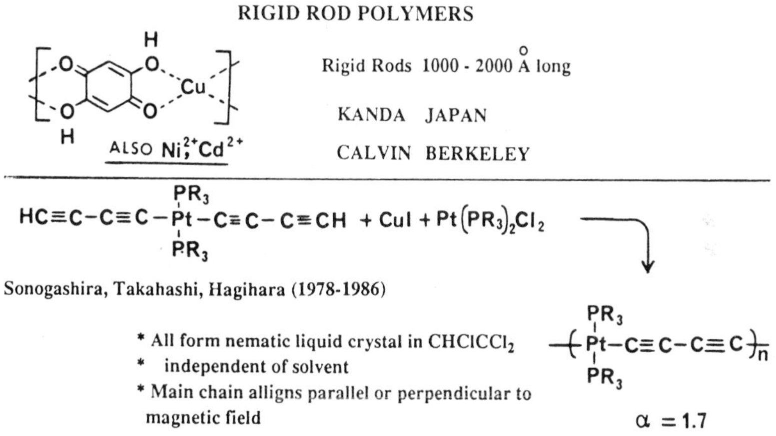

RIGID ROD POLYMERS

Rigid Rods 1000 - 2000 Å long

KANDA   JAPAN

CALVIN BERKELEY

Sonogashira, Takahashi, Hagihara (1978-1986)

* All form nematic liquid crystal in CHClCCl₂
* independent of solvent
* Main chain alligns parallel or perpendicular to magnetic field

$$-\!\!\left(Pt-C\equiv C-C\equiv C\right)_n$$

$\alpha = 1.7$

Scheme 5

Chain Orientation
in Magnetic Field

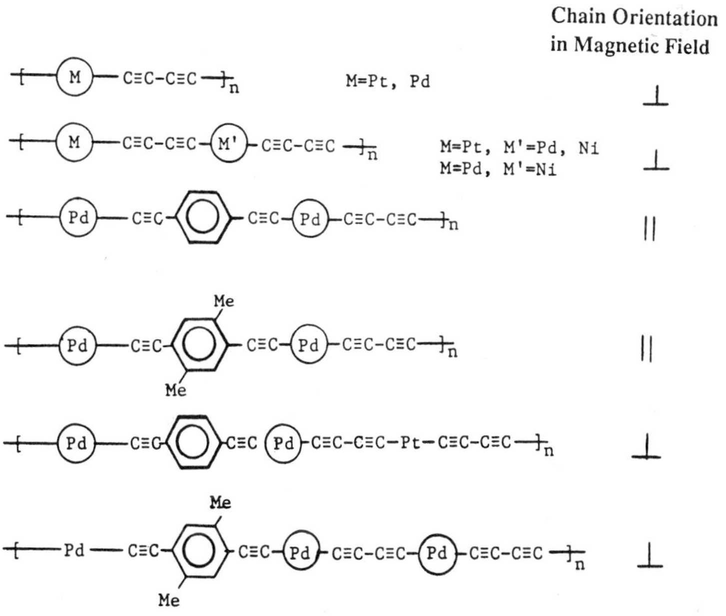

Scheme 6

5

In order to convey to the reader just a small portion of the vast diversity of known structures and applications a few examples have been selected. These examples are not necessarily representative of the field in any systematic way but they are selected to provide a flavor of what the field has to offer.

Organometallic condensation polymers have been used as dyes or pigments which resist bleeding, blooming or plateout.[61-63] An example, shown in Scheme 7, is the reaction product of fluorescein or sulfonphthaleim dyes with dicyclopentadienyldichlorotitanium or organostannanes. Degrees of polymerization or excess of 100 have been claimed for such polydyes. Such polymers should adhere to surfaces far better than the monomeric analogs.

POLYDYES AND COLORANTS

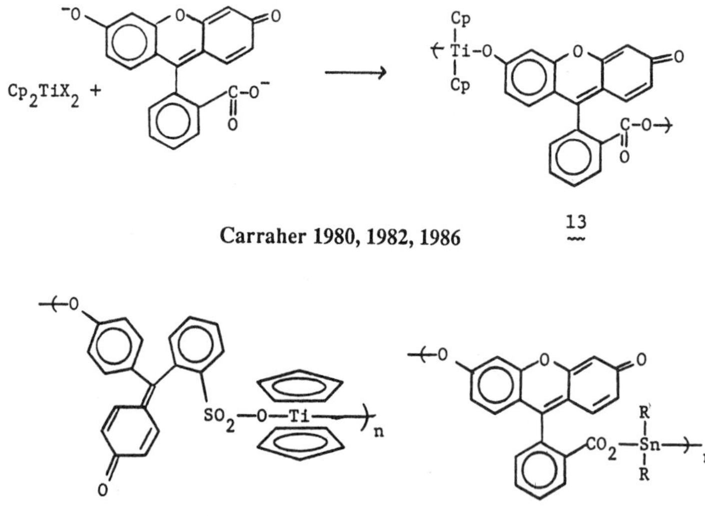

Carraher 1980, 1982, 1986

13

14

15

Scheme 7

Coordination complexes of platinum are under study as medicinal agents. For example, polyphosphazines have nitrogen atoms with unshared electron pairs in the main chain. These nitrogens can chelate platinum dichloride as shown in Scheme 8. The poly[bis(methylamine)phosphazene], shown in Scheme 8, has tumor-inhibiting activity against mouse P388 lymphocytic leukemia and was active in the Ehrlich ascites tumor-regression test.[61] The reaction of K_2PtCl_4 with pyrimidines, purines, hydrazines and diamines produces coordination polymers. An example of such a polymer which represses the replication of Poliovirus I and L RNA virus at a level of 10-20 μg/mL is shown in Scheme 8.[62]

Vinyl addition polymers of styrene, substituted in the para position with a europium chelate, have been made by Okamoto.[63] These polymers are being studied as active species in the electronic energy-transfer processes of lasers (see Scheme 8). Absorbed energy from xenon flashlamp pulsing is efficiently transferred to europium and the resulting polymers showed superfluorescence.[63] Even more bizarre was the use of poly-(vinylruthenocene) and poly(vinylosmocene) as preheat shields for targets in inertial-confinement nuclear fusion experiments.[64] Temperatures of 50×10^6 °C are needed and this is achieved by focusing high intensity laser beams (8×10^{12} W/cm.) on spheres of $^2H_2$ and $^3H_2$. The release of suprathermal electrons from the target is moderated by polymer layers having 1 to 4 atom % of high Z (Z=50-85) species and this permits higher fusion efficiencies. Poly(vinylruthencocene) and poly(vinylosomocene) were used in some of these experiments.

19

Tumor-inhibiting against
mouse P388 lymphocytic leukemia
and in Ehrlich ascites tumor-test

ALLOCK

20

Represses poliovirus I
and L rna virus replication
tumoral cell growth stopped

CARRAHER

LASER - RELATED

21

Laser electronic energy
transfer processes
OKAMOTO 1985
mma-copolymer superfluorescence
when pulsed by xenon flashlamps

22a M = Ru
22b M = Os

SHEATS
KOOL
PITTMAN

Preheat shields for laser
targets in inertial-confinement
nuclear fusion
8×10^{12}W/cm 50×10^6 °C

Scheme 8

A most interesting application of organometallic polymers involves the concept of "site isolation" in a polymer with "facilitated transport" through a membrane. In a series of papers the selective transport of oxygen or nitrogen through membranes containing organometallic polymers was examined.[65-69] Cyclopentadienyltricarbonylmanganese, upon irradiation with UV light, looses carbon monoxide and generates a coordinatively unsaturated complex which binds nitrogen reversibly (Scheme 9). However, the unsaturated species is unstable and rapid decomposition occurs by self-reaction in solution. However, when an octyl methacrylate copolymer of the 1-vinyl-3-methyl manganese complex is made and irradiated, the resulting coordinately unsaturated complex is stable since these sites are isolated from one another and can't react. New carbonyl bands are seen in the IR (Scheme 9). Exposure to nitrogen generates the corresponding nitrogen

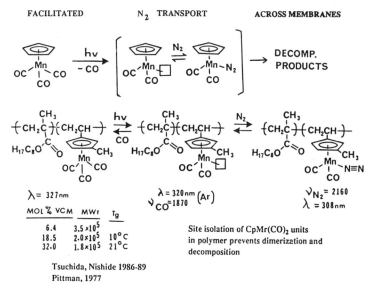

FACILITATED N_2 TRANSPORT   ACROSS MEMBRANES

DECOMP.
PRODUCTS

$\lambda = 327$nm

MOL % VCM	MWt	Tg
6.4	3.5×10^5	
18.5	2.0×10^5	10°C
32.0	1.8×10^5	21°C

$\lambda = 320$nm (Ar)
$\nu_{CO} = 1870$

$\nu_{N_2} = 2160$
$\lambda = 308$nm

Site isolation of CpMr(CO)₂ units
in polymer prevents dimerization and
decomposition

Tsuchida, Nishide 1986-89
Pittman, 1977

Scheme 9

7

complex (υ_{N_2} = 2160cm^{-1}) reversibly. A membrane composed of this polymer will facilitate the transport of nitrogen as shown in Scheme 10. Dual transport modes, both Henry and Langmuir modes, operate. Nitrogen reversibly jumps from one manganese atom, containing a vacant coordination site, to another. The copolymer compositions and properties can be varied over a range of properties.

DUAL-MODE TRANSPORT   (LANGMUIR AND HENRY)

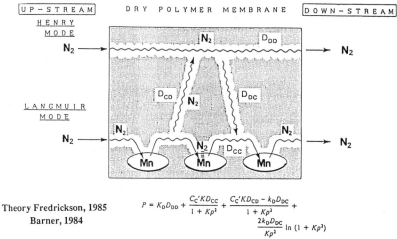

Theory Fredrickson, 1985
Barner, 1984

$$P = K_D D_{DD} + \frac{C_C' K D_{CC}}{1 + Kp^2} + \frac{C_C' K D_{CD} - k_D D_{DC}}{1 + Kp^2} + \frac{2k_D D_{DC}}{Kp^2} \ln (1 + Kp^2)$$

Experiments on this model Tsuchida, in press

Scheme 10

Oxygen transport has received considerable attention.[65-69] "Picket fence" cobalt porphyrin complexes are selectively blocked at the hindered face so that imidazole bases may only coordinate from the open side. As seen in Scheme 11, the imidazole-coordinated picket fence complex has a vacant coordination site at the top. The small linear oxygen molecule can traverse the "fence" to coordinate reversibly with cobalt from the top side. By dispersing such complexes in films one can build membranes which facilitate oxygen transport. This type of chemistry is one approach to making artificial blood.

OXYGEN BINDING AND TRANSPORT THROUGH MEMBRANE

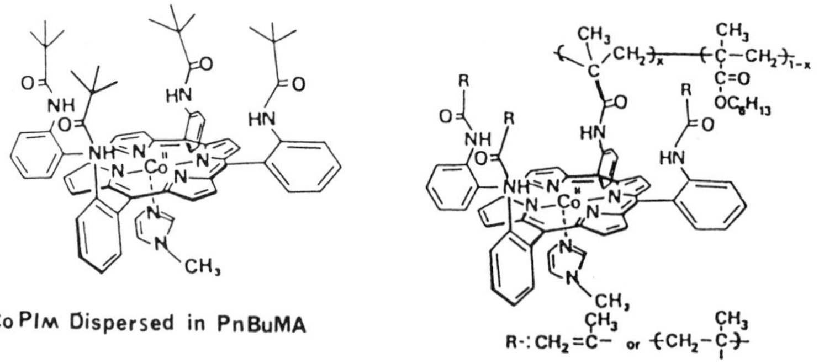

CoPIM Dispersed in PnBuMA

$P_{O_2}/P_{N_2} > 10$

$$CoPIM + O_2 \underset{k_{off}}{\overset{k_{on}}{\rightleftharpoons}} O_2\text{-}CoPIM$$

TSUCHIDA ETAL 1986, 1988

R-: $CH_2 = C-$ or $-CH_2-C-$

Bound

D= DIFFUSION COEFFICIENT OF O$_2$

D_b/D_d = 10

Scheme 11

Organometallic and inorganic polymers have frequently been investigated as possible conducting polymers. An early trick was to look for extended conjugation and then introduce mixed valence states. This most often led to charge-hopping type semiconductors. Thus, both poly(ferrocenylacetylene) and poly(vinylferrocene) were semiconductors which give highest values of conductivity at about 50/50 ferrocene: ferricenium ratios.[70]

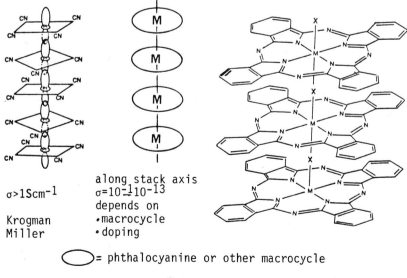

A large number of conjugated coordination polymers of regular structure have been made over many years. Though frequently insoluble amorphous powders, they have, in some cases exhibited, encouraging conductivities. An example of this class is the metal poly(benzodithiolene)s which have been prepared by the reaction of benzene-1,2,4,5-tetrathiol with Ni(II), Fe(II) and Cu(II).[71] They were paramagnetic conductors with conductivities ranging from 10^{-4} to 10^{-1}Scm^{-1}. A systematic study of structure versus conductivity in this area has proved very difficult so we will turn our attention to the stacked polymers where structure versus property advances have been forthcoming.

Example stacked polymers are shown in Scheme 12. These can be stacks giving metal chains or metal atoms may alternate with, for example, oxygen atoms. Tetracyanoplatinate stacked complexes are of interest because they have continuous metal-metal orbital overlap down the stack axis.[72-74] The stacks of square-planar Pt(CN)$_4$ units contain anions in spaces between the stacks. The cyano groups are staggered with respect to those coordinating adjacent Pt atoms. These, so-called Krogmann complexes, are one-dimensional conductors with conductivities greater than 1 Scm^{-1}. Work on the Krogmann complexes seems to have stimulated interest on phthalocyanine stacks also represented in Scheme 12. Here the ellipses represent the stacked macrocycles with direct metal-metal interactions down the chain. A variety of such stacked polymers have been made by the reaction of transition metal salts with tetranitriles[75,76] 1,2,4,5-tetracyanobenzene,[77] pyromellitic dianhydride[78] or tetracarboxylic acid derivatives.[79] These systems are semiconducting, typically, with pressed pellet conductivities ranging from 10^{-1} to 10^{-13}Scm^{-1}.

STACKED POLYMERS

$\sigma > 1$Scm^{-1}

Krogman
Miller

along stack axis
$\sigma = 10^{-1} - 10^{-13}$
depends on
• macrocycle
• doping

◯ = phthalocyanine or other macrocycle

Scheme 12

9

A more fruitful approach has been to make stacked phthalocyanine polymers with oxygen bridges between the metals. Dehydration of phthalocyanine complexes of Si, Ge and Sn produces a face-to-face oxygen bridged stacking pattern shown in Scheme 13.[80-86] Since electrical, optical and magnetic properties of "molecular metals" are a delicate function of complex architectural and electronic structural interaction, attempts to disentangle, understand and control properties has been almost impossible. Now, however, Marks and coworkers[80-86] have achieved significant advances in this effort.

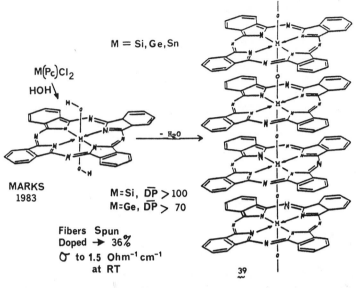

M = Si, Ge, Sn

M(Pc)Cl$_2$

HOH

MARKS
1983

M=Si, $\overline{DP}$ > 100
M=Ge, $\overline{DP}$ > 70

Fibers Spun
Doped → 36%
σ to 1.5 Ohm^{-1}cm^{-1}
at RT

- H$_2$O →

39

Scheme 13

The phthalocyanine polymer [Si(Pc)O]$_n$, where Pc = phthalocyanine, crystallizes naturally into an orthorhombic structure. For this neutral chain structure to become conducting it must be oxidized. This can be done electrochemically or by chemical oxidation. Scheme 14 illustrates the initial oxidation requires a significant over-potential. Minimal tunability of the resultant conductivity or oxidation is possible. During the oxidation counterions must move into the closely packed structure which causes a structural change from orthorhombic to the much more open tetragonal structure. Once in the tetragonal structure the counterions may easily distribute themselves in the vacant space between chains. Now reduction (undoping) or oxidation (doping) can be readily tuned continuously to counterion (BF$_4^-$) to Si ratios between 0 to 0.5. It is interesting to note that reduction of the tetragonal radical cation stacked structure gives back the neutral structure but it remains in the open tetragonal geometry. This is the first case where conduction band filling of a molecular metal is broadly tunable.

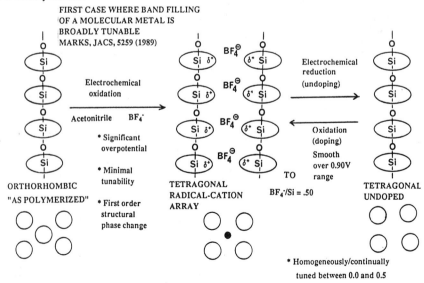

FIRST CASE WHERE BAND FILLING
OF A MOLECULAR METAL IS
BROADLY TUNABLE
MARKS, JACS, 5259 (1989)

Electrochemical
oxidation

Acetonitrile   BF$_4^-$

• Significant
overpotential

• Minimal
tunability

• First order
structural
phase change

ORTHORHOMBIC
"AS POLYMERIZED"

BF$_4^\ominus$

Electrochemical
reduction
(undoping)

Oxidation
(doping)

Smooth
over 0.90V
range

TO

BF$_4^-$/Si = .50

TETRAGONAL
RADICAL-CATION
ARRAY

TETRAGONAL
UNDOPED

• Homogeneously/continually
tuned between 0.0 and 0.5

Scheme 14

In tetrahydrofuran, tetragonal [Si(Pc)O]$_n$ can be reversibly n-doped to yield [Bu$_4$N$^+$]$_{0.09}$[Si(Pc)O]$_n$ (Scheme 15). Doping stoichiometry is largely a function of anion size due to packing restrictions within the tetragonal crystal structure. In contrast, the [Ni(Pc)]$_n$ polymer (without oxygen bridges) has poor tunability and a large overpotential. Upon oxidation, its native monoclinic slipped-stack β-phase structure undergoes a phase transition to a tetragonal structure. However, upon undoping, spontaneous conversion back to the monoclinic slipped-stack β-phase occurs. The oxygen-bridged germanium polymer [Ge(Pc)O]$_n$ undergoes irreversible oxidation with Ge-O bond cleavage.

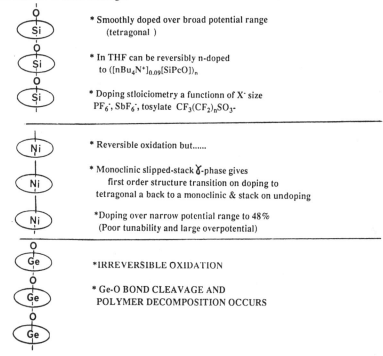

* Smoothly doped over broad potential range
   (tetragonal)

* In THF can be reversibly n-doped
   to ([nBu$_4$N$^+$]$_{0.09}$[SiPcO])$_n$

* Doping stloiciometry a functionn of X$^-$ size
   PF$_6^-$, SbF$_6^-$, tosylate CF$_3$(CF$_2$)$_n$SO$_3$-

* Reversible oxidation but......

* Monoclinic slipped-stack ɣ-phase gives
   first order structure transition on doping to
   tetragonal a back to a monoclinic & stack on undoping

*Doping over narrow potential range to 48%
   (Poor tunability and large overpotential)

*IRREVERSIBLE OXIDATION

* Ge-O BOND CLEAVAGE AND
   POLYMER DECOMPOSITION OCCURS

Scheme 15

Thin films of porphyrin-metal polyamides have been prepared by the interfacial polymerization of tetrakis chloride derivatives with either aliphatic dianines or with tetrakis amino derivatives of the porphyrin-metal complexes.[87] Films with thicknesses in the 0.01-10μm range, display unique chemical asymmetry. Opposite surfaces show different concentrations of functional groups. When placed between identical semitransparent electrodes and irradiated with broad-band or pulsed laser light the films developed directional photopotential. Photopotentials of 25mV were seen. The directionality of the photopotentials is the result of electron transfer toward the acid surface of the asymmetric film.

<u>Catalytic Applications</u>. A large variety of organic polymers have been derivatized with ligands to which metal complexes have been bound. These metal complexes were then used as "immobilized or anchored homogeneous catalysts" to promote organic reactions. Alternatively, well defined metal-ligand complexes,

11

functionalized to behave as monomers, have been polymerized to give polymer-bound catalysts. This topic has resulted in large literature and it has been extensively reviewed.[88-91] A few examples will be cited from what now is a very extensive literature.

Phosphinated polystyrene resins have been functionalized with both bisphosphine nickel carbonyl and tristriphenylphosphinehydridocarbonylrhodium. Using this system, butadiene can be cyclodimerized to vinylcyclohexane and selectively hydroformylated at the terminal double bond in sequence.[92-94] Similarly, the use of a NiBr$_2$ chelate, reduced by NaBH$_4$, together with the rhodium polymer permitted the linear oligomerization-selective hydroformylation of butadiene.[95] These were both one-pot processes.

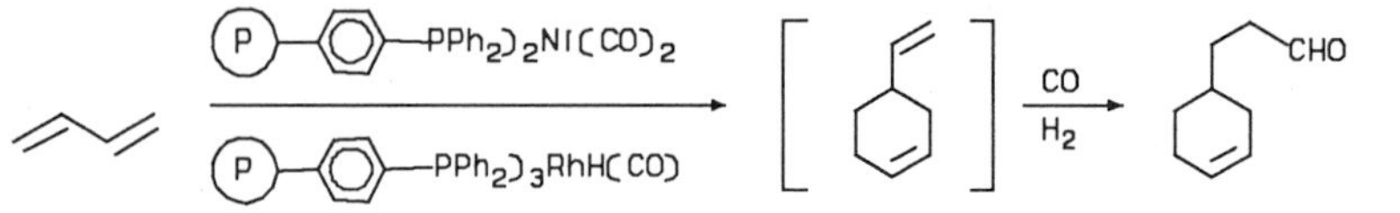

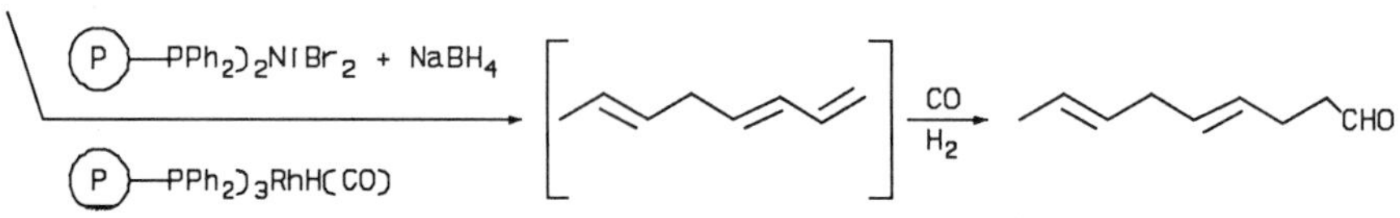

The use of asymmetric catalysts in chiral syntheses is taking on increasing importance. Asymmetric ligands or asymmetric metal complexes used in these transformations are quite expensive and need to be efficiently separated from reaction mixtures and recycled. Scheme 16 shows the preparation of a polymer-anchored dibenzophosphole-DIOP platinum-tin catalysts system.[94-96] The asymmetric ligand places the Pt-SnCl$_3$ system in a chiral environment. This catalyst has given the highest enantiometric excesses ever observed in catalytic hydroformylation. The initially achieved 70-83% e.e. values[94,95] were improved to >95% by the use of triethylorthoformate (TEOF) as the solvent.[96]

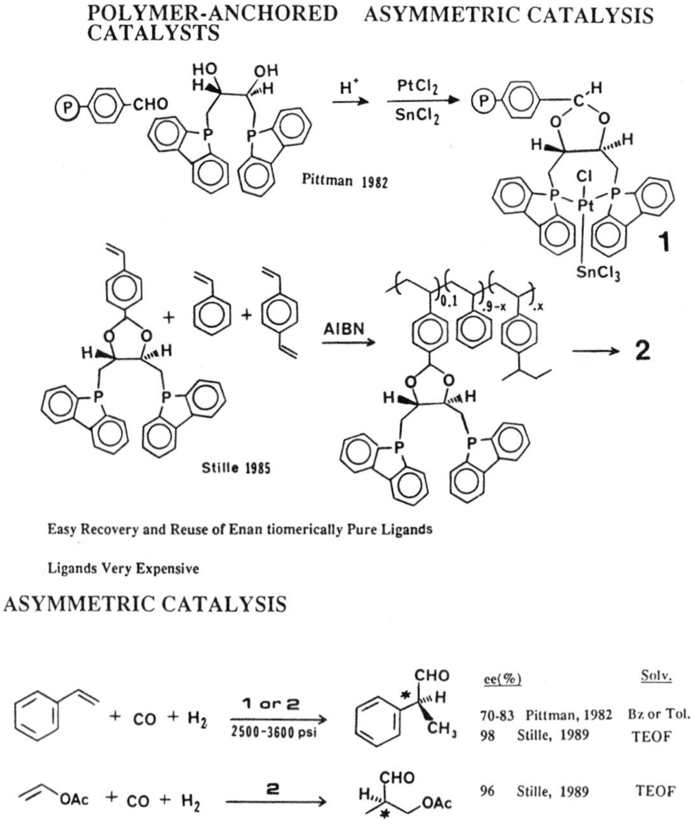

POLYMER-ANCHORED ASYMMETRIC CATALYSIS CATALYSTS

Pittman 1982

Stille 1985

1

2

Easy Recovery and Reuse of Enan tiomerically Pure Ligands

Ligands Very Expensive

ASYMMETRIC CATALYSIS

	ee(%)		Solv.
	70-83	Pittman, 1982	Bz or Tol.
	98	Stille, 1989	TEOF
	96	Stille, 1989	TEOF

Scheme 16

INORGANIC POLYMERS

For convenience, inorganic polymers[98] are defined as those without carbon in the backbone. Thus, polysiloxanes would be classified as inorganic polymers. Given the extensive commercial development of poly-siloxanes, no more will be said here about this class of materials and the reader is referred elsewhere.[99-102] Polyphosphazines, with alternating phosphorous and nitrogen atoms in the backbone, are defined as inorganic polymers using our definition as are poly(boron nitride), poly(sulfur nitride), polysulfur, poly(silanes), poly(germanes), polymeric metal phosphinates (see Scheme 3), metal oxide polymers such as silica, alumina and related systems and what are commonly called inorganic fibers such as boron-tungsten, alumina-zirconnia, silicon nitride and silicon carbide. The latter, of course has carbon in the backbone but it clearly is an "inorganic" polymer. Similarly, poly(carborane-siloxanes) are considered inorganic despite the fact the polymer chains contain the carbons of the carborane cages. To illustrate some of the recent activity and some of the structural diversity and charm in this field a few classes of these polymers are briefly mentioned.

poly(siloxane)   poly(phosphazine)   poly(boron nitride)   poly(sulfur nitride)

ring structures

poly(sulfur)   poly(silane)   poly(germane)   poly(titanosiloxane)

silicon carbide   silicon nitride

Poly(phosphazenes)[103-105] contain alternating nitrogen and phosphorus atoms and alternating double bonds in the backbone. Phosphorus is pentavalent with two substituents which can be varied widely (alkyl, aryl, alkoxy, aryloxy, arylamino, halogen or pseudohalogen) which leads to a wide variety of material properties for diverse applications. Poly(phosphazenes) are most readily prepared by the ring opening polymerization of hexachlorocyclotriphosphazene to give degrees of polymerization up to 15×10^3. The chlorines are readily replaced by nucleophiles to lead to the large diversity of polymers in this class known today. The properties

POLY(PHOSPHAZENES)

H. Allcock, Chemtech
5,552 (1975)

are sharply dependent upon the subtituent as illustrated by the sharp changes in glass transition temperatures, Tg, for the following substituents: CH_3O, -76°C; C_6H_5O, 6°C; C_6H_5NH-, 105°C. The thermal stability of several classes are outstanding with onset of weight loss above 300°C (via TGA) and good flame-retardant properties are exhibited.[103-105] Fluoroalkoxyphosphazene vulcanizates have excellent solvent resistance.[106] These properties as well as their high elastomer elongations between -60 to 200°C have resulted in their use in gaskets, damping materials and petroleum piping for arctic applications.[107]

Poly(carborane-siloxanes) have a linear structure in which R′ and R″ can be alkyl, fluoroalkyl or aryl groups.[108] The main chain contains the carborane polyhedra. Most commonly $C_2B_{10}H_{10}$ is the carborane used,

13

although the closo-carboranes $C_2B_5H_7$ and $C_2B_{10}H_{12}$ have also been used.[109,110] Interest in these polymers centers on their high thermal and oxidative stability and flame resistance. Rapid weight loss begins at 400°C or above. When R' and R'=CH$_3$, the value of Tg decreases from -42 to -88°C as n increases from 1 to 5.[·] These properties have resulted in the application of poly(carborane-siloxanes) in O-rings, gaskets, wire coatings and gas chromatography stationary phases.

POLY(CARBORANE-SILOXANES)

R. Williams, Pure Appl. Chem. 29, 569 (1972)
- High Thermal and Oxidative Stability
- R$_2$ Ge and borazine bridges also made

Poly(sulfur nitride) was first prepared in 1910 by Burt who passed vapors of cyclic tetrasulfur tetranitride, S_4N_4, over heated silver gauze.[111] In the 1970s purer products were obtained and strikingly lustrous golden poly(sulfur nitride) films, made of a malleable crystalline fibrous material, were obtained.[112] Parallel fiber layers are made of aligned chains of alternating sulfur and nitrogen where all S-N bond lengths are similar and correspond to a sulfur-nitrogen bond order intermediate between single and double bonds. Poly(sulfur nitride) exhibits a very high and anisotropic electrical conductivity (3700 Scm^{-1} along the crystalline axis) which increases on cooling to 4.2°K.[113,114] The polymer becomes superconducting at 0.26°K.[115] The conductivity increases on doping with bromine $(SNBr_{0.4})_n$ to a value of 3.8×10^4 Scm^{-1} at room temperature parallel to the fibers and only 8 Scm^{-1} perpendicular to the fiber axis.

Burt, J. Chem. Soc. 1171 (1910)
Heeger, MacDiarmid, JACS, 97, 6358 (1975)
Baughman, Chance, J. chem. Phys. 66, 401 (1977)

The delocalized chalcogen-based polymers, poly(carbon disulfide) and poly(carbon diselenide) were of special interest after the observation of superconductivity in $(SN)_x$. Carbon disulfide can be polymerized at high pressures (>4.5G Pa) at 175°C to give a black semiconducting solid[116] which was assigned a head-to-tail linear structure.[117] However, the head-to-head structure is thought to be 16.3 kJ mol.$^{-1}$ more stable so this might be more likely.[118] This polymer has been prepared by anionic initiation,[118,119] plasma deposition,[120] photolysis[121] and irradiation[122] to give structures that apparently vary or give rise to morphologically heterogeneous mixtures.[123] CSe$_2$ can be polymerized at ambient or high pressures[124,125] to give a variety of materials ranging from amorphous insulators to semiconductors to a crystalline product with a room temperature conductivity of 50 Scm^{-1}. A two dimensional sheet structure[126,127] with head-to-head bonding[124] is said to be more stable than the head-to-tail structure.[118]

14

CSe$_2$ and CS$_2$ Polymers

CS$_2$ $\xrightarrow{\Delta}$ head–to–tail   .head-to-head

CSe$_2$ $\xrightarrow{\Delta}$ head–to–head $\xrightarrow[\text{pressure}]{\text{anneal}}$ SHEET STRUCTURE $\sigma = 50\,\Omega^{-1}\,cm^{-1}$

CS$_2$ + CSe$_2$ $\xrightarrow{\;/\!/\;}$ COPOLYMER

Bridgman (1941)
Baughman (1988) J. Macromol. Sci. Chem. 799
Kobayashi Chem. Lett. 1407 (1983)
        Bull. Chem. Soc. Jpn. 3821 (1986)

Poly(silanes)

Poly(silanes) are a new class of inorganic polymers having an all-silicon backbone. For 60 years polymers containing all-silicon backbones remained a dream. High school and freshman chemistry courses taught that only carbon atoms could catenate together to form long linear chains. Despite the success in preparing poly(siloxanes), long chains of other single atoms besides carbon did not exist. Poly(dimethylsilane) was finally prepared by reacting dichlorodimethylsilane with sodium but the materials were insoluble and intractable.[128] The key developments leading to soluble and characterizable poly(silanes) were, (a) the introduction of large substituent groups which decrease order and (b) the use of two different substituents.[129,130] This is represented below for methylcyclohexydichlorosilane polymerization and for a copolymerization. High molecular weights (>1,000,000) were achieved of soluble, tractable, readily purified polymers. These polymers could be molded and cast into films. Improved synthetic procedures have been developed and reviewed.[131-133] The rate of research on poly(silanes) has accelerated to a furious pace in synthesis, structural studies, photochemistry, theoretical studies, electronic, lithographic, rheological, piezoelectric and piezochromic properties, solid state phase transitions, photophysics and molecular dynamics. In the single issue of *Polymer Preprints*, **31** (2) August 1990 some 143 papers appeared!

POLYSILANES

West etal. J. Polym. Sci. Polym. Chem. Ed. **22**, 59 and 225 (1984).
  Mol. wts. above 400,000
  UV depends on n
  Main chain scission on absorption (polymerization initiators)
  Contrast enhancing lithographic resists (bleaching latency)

15

$$\text{CH}_2\text{CH}_2\text{Ph}$$

Cl—Si—Cl   +   (cyclopentane ring with SiCl₂)   →[Na, Δ, toluene]   poly[(Si(CH₂CH₂Ph)(CH₃))—Si(ring)]ₙ

Poly(silanes) display UV absorption due to σ-delocalization along the all silicon backbone of the polymer which has no parallel for carbon chain polymers or siloxanes. For example, poly(cyclohexylmethylsilane) exhibits an absorption band at 326mm[130] while poly(dihexylsilane) has a λmax at 317.[132] Poly(silanes) with phenyl substituents absorb strongly around 330nm resulting from an interaction between the phenyl groups and the silicon backbone which acts as a σ→σ* or σ→π* chromophore. The conformational structure of many symmetrically substituted poly(di-n-alkylsilanes) have been determined.[134-136] The silicon bond conformation is *trans* planar when the alkyl substituent has one, two, three, six, seven, or eight carbons but it becomes a 7/3 helix with substituents with four or five carbons. The electronic properties show a weak correlation with the silicon bond conformation.[134-138] In a conformationally discolored state or in the regular 7/3 helical conformation the λmax values are ca. 315nm whereas for the all *trans* conformation the λmax values are observed from 332 to 374nm. Some homopolymers exhibit strong thermochromic[134,135] and piezochromic[136] properties due to conformational changes. For example, the λmax of poly(di-n-hexylsilane) in solution or after baking at 100°C is 317nm but after 3 hrs. at 21°C films exhibit a λmax = 371nm.

Poly(silanes) have a high sensitivity to chain cleavage on exposure to UV light.[139,140] Chain cleavage occurs with quantum yields of up to 0.97 in solution though they are much lower in the solid state.[141,142] As cleavage occurs a progressive blue shift and continuous bleaching of the UV spectrum occurs, both in the solid state and solution. As the molecular weight decreases both the λmax and ε values decrease. Photodecomposition produces substituted silylenes and silyl radicals. Thus, small amounts poly(silanes) can serve as photoinitiators, which are relatively oxygen insensitive, for a variety of monomers.[143,144] The 3M Company has commercialized this process.

$$\left(\text{Si}_R\text{-Si}_R\text{-Si}_R\right)_n \xrightarrow{h\nu} 2 \sim\!\!\sim\!\dot{\text{Si}}R_2 + R_2\text{Si:}$$

$$\sim\!\!\sim\!\dot{\text{Si}}R_2 + CH_2\!\!=\!\!CHX \longrightarrow \sim\!\!\sim\!\text{Si}R_2\!-\!CH_2\!-\!\dot{C}HX$$

The unusual ability of poly(silanes) to absorb UV light led to the demonstration that AsF₅-doped poly(silastyrene) is a semiconductor.[139] The band gap in common silane polymers is about 4eV in contrast to a value of almost 8eV for carbon chains. Conducting and photoconducting properties have been studied.[145-147] Only holes were found mobile (mobilities of ~ 10^{-4} cm² V⁻¹s⁻¹ at 20°C) and it appears hole transport occurs via the σ electronic states of the silicon backbone rather than hopping between chain substituents.

High solubilities, ready formation of high optical quality films, and both bleaching of the absorption spectrum and chain scission on exposure to UV irradiation have led IBM and others to examine poly(silanes) as resists for bilayer UV lithography and oxygen reactive ion etching (O₂-RIE) barriers.[148-150] Upon exposure to an oxygen plasma polysilanes oxidize to an SiO₂ layer which resists further etching. Once a pattern is developed in the exposed resist, RIE can be used to cut through underlying layers and image the wafer. Clean submicron images have been generated in a bilayer process using a 248-nm excimer laser exposure tool.[151] The use of poly(silanes) in contrast enhancement lithography is especially clever.[152,153] A thin layer of poly(silane) on top of the photoresist sharpens the diffraction-distorted mask image giving improved resolution. This occurs because the diffracted light is less intense and the edge areas of the image bleach less rapidly.

Pyrolysis of Polysilanes to Silicon Carbide

Pioneering work by Yajima[154,155] showed that poly(dimethylsilane) can be converted into β-SiC fibers of very high tensile strength (350kg mm⁻²). Thermolysis at 400-480°C gives a poly(carbosilane) which is

16

fractionated and then melt-spun into fibers. The fibers are oxidized, at their surfaces, by air at 350°C to provide rigidity and then pyrolyzed at 1200°C under nitrogen to give crystalline β-SiC.

$$\left(\begin{array}{c} CH_3 \\ | \\ Si \\ | \\ CH_3 \end{array}\right)_n \xrightarrow{470°C} \left(\begin{array}{c} H \\ | \\ Si-CH_2 \\ | \\ CH_3 \end{array}\right)_n \xrightarrow[\text{to fibers}]{\text{melt-spun}}$$

$$\xrightarrow[\text{2. } N_2, \text{ 1200°C}]{\text{1. Air, 190°C}} \beta\text{-SiC} + H_2 + CH_4$$

The pyrolysis of polymers containing η^4-dienetricarbonyliron moieties was shown to produce Cr_2O_3 and iron oxides, respectively, dispersed as tiny particles throughout the material.[154] Yasuda has followed the progress of a high temperature thermal decomposition of a η^4-dienetricarbonyliron-containing copolymer under an Ar atmosphere by [57]Fe-Mössbauer spectroscopy. As seen in Scheme 17 the decomposition at 900°C gives Fe_3C and continued pyrolysis to 1400°C gives a mixture of α-Fe and δ-Fe. Yasuda also pyrolyzed a variety of fibers containing organometallic units and achieved both metal particles or metal phosphide particles of uniform sizes with the resulting carbon fibers. An example is shown in Figure 1 which pictures carbon fibers with uniform Pd_5P_2 and PdP_2 particles of uniform size distributed within the body of the fiber.

$$\left(\begin{array}{c} CH-CH_2 \\ | \\ O=C \\ \quad OCH_2CH_2- \end{array}\right)_n \bigcirc Cr(CO)_3 \xrightarrow{\Delta \text{ or } h\nu} \begin{array}{c} \text{crosslinked} \\ \text{polymer} \end{array} + Cr_2O_3$$

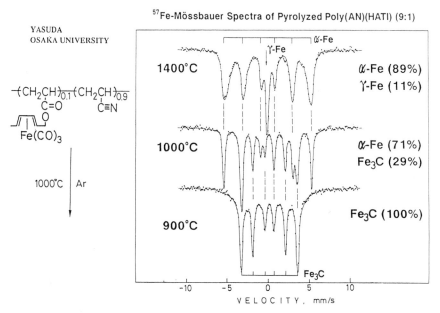

[57]Fe-Mössbauer Spectra of Pyrolyzed Poly(AN)(HATI) (9:1)

YASUDA
OSAKA UNIVERSITY

$$\left(CH_2CH\right)_{0.1}\left(CH_2CH\right)_{0.9}$$
$$\begin{array}{cc} C=O & C\equiv N \\ | \\ O \\ Fe(CO)_3 \end{array}$$

1000°C | Ar

1400°C α-Fe (89%) γ-Fe (11%)

1000°C α-Fe (71%) Fe_3C (29%)

900°C Fe_3C (100%)

VELOCITY, mm/s

Scheme 17

17

Poly(germanes) and Poly(silynes)

The successful preparation of high molecular weight poly(silanes) suggested that catenated germanium analogs might also be stable. Like silicon, germanium derivatives form cyclic and acyclic catenates. Now the first soluble, high molecular weight, substituted germanium homopolymers and silicon-germanium copolymers have been made by the Wurtz-type reaction of dichlorogermanium precursors with sodium dispersions. As frequently found in poly(silane) synthesis, poly(germanes) seem to exhibit binodal molecular weight distributions.

$$
\underset{\overset{|}{R}}{\overset{\overset{R}{|}}{Cl-Ge-Cl}} \;+\; 2Na \;\xrightarrow[\Delta]{\text{toluene}}\; \underset{\overset{|}{R}}{\left(\!\overset{\overset{R}{|}}{Ge}\!\right)_{\!n}} \;+\; 2NaCl
$$

$$
\underset{\overset{|}{R}}{\overset{\overset{R}{|}}{Cl-Ge-Cl}} \;+\; \underset{\overset{|}{R}}{\overset{\overset{R}{|}}{Cl-Si-Cl}} \;\xrightarrow[\Delta]{\underset{\text{toluene}}{2Na}}\; \underset{\overset{|}{R}\;\;\overset{|}{R}}{\left(\!\overset{\overset{R}{|}\;\;\overset{R}{|}}{Ge-Si}\!\right)_{\!n}} \quad \text{random}
$$

YASUDA
OSAKA UNIVERSITY

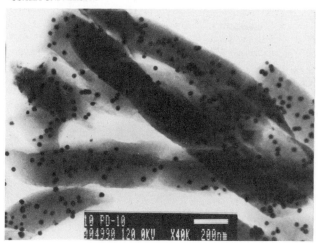

Pd$_5$P$_2$, PdP$_2$ / C Composite

Figure 1

The absorption maxima of the germanium homopolymers in solution were about 20nm red shifted from their corresponding silicon derivatives. In the solid state, poly(di-n-hexylgermane) and poly(di-n-octylgermane) are strongly thermochromic. This effect has been attributed to the conformational locking of the backbone caused by crystallization of the side groups as the temperature is lowered. The germanium-silicon copolymers are also strongly thermochromic and their long wavelength λmax is blue shifted about 8nm from the respective homopolymers. The homopolymers and copolymers undergo scission to lower molecular weight materials when subjected to light.

What structure would be obtained if RSiCl$_3$ (R = alkyl) was subjected to the Wurtz-like reaction used in poly(silane) synthesis? Would poly(silynes) be formed having conjugated double bonds or would three dimensional networks of sp^3-hybridized silicon result? The use of Na/K alloys and the application of high intensity ultrasound permitted the smooth homogeneous Wurtz-type conversion of hexyltrichlorosilane into a polymer.[158,159] Yellow, hexane soluble polymers with Mw = 2-5x10^4 (GPC) and monodal envelopes of polydispersities of about 2 were isolated.

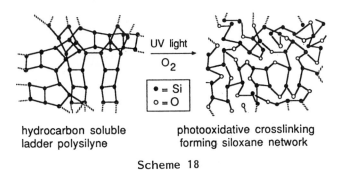

This polymer had no crystallinity or structural regularity (DSC, x-ray). Its ^{29}Si NMR spectrum demonstrated the silicon atoms were sp^3-hybridized with three silylsubstituents and eliminated the presence of silicon-silicon double bonds as a primary structural feature. The material is constructed of sp^3-hybridized monoalkylsilane moieties assembled into rigid but irregular networks of Si-Si σ-bonds consisting of a sheet like connection of ring structures. The poly(alkylsilynes) exhibit intense σ→σ* near-UV absorption bands tailing into the visible. The breadth and intensity of these bands are the result of the extension of Si-Si σ-conjugation into additional dimensions across the silicon frame network.

Poly(n-hexylsilyne) is far more stable to photo-degradation in inert atmospheres than are linear poly(silanes) due to the delocalized nature of excitations and the greater propensity of the network to enforce recombination of the photogenerated radicals. Photooxidation occurs upon irradiation in air as crosslinks are formed to give polymeric silicon networks. This photooxidation bleaches the UV absorption and is accompanied by a large decrease in refractive index (from 1.65 to 1.45 as Si-Si bonds are replaced by Si-O-Si). This suggests the use of poly(silanes) as the basis of a new approach to thin film optical waveguide fabrication[160] and as negative photomasks.

hydrocarbon soluble
ladder polysilyne

photooxidative crosslinking
forming siloxane network

Scheme 18

Polymers with Metal Oxygen Backbones

A large variety of aluminum and silicon polymers with metal oxygen backbones have been made besides the poly(siloxanes). Poly(aluminosiloxanes) contain an Si-O-Al-O backbone. A typical example[161] results from the reaction of sodium salts of dimethylsiloxane oligomers with aluminum chloride. Polymers with Si/Al ratios of 0.8 to 23 have been made. Low Si/Al ratios are brittle and insoluble having a 3-dimensional structure while those with Si/Al ratios of 7 to 23 are soluble.

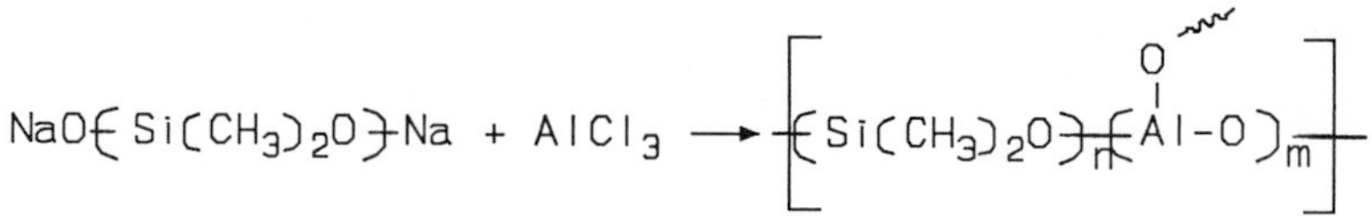

$$NaO\!\!\left[Si(CH_3)_2O\right]\!\!Na + AlCl_3 \longrightarrow \left[\left[Si(CH_3)_2O\right]_n\!\!\left[\underset{\underset{\displaystyle \sim\!\!\sim}{\overset{\displaystyle O}{|}}}{Al}-O\right]_m\right]$$

Poly(aluminophenylsiloxanes), having a Si/Al ratio of 4, is completely infusible but very soluble in organic solvents. It can be plasticized. A ladder structure has been assigned.

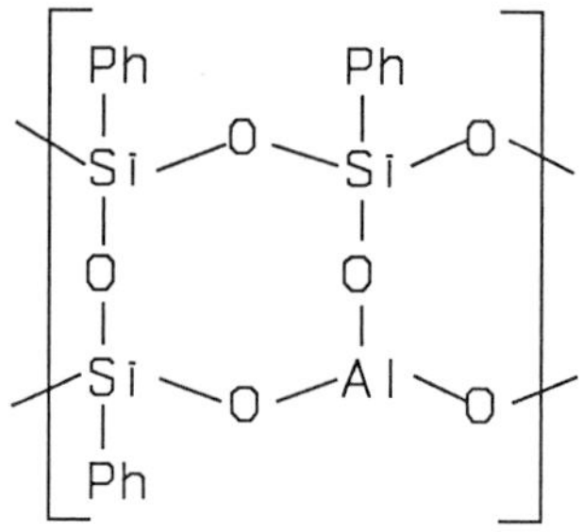

Poly[(acyloxy)aloxanes] can be readily prepared by the one-pot sequential reaction of either trialkyl-[162] or trialkoxyaluminums[163] with carboxylic acids and then water. Water promotes the hydrolytic polymerization. Some gelation occurs after 1.8 equiv. of ethane are evolved. The aluminum can be properly coordinated by the addition of another mole of carboxylic acid. After this step the polymers become more soluble, achieve higher molecular weights and exhibit better solution processability than the original polymer. This "preceramic polymer" was used in the preparation of high performance alumina fibers but, unlike other preceramic alumina polymers,[164] this group was not melt-processable. When RCOOH = n-dodecanoic acid and R'COOH = 3-ethoxypropanoic acid a series of polymer (mol. wt. >50,000) was made which could be melt spun into thin fibers at above 200°C.[165]

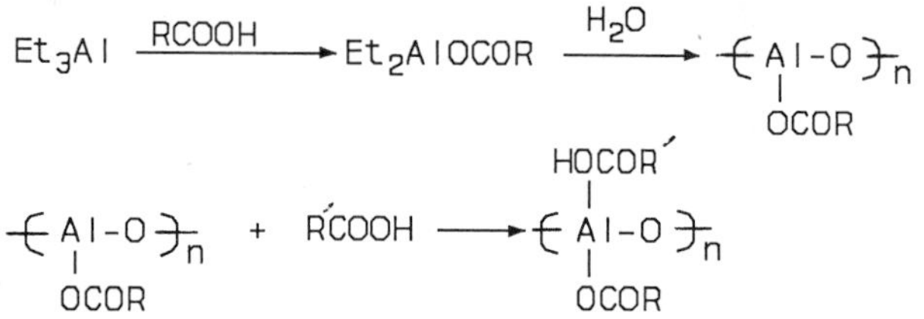

$$Et_3Al \xrightarrow{\text{RCOOH}} Et_2AlOCOR \xrightarrow{\text{H}_2O} \underset{\underset{\displaystyle OCOR}{|}}{\left[Al-O\right]_n}$$

$$\underset{\underset{\displaystyle OCOR}{|}}{\left[Al-O\right]_n} + \acute{R}COOH \longrightarrow \underset{\underset{\displaystyle OCOR}{|}}{\overset{\overset{\displaystyle HOCOR'}{|}}{\left[Al-O\right]_n}}$$

The long dodecanoic side chain imparts thermoplasticity while the 3-ethoxypropanoic group gives good dry-spinability. Since carboxylate ligands are bidentate, the six coordination sites at aluminum can be completely filled but the exact ratios of acids used (to each other and to Al) is important. An idealized structure in which coordination of the ethoxy groups to an adjacent aluminum, having only one carboxylate group attached, is shown below.

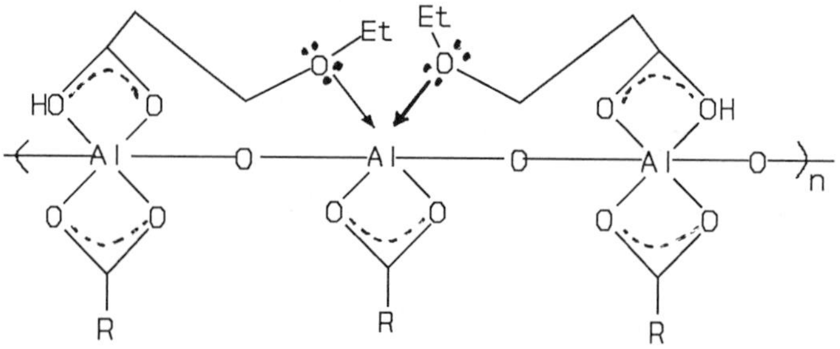

Random -Si-O-Ge-O- and -Si-O-Sn-O- polymers have been made by cohydrolysis of dialkyldichlorosilanes with dialkyldihalogermanes (or stannanes).[166] Further exposure to sulfuric acid gave rubber-like materials. Condensation of dimethylsilanediol with titanocene dichloride results in the incorporation of

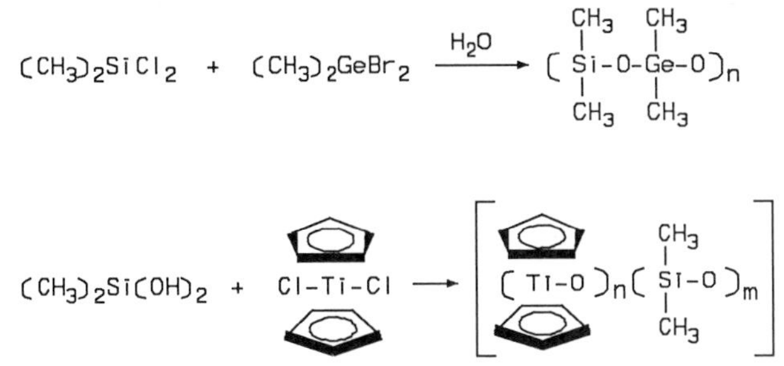

$$(CH_3)_2SiCl_2 \ + \ (CH_3)_2GeBr_2 \ \xrightarrow{H_2O} \ \begin{matrix} CH_3 & CH_3 \\ | & | \\ (\ Si-O-Ge-O \ )_n \\ | & | \\ CH_3 & CH_3 \end{matrix}$$

titanocene units. An interesting "oxidation-reduction copolymerization" of bis[bis(trimethylsilyl)amido]germanium (divalent germanium) with various p-benzoquinones resulted in alternating germanium(IV) and p-hydroquinone units.[167] Copolymers of relatively high molecular weights (M_w >3x10⁴ for p-benzoquinones and M_w = 3.6x10⁵ for 1,4-naphthoquinone) were obtained which were soluble in hexane, benzene and chloroform. The copolymerizations occurred readily at -78°C in toluene even when the sterically hindered 2,5-di-t-butyl-1,4-benzoquinone monomer was employed. The first step of the reaction probably involves the formation of the zwitterion or the diradical shown below. Two of these intermediates then react to give a dimeric zwitterion or diradical. Successive reactions leads to polymer.

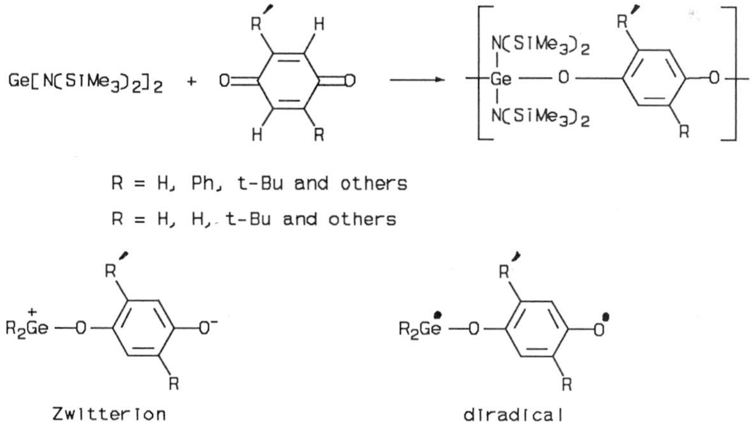

R = H, Ph, t-Bu and others

R = H, H, t-Bu and others

Zwitterion

diradical

Sol-Gel Inorganic Polymers

Metal-oxygen backbone polymers and networks are often used as precursors to ceramics. The preparation of such preceramic polymers by the sol-gel process has witnessed explosive growth. A true marriage between inorganic polymer chemistry and ceramic materials is traced by several symposia highlighting recent developments.[168-170] In sol-gel processes, ceramic precursor polymers are formed in solution at ambient temperature by classic chemistry.[171,172] They are shaped by casting, film formation or fiber drawing and then consolidated into dense glasses or polycrystalline ceramics.[171,172]

Commonly, sol-gel processes employ alkoxides of silicon, boron, titanium and aluminum. In alcohol-water solutions, the alkoxide groups are removed stepwise by either acid- or base-catalyzed hydrolysis and replaced by hydroxyl groups. Next -M-OH moieties form -M-O-M- linkages and water. Thus, polymeric chains branch, grow and interconnect as shown in the equations outlined in Scheme 19 for a silicate sol and a silicate-zirconate sol. Gelatin occurs, eventually, as growing polymers form networks which grow to span the entire volume of the reacting solution. At this point (the gel point) both the viscosity and elastic modulus rapidly increase. The rate of sol and gel formation is exceptionally sensitive to pH, H_2O/alcohol ratio and temperatures.

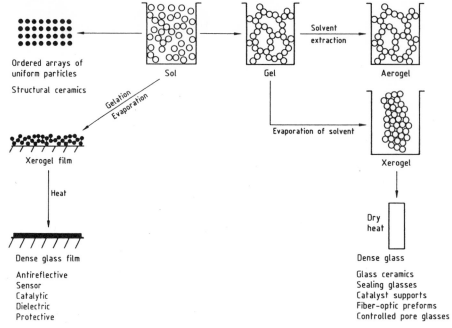

Scheme 19

The gel is a viscoelastic material composed of interpenetrating liquid and solid phases. Scheme 20 illustrates the overall outline of a sol-gel process from sol formation to product. The gel structure retards escape of liquid, prevents structure collapse and it freezes in shapes. The sudden increase in viscosity locks in place shapes formed by casting, drawing of fibers or film formation. Consolidation to dense glasses or ceramics occurs during heat treatment and sintering. Silica networks can be modified homogeneously at the molecular/atomic level by addition of alkoxides of B, Al, Ti and Zr allowing enormous flexibility to ceramic design. For example, Scheme 19 illustrates sol-gel chemistry for SiO_2/ZrO_2 and $SiO_2/ZrO_2/Al_2O_3$ ceramics. Obviously, products produced by the sol-gel route will be different than those formed by mixing oxide powders together and sintering them.

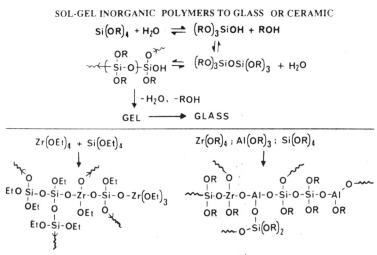

Brinker, Ulrich (Eds.), "Better Ceramics through Chemistry II", 1986

Scheme 20

Since the gels are tailored by low temperature chemical reactions it should be possible to incorporate organic units into the silicate gels. This has been done and resulted in a new family of materials called ceramers.[173] Thus, vinyl groups, epoxides and methacrylates have been polymerized to give interpenetrating networks with a wide variety of structures and properties.[174-176] Scheme 21 summarizes a few such ideas. The incorporation of organic polymeric units of various sizes could lead to tougher more impact resistant ceramics.

Short alkoxy oligomeric chains, terminated with organic epoxides, have been polymerized to give chemically bound poly(ethyleneoxide) chains which are then locked by further hydrolyzing the alkoxides to form the metal oxide gel structure. The incorporation of end-capped poly(tetramethylene oxide) (PTMO) blocks (molecular mass 650-2000) into tetraethoxysilane sol-gel glasses has been reported.[177] This is illustrated in Scheme 21. These materials exhibit high extensibility and contain interdispersed organic polymer and inorganic polymer regions. The PTMO blocks are well dispersed but some degree of local phase separation occurs. Will we see "stretchy glass" one day? Applications of ceramers now include adhesives for glass surfaces,[176] protective coatings for medieval stained glass[178] and scratch-resistant coatings for eyeglass lenses.

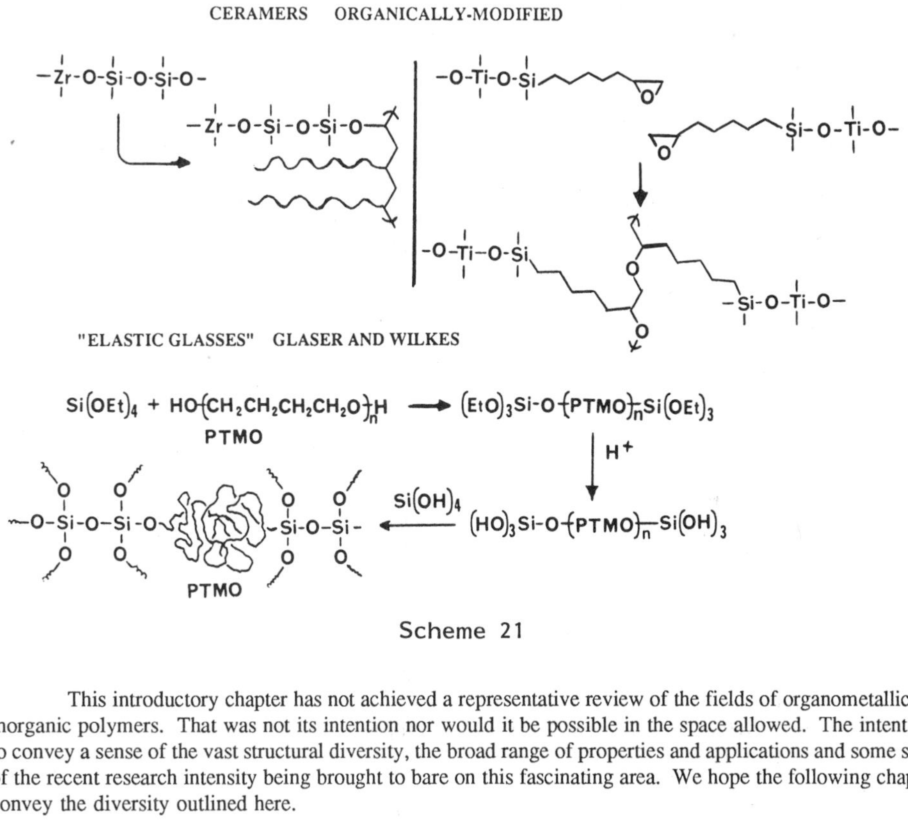

CERAMERS   ORGANICALLY-MODIFIED

"ELASTIC GLASSES"   GLASER AND WILKES

Scheme 21

This introductory chapter has not achieved a representative review of the fields of organometallic and inorganic polymers. That was not its intention nor would it be possible in the space allowed. The intent was to convey a sense of the vast structural diversity, the broad range of properties and applications and some sense of the recent research intensity being brought to bare on this fascinating area. We hope the following chapters convey the diversity outlined here.

REFERENCES

1.   C. U. Pittman, Jr., C. E. Carraher, Jr. and J. R. Reynolds, "Organometallic Polymers" in the "Encyclopedia of Polymer Science and Engineering," H. Mark, N. Bikales, C. Overberger and J. Menges, eds., Vol. 10, 541-594, John Wiley and Sons (1987).

2.   A. D. Pomogailo and V. S. Savostianov, Advances in the Synthesis and Polymerization of Metal-Containing Monomers, *J. Macromol. Sci. Rev. in Macromol. Sci. Chem. Phys.* C25(3), 375-479 (1985).

3.   J. E. Sheats, C. E. Carraher, Jr. and C. U. Pittman, Jr., eds., "Metal-Containing Polymer Systems," Plenum Publishing Corp., New York (1985) pp. 1-523.

4.   C. E. Carraher, Jr., J. E. Sheats and C. U. Pittman, Jr., eds., "Advances in Organometallic and Inorganic Polymer Science," Marcel Dekker, Inc., New York (1982) pp. 1-449.

5. C. E. Carraher, Jr., J. E. Sheats and C. U. Pittman, Jr., "Organometallic Polymers," Academic Press, Inc., Orlando, FL (1978) pp. 1-346.

6. K. A. Andrianov, "Metallorganic Polymers," John Wiley and Sons, Inc., New York (1965).

7. E. W. Neuse and H. Rosenberg, "Metallocene Polymers," Marcel Dekker, Inc., New York (1970).

8. M. Zeldin, K. J. Wynne and H. R. Allcock, eds., "Inorganic and Organometallic Polymers," *ACS Symposium Series*, **360** (1988) pp. 1-449.

9. C. E. Carraher, Jr., and C. U. Pittman, Jr., "Inorganic Polymers," in "Ullmann's Encyclopedia of Industrial Chemistry," Fifth Edition, VCH Press, Vol. A14 (1989) pp. 241-262.

10. S. N. Borison, M. G. Voronkov and E. Lukevits, "Organosilicon Heteropolymers and Hetero-compounds," Plenum Press, New York (1970).

11. H. R. Allcock, "Heteroatom Ring Systems and Polymers," Academic Press, New York (1967).

12. A. L. Rheingold, "Homoatomic Rings, Chains and Macromolecules of Main Group Elements," Elsevier Scientific Publishing Co., New York (1977).

13. F. G. A. Stone and W. A. G. Graham, "Inorganic Polymers," Academic Press, New York (1962).

14. B. P. Block, *Inorganic Macromol. Rev.*, **1** (2): 115 (1970).

15. C. J. Brinker, et al, "Inorganic and Organometallic Polymers," *ACS Symposium Series*, **360**, (1988).

16. L. L. Hench and D. R. Urich, eds., "Science of Ceramic Chemical Processing," Wiley-Interscience, New York (1986).

17. N. H. Ray, "Inorganic Polymers," Academic Press, New York (1978).

18. M. F. Lappert and G. J. Leigh, eds., "Developments in Inorganic Polymer Chemistry," Elsevier, Amsterdam (1962).

19. H. R. Allcock, "Rings, Clusters and Polymers of the Main Group Elements," *ACS Symposium Series*, **232** (1983).

20. C. U. Pittman, Jr., C. C. Lin and T. D. Rounsefell, *Macromolecules*, **11**, 1022 (1978).

21. C. U. Pittman, Jr., G. V. Marlin and T. D. Rounsefell, *Macromolecules*, **6**, 1 (1973).

22. C. E. Carraher, Jr. and S. Bajah, *Br. Polym. J.*, **15**, 9 (1974).

23. C. E. Carraher, Jr. in P. Millich and C. E. Carraher, Jr., eds., "Interfacial Synthesis," Vol. II, Marcel Dekker, Inc., New York (1977), chapter 20.

24. C. E. Carraher, Jr. and R. Nordin, *J. Appl. Polym. Sci.*, **18**, 53 (1974).

25. C. E. Carraher, Jr. and S. T. Bajah, *Br. Polym. Jr.*, **14**, 42 (1973).

26. C. E. Carraher, Jr. et al, *J. Macromol. Sci. Chem.*, **17**, 1121 (1983).

27. Y. Sasaki, L. L. Walker, E. L. Hurst and C. U. Pittman, Jr., *J. Polym. Sci., Polym. Chem. Ed.*, **11**, 1213 (1973).

28. M. H. George and G. F. Hayes, *J. Polym. Sci. Polym. Chem. Ed.*, 1049 (1975); **14**, 475 (1986).

29. C. U. Pittman, Jr. and C. C. Lin, *J. Polym. Sci. Polym. Chem. Ed.*, **17**, 271 (1979).

30. C. U. Pittman, Jr. and T. D. Rounsefell, *Macromolecules*, **9**, 937 (1976).

31. M. D. Rausch, C. U. Pittman, Jr., et al, *J. Organometal. Chem.*, **137**, 199 (1977).

32. C. U. Pittman, Jr., M. D. Rausch, et al, *Macromolecules*, **11**, 560 (1978).

33. M. D. Rausch and C. U. Pittman, Jr., *J. Organometal. Chem.*, **205**, 353 (1981).

34. C. U. Pittman, Jr., M. D. Rausch, et al, *Macromolecules*, **14**, 237 (1981).

35. M. D. Rausch and C. U. Pittman, Jr., *J. Amer. Chem. Soc.*, **104**, 984 (1982).

36. C. U. Pittman, Jr. and B. Surynarayanan, *J. Amer. Chem. Soc.*, **96**, 7916 (1974).

37. D. W. Macomber and M. D. Rausch, *J. Organometal. Chem.*, **250**, 311 (1983).

38. C. U. Pittman, Jr. et al, *Macromolecules*, **3**, 105, 746 (1970); **4**, 155 (1971).

39. C. U. Pittman, Jr., O. E. Ayers and S. P. McManus, *J. Macromol. Sci. Chem.*, **7**, (8) 1563 (1973).

40. V. V. Korshak, et al, *Vysokomol. Soedin*, **5**, 1284 (1963).

41. M. D. Rausch, C. U. Pittman, Jr. et al in "New Monomers and Polymers," B. M. Culbertson and C. U. Pittman, Jr., eds., Plenum Publishing Corp. New York (1984), pp. 243-267.

42. C. U. Pittman, Jr. in E. I. Becker and M. Tsutsui, eds., "Organometallic Reactions," Vol. 6, Marcel Dekker, Inc., New York (1977) pp. 1-62.

43. C. U. Pittman, Jr. and C. C. Lin, *J. Polym. Sci. Polym. Chem. Ed.*, **17**, 271 (1979).

44. Y. Morita, et al, *Kobunshi Ronbunshu*, **37**, 677 (1980).

45. K. Gonsalves, L. Zhan-ru, R. W. Lenz and M. D. Rausch, *J. Polym. Sci. Polym. Chem. Ed.*, **23**, 1707 (1985).

46. S. L. Sosin, V. V. Korshak, et al, *Vysokomol Soedin*, 22 (9), 699 (1970).

47. C. U. Pittman, Jr. and A. Hirao, *J. Polym. Sci. Polym. Chem. Ed.*, **16**, 1197 (1978); **15**, 1677 (1977).

48. J. T. Wrobleski and D. B. Brown, *Inorg. Chem.*, **18**, 498, 2738 (1979).

49. S. Kanda, *Nippon Kagaku Zasshi*, **83**, 560 (1962).

50. T. J. Marks, et al, *J. Amer. Chem. Soc.*, **106**, 3207 (1984).

51. L. Richards, I. Koufis, C. S. Chan, J. L. Richards and C. Cotter, *Inorg. Chem.*, **105**, L21 (1985).
52. S. Kanda and Y. Saito, *Bull. Chem. Soc. Jpn.*, **30**, 192 (1957).
53. R. H. Bailes and M. Calvin, *J. Amer. Chem. Soc.*, **69**, 1886 (1947).
54. S. Takahashi, H. Morimoto, E. Murata, S. Kataoka, K. Sonogashira and N. Nagihara, *J. Polym. Sci. Polym. Chem. Ed.*, **20**, 565 (1982).
55. S. Takahashi, et al, *J. Chem. Soc. Chem. Commun.*, 3 (1984).
56. S. Takahashi, et al, *Mol. Cryst. Liq. Cryst.*, **82**, 139 (1982).
57. K. Sonogashira, S. Takahashi and Hagihara, *Macromolecules*, **10**, 879 (1977).
58. K. Sonogashira, et al, *J. Organometal. Chem.*, **145**, 101 (1978).
59. S. Takahashi, et al, *Macromolecules*, **11**, 1063 (1978).
60. S. Takahashi and K. Sonogashira, *Kobunshi*, **29** (5), 395 (1980).
61. H. Allcock in "Organometallic Polymers," C. E. Carraher, J. E. Sheats and C. U. Pittman, Jr., eds., Academic Press, Orlando, FL (1978) pp. 283.
62. C. E. Carraher, Jr., W. J. Scott, J. A. Schroeder and D. J. Giron, *J. Macromol. Sci. Chem. Part A*, **15** (4), 625 (1981).
63. Y. Okamoto, S. S. Wang, K. H. Zhu, E. Banks, B. Garetz and E. K. Murphy in Ref. 3, pp. 425.
64. J. E. Sheats, F. Hessel, L. Tsarcuhas, K. G. Podejko, T. Porter, L. B. Kool and R. L. Nolan, Jr. in Ref. 3, pp. 137-147. Also see same authors, *Polym. Mater. Sci. Eng.*, **49** (2), 363 (1983).
65. E. Tsuchida and H. Nishide in "Liposomes as Drug Carriers," G. Gregoriadis, ed., John Wiley and Sons, New York (1988), chapter 40.
66. E. Tsuchida, H. Nishide, M. Ohyanagi and O. Okada, *J. Phys. Chem.*, **92**, 6461 (1988).
67. H. Nishide, M. Ohyanagi, O. Okada and E. Tsuchida, *Macromolecules*, **19**, 494 (1986).
68. E. Tsuchida, et al, *Macromolecules*, **22**, 2103 (1989).
69. H. Nishide, M. Ohyanagi, O. Okada and E. Tsuchida, *Macromolecules*, **21**, 2910 (1988).
70. C. U. Pittman, Jr. and Y. Sasaki, *Chem. Lett. Jpn.*, 383 (1975).
71. C. W. Dirk, M. Bousseau, P. H. Barrett, F. Moraes, F. Wudl and A. J. Heeger, *Macromolecules*, **19**, 266 (1986).
72. K. Krogmann and H. D. Hausen, *Z. Anorg. Allg. Chem.*, **358**, 67 (1968).
73. K. Krogmann, *Angew. Chem. Int. Ed. Engl.*, **8**, 35 (1969).
74. L. S. Miller and A. J. Epstein, eds., *Ann. N. Y. Acad. Sci.*, 313 (1978); L. S. Miller, ed., "Extended Linear Chain Compounds," Vol. 1-3, Plenum Press, New York (1982).
75. G. Kossmehl and M. Rohde, *Macromol. Chem.*, **178**, 715 (1977).
76. W. D. Bascom, R. L. Cottington and T. Y. Ting, *J. Mater. Sci.*, **15**, 2097 (1980).
77. C. J. Norrel, et al, *J. Polym. Sci. Polym. Chem. Ed.*, **12**, 913 (1974).
78. L. Kreja and A. Plewka, *Electrochim Acta*, **25**, 1283 (1980).
79. C. S. Marvel and J. H. Rassweiler, *J. Amer. Chem. Soc.*, **80**, 1197 (1958).
80. J. L. Petersen, C. S. Schramm, D. R. Stojakovic, B. M. Hoffman and T. J. Marks, *J. Amer. Chem. Soc.*, **99**, 286 (1977).
81. B. M. Hoffman and T. J. Marks, *J. Amer. Chem. Soc.*, **99**, 286 (1977); T. J. Marks, et al, *ibid*, **106**, 3207 (1984).
82. T. J. Marks, et al, *ACS Div. Org. Coat. Plast. Chem. Papers*, **41**, 127 (1979); T. J. Marks, et al, *Synth. Met.*, **15**, 115 (1986).
83. T. J. Marks, *Science*, **227**, 881 (1985).
84. T. J. Marks, et al, *J. Amer. Chem. Soc.*, **108**, 7595 (1986).
85. T. J. Marks, et al, *Mol. Cryst. Liq. Cryst.*, **118**, 349 (1985); *ibid*, **93**, 355 (1983).
86. J. G. Gaudiello, G. E. Kellogg, S. M. Tetrick and T. J. Marks, *J. Amer. Chem. Soc.*, **111**, 5259 (1989).
87. C. C. Wamser, et al, *J. Amer. Chem. Soc.*, **111**, 8485 (1989).
88. C. U. Pittman, Jr. in G. Wilkinson, F. G. A. Stone and E. W. Abel, eds., "Comprehensive Organometallic Chemistry," Vol. 8, Pergamon Press, Oxford, UK (1982), Chapter 55, pp. 553-608.
89. F. R. Hartley, "Supported Metal Complexes," D. Reidel Publishers, Dordrecht, The Netherlands (1985), pp. 1-318.
90. Y. Chauvin, D. Commereuc and F. Dawans, *Prog. Polym. Sci.*, **5**, 95 (1977).
91. C. U. Pittman, Jr. in P. Hodge and D. C. Sherrington, eds., "Polymer-Supported Reactions in Organic Synthesis," John Wiley and Sons, Inc., Chichester, UK (1980), pp. 249-291.
92. C. U. Pittman, Jr. and L. R. Smith, *J. Amer. Chem. Soc.*, **97**, 1749 (1975); *ibid*, **97**, 1742 (1975).
93. S. E. Jacobson and C. U. Pittman, Jr., *J. Chem. Soc. Chem. Commun.*, 187 (1975).
94. C. U. Pittman, Jr. and L. R. Smith, *J. Amer. Chem. Soc.*, **97**, 341 (1975).
95. C. U. Pittman, Jr., Y. Kawabata and L. I. Flowers, *J. Chem. Soc. Chem. Commun.*, 473 (1982).
96. G. Consiglio, P. Pino, L. I. Flowers and C. U. Pittman, Jr., *J. Chem. Soc. Chem. Commun.*, 612 (1983).

97. J. K. Stille, *Reactive Polymers*, **10**, 165 (1989).
98. C. E. Carraher and C. U. Pittman, Jr., "Inorganic Polymers" in "Ullmann's Encyclopedia of Industrial Chemistry," Vol. A14, Fifth Edition, VCH Press (1989) pp. 241-262.
99. W. Noll, "Chemistry and Technology of Silicons," Academic Press, New York (1968).
100. W. Lynch, "Handbook of Silicon Rubber Fabrication," Van Nostrand, New York (1978).
101. M. Ranney, "Silicons," Noyes Data Corp., Park Ridge, NJ (1977).
102. B. Arkles, *Chemtech*, **13**, 542 (1983).
103. H. R. Allcock, "Phosphorus-Nitrogen Compounds," Academic Press, New York (1972).
104. H. R. Allcock, *Chemtech*, **5**, 552 (1975).
105. H. R. Allcock, *Angew. Chem. Int. Ed. Engl.*, **16**, 147 (1977).
106. J. C. Vicic and K. A. Reynard, *J. Appl. Polym. Sci.*, **21**, 3185 (1977).
107. D. F. Lohr and J. A. Beckman, *Rubber and Plastics News*, **16** (1982).
108. K. O. Knollmueller, R. N. Scott, H. Kawasnik and J. R. Sieckhaus, *J. Polym. Sci. Polym. Chem. Ed.*, **9**, 1071 (1971).
109. E. N. Peters, et al, *Rubber Chem. Technol.*, **48**, 14 (1979).
110. R. W. Williams, *Pure Appl. Chem.*, **29**, 569 (1972).
111. F. P. Burt, *J. Chem. Soc.*, 1171 (1910).
112. A. Heeger, A. MacDiarmid, et al, *J. Amer. Chem. Soc.*, **97**, 6358 (1975).
113. R. H. Baughman, P. A. Apgar, R. R. Chance, A. G. MacDiarmid and A. F. Garito, *J. Chem. Phys.*, **66**, 401 (1977).
114. A. G. MacDiarmid in R. B. King, ed. "Inorganic Compounds with Unusual Properties," Chapt. 6, ACS Publishers, Washington, D. C. (1976).
115. R. L. Green, et al, *Lect. Notes Phys.*, **65**, 603 (1977).
116. P. W. Bridgeman, *Proc. Am. Acad. Arts Sci.*, **74**, 399 (1941).
117. E. Whalley, *Can. J. Chem.*, **38**, 2105 (1960).
118. Y. Okamoto, Z. Iqbal and R. H. Baughman, *J. Macromol. Sci. Chem.*, **A25**, 799 (1988).
119. J. Tsukamoto and A. Takahashi, *Jpn. J. Appl. Phys.*, **25**, L338 (1986).
120. L. A. Wall and D. W. Brown, *J. Polym. Sci. Part C*, **4**, 1151 (1964).
121. M. Berthelot, *Ann. Chim. Phys.*, **11**, 15 (1936).
122. Y. Asano, *Jpn. Appl. Phys.*, **22**, 1618 (1983).
123. Z. Iqbal, et al, *J. Chem. Phys.*, **85**, 4019 (1986).
124. A. J. Brown and E. Whalley, *Inorg. Chem.*, **7**, 1254 (1968).
125. Y. Okamoto and P. S. Wojciechowski, *J. Chem. Soc. Chem. Commun.*, 386 (1982).
126. H. Kobayashi, A. Kobayashi and Y. Sasaki, *Mol. Cryst. Liq. Cryst.*, **118**, 427 (1985).
127. A. Kobayashi, et al, *Bull. Chem. Soc. Jpn.*, **59**, 3821 (1986).
128. J. P. Wesson and T. C. Williams, *J. Polym. Sci. Polym. Chem. Ed.*, **19**, 65 (1981).
129. X. H. Zhang and R. West, *J. Polym. Sci. Polym. Chem. Ed.*, **22**, 159, 225 (1984).
130. R. West, et al, *Am. Ceram. Soc. Bull.*, **62** (8), 899 (1983).
131. R. West in "Chemistry of Organic Silicon Compounds," S. Patai and Z. Rappoport, eds., John Wiley, New York (1989) p. 1207.
132. R. D. Miller and J. Michl, *Chem. Rev.*, **89** (6), 1359 (1989).
133. R. D. Miller, *Agnew. Chem. Int. Ed. Engl. Adv. Mater.*, **28**, 1733 (1989).
134. F. C. Schilling, F. A. Bovey, A. J. Lovinger and J. M. Zeigler in "Silicon-Based Polymer Science," J. M. Zeigler and F. G. Fearon, eds., Advances in Chemistry Series 224, ACS, Washington, D. C. (1990) pp. 341-378.
135. F. C. Schilling, A. J. Lovinger, J. M. Zeigler, D. D. Davis and F. A. Bovey, *Macromolecules*, **22**, 3055 (1989).
136. F. C. Schilling, et al, *Macromolecules*, **22**, 4645 (1989).
137. R. D. Miller, et al, "Inorganic and Organometallic Polymers," *ACS Symposium Series*, **360** (1988).
138. J. Michl, et al, "Inorganic and Organometallic Polymers," *ACS Symposium Series*, **360** (1988).
139. R. West, L. D. David, P. I. Djurovitch, K. L. Stearley, K. S. Strinivasan and H. Yu, *J. Amer. Chem. Soc.*, **103**, 7352 (1981).
140. P. Trefonas, R. West, R. D. Miller and D. Hofer, *J. Polym. Sci. Polym. Lett. Ed.*, **21**, 823 (1983).
141. R. D. Miller, J. E. Guillet and J. Moore, *Polym. Preprints*, **29** (1), 552 (1988).
142. H. Ban and K. Sukegawa, *J. Polym. Sci. Polym. Chem. Ed.*, **26**, 521 (1988).
143. R. West, A. R. Wolff and D. J. Peterson, *J. Radiat. Curing*, **13**, 35 (1986).
144. A. Wolff and R. West, *Appl. Organomet. Chem.*, **1**, 7 (1987).
145. H. Naarman, et al, *German Patent* DE 3634281 (1988).
146. R. G. Kepler, J. M. Zeigler, L. A. Harrah and S. R. Kurtz, *Phys. Rev. B.*, **35**, 2818 (1987).

147. M. Fujino, *Chem. Phys. Lett.*, **136**, 451 (1987).
148. H. Hiroka (to IBM Corp.), *U. S. Patent* 4,464,460 (1984).
149. D. C. Hofer, R. D. Miller and C. G. Willson, *J. Soc. Photo. Opt. Instrum. Eng.*, **469**, 16 (1984); *ibid*, **469**, 108 (1984).
150. D. C. Hofer, R. D. Miller and C. G. Willson, *Proc. SPIE*, **469**, 16 (1984).
151. R. D. Miller, et al, *J. Polym. Eng. Sci.*, **29** 882 (1989) and *J. Polym. Mater. Sci. Eng.*, **60**, 49 (1989).
152. R. D. Miller, et al, *J. Polym. Eng. Sci.*, **26**, 1129 (1986).
153. D. C. Hofer, et al, *SPIE*, **469**, 108 (1984).
154. C. U. Pittman, Jr. et al, *J. Polym. Sci. Chem. Ed.*, **11**, 2753 (1973); *J. Macromol. Sci. Chem.*, **A7** (8), 1563 (1973).
155. Y. Yasuda, personal communication.
156. P. Trefonas and R. West, *J. Polym. Sci. Polym. Chem. Ed.*, **23**, 2099 (1985).
157. R. D. Miller and R. Sooriyakumaran, *J. Polym. Sci. Polym. Chem. Ed.*, **25**, 111 (1987).
158. P. A. Bianconi and T. W. Weidman, *J. Amer. Chem. Soc.*, **110**, 2342 (1988).
159. P. A. Bianconi, F. S. Schilling, T. W. Weidman, *Macromolecules*, **22**, 1697 (1989).
160. L. A. Hornak, T. W. Weidman and E. W. Kwock, *J. App. Physics*, **67** 2235 (1990).
161. D. C. Bradley, "Polymeric Metal Alkoxides, Organometalloxanes and Organometalloxanosiloxanes" in "Inorganic Polymers," F. G. A. Stone and W. A. G. Graham, eds., Academic Press, New York (1962), chapter 7.
162. Y. Kimura, S. Sugaya, T. Ichimura and I. Taniguchi, *Macromolecules*, **20**, 2329 (1987).
163. Y. Kimura, et al, *Macromol. Chem. Rapid Commun.*, **6**, 247 (1985).
164. E. Ichiki, *Kagaku to Kogyo*, **31**, 706 (1978); *Philos. Trans. R. Soc. London A.*, **No. 294**, 407 (1980).
165. Y. Kimura, M. Furukawa, H. Yamane and T. Kitao, *Macromolecules*, **22**, 79 (1989).
166. I. K. Stavitskii, *Vysokomol. Soedin*, **1**, 1502 (1959).
167. S. Kobayashi, S. Iwata, M. Abe and S. Shoda, *J. Amer. Chem. Soc.*, **112**, 1625 (1990).
168. L. L. Hench and D. R. Ulrich, eds., "Science of Ceramic Chemical Processing," Wiley-Interscience, New York (1986).
169. C. J. Brinker, D. Clark and D. R. Ulrich, eds., "Better Ceramics Through Chemistry II," Materials Res. Soc., Pittsburgh, PA (1986).
170. "Proceedings of the IIIrd Internat. Conference on Ultrastructure Processing of Glasses, Ceramics and Composites," Feb. 24-27 (1987).
171. C. J. Brinker, et al, in "Inorganic and Organometallic Polymers," M. Zeldin, K. J. Wynne and H. R. Allcock, eds., *ACS Symposium Series*, **360** (1988).
172. D. R. Ulrich, *Chemtech*, **18**, 242 (1988).
173. H. K. Schmidt in "Inorganic and Organometallic Polymers," M. Zeldin, K. J. Wynne and H. R. Allcock, eds., *ACS Symposium Series*, **360** (1988).
174. H. Schmidt, H. Scholze, "Springer Proceedings in Physics," Vol. 6, Aerogels, Springer Publ. Co. (1986), p. 49.
175. H. Schmidt, H. Scholze and G. Tunker, *J. Non. Cryst. Solids*, **80**, 557 (1986).
176. G. Philipp and H. Schmidt, *J. Non. Cryst. Solids*, **63**, 261 (1984).
177. H. Huang, R. H. Glaser and G. L. Wilkes, in "Inorganic and Organometallic Polymers," M. Zeldin, K. Wynne and H. Allcock, eds., *ACS Symposium Series*, **360** (1988).
178. G. Tunker, H. Patzelt, H. Schmidt and H. Scholze, *Glastechn Ber.*, **59**, 272 (1986).

METAL-CONTAINING POLYMERS IN THE SOVIET UNION

A.D. Pomogailo

Institute of Chemical Physics, USSR Academy of Sciences
142432 Chernogolovka, Moscow Region, USSR

INTRODUCTION

A combination of polymers with metals was found to show a number
of novel useful properties in the synthesized products. First, they
turned out to be efficient and selective catalysts for various
reactions. Second, metal-containing polymers acquire specific
physico-mechanical and operating features. Finally, the introduction of
metal ions into a polymer structure may give rise to biocidal activity
of these materials.

Immobilization of metal-containing complexes (MX_n) on polymers is
a new trend in chemistry. It was not until 20 years ago that this
branch acquired a reliable foundation for its development. The
fundamental principles of the MX_n-polymer binding mechanism have been
formulated recently and a deep insight into the nature of the MCM
structural arrangement gained. Three approaches are most widely used in
the synthesis of metal-containing polymers: polycondensation of
metal-containing compounds (the synthesized products containing a metal
in the main chain), polymer-analogous conversions and copolymerization
of metal-containing monomers (MCM) (metal ions are in the side polymer
chains). The chemical binding of metals to polymers may be effected via
any bonds such as: $\sigma-$, $n\sqrt{-}$, $\pi-$,
ionic and chelate types, depending on the nature of the reagents
involved.

Inorganic and Metal-Containing Polymeric Materials
Edited by J. Sheats *et al.*, Plenum Press, New York, 1990

To our mind, a considerable contribution to the development of this branch has also been made by the Soviet scientists. V.A. Kargin, V.A. Kabanov, and V.V. Korshak, to name a few, were pioneers in this field. These problems are the focus of various conferences and symposia dealing with the chemistry of high molecular and coordination compounds, catalysis, etc. (cf., e.g. (1-3).

The present review presents a brief analysis of this contribution.

STUDIES OF METAL-ION-POLYMER BINDING IN SOLUTIONS

Polymers as ligands behave differently from their low molecular analogues in diluted solutions. This can be assigned to the presence of a variety of reaction sites as well as changes of conformation and macromolecule shape in solution during a reaction. Different approaches can be chosen to analyze interactions in the MX_n-polymer systems, depending on what is considered to be a central particle - a macroligand or a metal ion (4-8). Chain molecule equilibrium constants (K) can be calculated by analyzing a consecutive addition of a transition metal ion (M) as a central particle to the functional polymer group (L):

$$LM_{i-1} + M \overset{K_i}{\rightleftharpoons} LM_i$$

where LM_i is the polymer chain with i M units added M

$$K_i = \frac{[LM_i]}{[LM_{i-1}][M]}$$

The total constant of complex formation, expressed in terms of the running concentrations L and M is:

$$\bar{K} = \frac{[LM_i]}{[L][M]^i} = \prod_{j=1}^{j=i} K_j$$

The function of formation $(\tilde{n})$ can be determined as:

$$\tilde{n} = \frac{\sum_{i=1}^{i=N} iK_i[M]^i}{1 + \sum_{i=1}^{i=N} K_i[M]^i}$$

Complex formation with linear polymers proceeds at high rates $(K = 10^6 - 10^9 \text{ l.mol}^{-1}.\text{s}^{-1})$.

Chain molecular reactions are distinguished by the formation of a variety of complexes of identical chemical composition but different distribution of the reacted MX_n along the chain. This creates a compositional inhomogeneity in the distribution of the macrocomplexes over the lengths of reacted and unreacted blocks. For various reasons (e.g., "chain effect" and "neighbour" effects) not all of the reaction sites participate in MX_n binding. Therefore, in calculating the constants of formation, real concentrations of the fragments interacting with MX_n should be taken into account. Thus, in the formation of identical macrocomplexes in the Cu(2+)-poly(4-vinylpyridine) (P4VP) system, an effective constant (K_{eff}) is calculated as follows (9):

$$K_{eff} = \frac{\left[Cu(4VP)_4^{2+}\right]}{\left\{\frac{\beta[4\text{-}VP]_o}{4} - \left[Cu(4\text{-}VP)_4^{2+}\right]\right\}\left\{[Cu]_o - \left[Cu(4\text{-}VP)_4^{2+}\right]\right\}}$$

where B is the factor defining the maximum number of the 4-VP units involved in the formation of the Cu(4-VP) complexes.

The attachment of MX_n to the polymer chain is accompanied not only by a direct chemical act but changes in the "local rigidity" of the chain, as well, to increase polymer reactivity (as exemplified by polyoxyethylene (10, 11)):

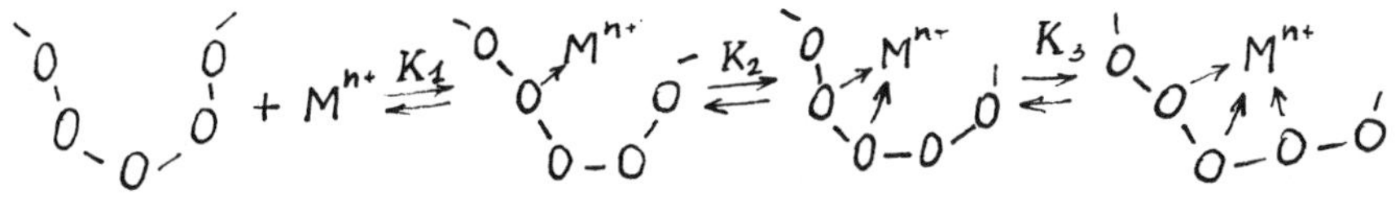

In other words, coordination causes the bending of the polymer chain and renders its conformation favourable for further reactions.

The constant of macrocomplex formation can be expressed as follows:

$$\overline{K} = \prod_{i=1}^{i=N} K_i = \sigma K_i$$

where $\delta = K_i/K_{i-1}$ is the cooperativity parameter showing a factor by which the constant of formation increases after each successive attachment as compared with the previous one. Usually $K_1/K_2 \cdot \delta = 10^{-4}$ - 10^{-8}. Therefore, a given interaction with one chain is completed after all the potential interaction centers have been exhausted.

Thus, it is thermodynamically more favourable for macromolecules to form coordinatively saturated complexes (even if in a lesser amount) than unsaturated ones. This leads to a simultaneous coexistence within the reaction volume of macrocomplexes with a maximum amount of bound L and uncoordinated macroligands, i.e., an "everything or nothing" principle is realized in this case.

MX_n IMMOBILIZATION BY CROSS-LINKED POLYMERS

On analyzing the MX_n reactions involving insoluble polymers, two types of restrictions should be considered, first, diffusional (transfer of the functional groups after the system has crossed a vitrification point) and topological ones (a virtually complete absence of the translational diffusion in these groups grafted onto the polymer backbone). The diffusional process results primarily in the reaction of the surface functional groups; therefore, the penetration of the MX_n into the polymer block is restricted.

The interactions in such systems are evaluated by binding isotherms described by a modified Langmuir equation which can be used for finding the constants of formation:

$$ [M]/[M]_b = 1/\bar{K} + \left(1/f_{max}\right)[M] $$

where M_b is the concentration of the bound metal: f_{max} is the constant characterizing the limiting binding (absorption) of the metal by the functional polymer groups (i.e., $[M]_b / [L]_o$).

This approach was used to estimate the constants of complex formation for Cu(2+) and Ni(2+), the complexes incorporating with polyethylene (PE)-grafted - poly(acrylic) acid (12, 13). For cross-linked polymers K is one-two orders of magnitude higher than that for soluble macroligands. However, the rate of metal complex formation is markedly lower than that of linear polymers and is controlled by the rate of MX_n diffusion towards the functional groups. In this case, the formation of metal complexes is preceded by that of outer-sphere complexes with their subsequent conversion to the inner sphere species and diffusion to the second polymer ligand, etc. Therefore, due to the lower MX_n concentrations at the reaction site, the gross composition of the complexes may differ from that of the soluble polymers. Thus, macrocomplexes with a coordinatively unsaturated transition metal can be formed just in these cases, which is important for many metal-containing polymer applications. Also, the composition and stability of the resultant complexes are affected by carrier porosity. As a rule, transition from a nonporous (gel-like) to a porous structure is accompanied by a decrease in the average constant of complex formation. This is attributed to a closer packing of the chains and a lower mobility of their functional groups.

In most cases, the average constants of formation for such systems are calculated as follows:

$$\overline{K} = [M]_b \Big/ [M][L]^m$$

where m is the number of polymer functional groups entering into the first coordination sphere of a transition metal. All functional groups bound to the MX_n molecule were considered as single n-dentate ligands. This situation is similar to MX_n binding in polymer solutions, consisting of formation of the first bond (second order reaction) and subsequent ones (first order reactions), all the back reactions being of the first order. The constancy of the mean metal center composition was experimentally exemplified by Cu(2+) and Mo(6+) complexes, immobilized on carboxylmethyl cellulose and Mo(6+) on cellulose phosphate (14, 15).

THE INTERACTION OF MX_n WITH GRAFTED POLYMERS

One of the disadvantages of traditional macroligands is a relatively poor utilization of their functional groups. Therefore, extensive studies are being carried out to find new types of macroligands (16). The most widely used methods for these purposes consist in the chemical modification of the carrier polymer surface (i.e., oxidation, chlorination, amination, phosphorylation, etc.). The difficulties which are encountered in doing so are diversity of the reaction routes and uncontrolled transformations both through the layer

depth and types of the functional groups. Therefore, localization of
the reaction centers on the polymer surface is of particular interest
for many purposes. A method of graft polymerization of monomers with
ligand-supported functional groups (17-19) is considered optimal to this
end.

A general scheme for the production of such functionalized
polymers can be presented as follows: The polymer support (polyethylene
(PE), polypropylene (PP), polystyrene (PS), polytetrafluoroethylene
(PTFE), ethylene-propylene copolymers (CEP), etc.) are subjected to
mechanical, chemical, radiation-chemical (-irradiation or with
accelerated electrons) or high-frequency (HF), UV-irradiation treatment
with a subsequent grafting of the appropriate monomers:

$$\left] \xrightarrow{\text{initiation}} \left] \cdot \right] + nCR_1R_2{=}CR_3X \rightarrow \left[\begin{array}{cc} R_1 & R_3 \\ | & | \\ {+}C & {-} C) \\ | & | \\ R_2 & X \end{array} \right]_n$$

Polymer
surface

where R_1, R_2, R_3 - H, CH_3, C_2H_5, $CH_2{=}CH-$, C_6H_5, Hal and
    X - functional groups or heteroatom.

Gas-phase graft polymerization induced by ionizing radiation
(fundamentals are generalized in another review (20)), accelerated
electrons, low-temperature low-pressure crown discharge, etc., is an
efficient technique for polymer modification. Table 1 gives
experimental data on grafting allyl sulfide (AS), allyl alcohol (AAl)
and other monomers, together with radiation-chemical yields (Ggr)
resulting from intensity (I) and irradiation dose (D).

To produce composite materials with improved properties, a high
(tens of percent) degree of grafting is necessary; whereas for the
synthesis of metal-containing polymers a few percent grafting is
sufficient. For example, to obtain a 100-300 $\overset{o}{A}$ thick cover, it is
sufficient to graft 1-2% monomer (Fig. 1).

Table 1. Parameters of Radiation-Induced Vapor-Phase Grafted Polymerization

Polymer support	Grafted monomer	T, °C Support	T, °C Monomer	I, rad/sec	D, Mrad	Amount of grafted monomer, %	G_{gr}
PE	Allyl sulfide (AS)	40	20	80	2.4	0.4	23
PE	Diallyl sulfide (DAS)			220	10.0	1.2	11
PE	Methyl vinylketone (MVK)			63	1.0	3.6	510
PE	Vinyl acetate (VA)			220	6.0	12.5	200
CEP	VA				6.0	5.0	100
PP	VA				5.0	5.0	115
PS	VA				4.0	1.3	35
PE	Methyl methacrylate (MMA)	50	20	60	1.0	13.0	1300
PE	2-vinylpyridine (2-VP)	50	30		5.0	2.0	20
PE	4-vinylpyridine (4-VP)	70	50	80	5.0	2.0	30
PE	Propargyl alcohol	40	20	70	60.0	1.8	8
PE	Dibutyl ether of vinyl-phosphoric acid (DEVPAc)	40	40	80	10.0	5.0	
PE	Allyl alcohol (AAl)	20	20	11	12.0	2.0	25

[a]Liquid phase grafting on PE suspension in 5% dibutyl ether of vinylphosphonic acid in benzene

Thus, a method of functional cover grafting opens a unique possibility to create macroligands of practically any type: Mono- and polyfunctional, including those suitable for initiating numerous polymer-analogous conversions. MX_n binding by these polymers is governed by the same rules as in the case of linear macroligands. The composition, structure, and topology of the immobilized metal-containing

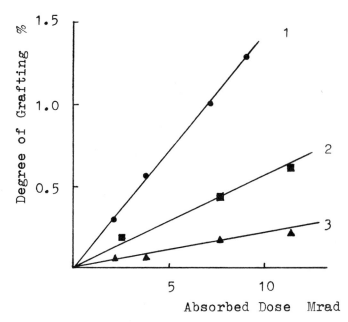

Fig. 1. Amount of grafted polymer PAA1 (1), PAA (2), and PDAA (3) onto PE powder vs. dose: temp. $20^{\circ}C$, dose rate 3 krad/min.

complexes depend on the nature of the reacting groups and reaction conditions (21, 22). It is important that this leads to formation of both isolated metal centers and cluster-type aggregations - regions of a high local MX_n concentration in the polymer surface cover. Distances between the isolated centers are over 22 Å and those between the aggregations from 7 to 9 Å. Characteristics of MX_n binding by polymers are given in Table 2.

Table 2. Immobilization of Complex Catalyst Components on Polymers with Grafted Layer

Polymer carrier	Amount of grafted component, wt%	Immobilized compound	Bound metal content (mole/g).10^4	Functional groups, mole/mole
PE-gr-PDAA	2.5	$TiCl_4$	2.5	0.96
	3.1	VCl_4	3.8	1.19
	3.1	$Ti(C_5H_5)_2Cl_2$	0.1	0.031
	4.0	VCl_4	2.5	0.36
PE-gr-PAA	4.0	$Ti(OC_4H_9)_4$	0.1	0.014
	4.0	$Ti(C_5H_5)_2Cl_2$	0.1	0.014
	3.6	$TiCl_4$	0.6	0.10
PE-gr-PAAl	3.6	VCl_4	3.0	0.50
	3.6	$Ti(OC_4H_9)_4$	0.15	0.025
PS-gr-PAAl	1.8	$VOCl_3$	1.6	0.52
PP-gr-PAAl	1.6	$TiCl_4$	0.8	0.29
	6.5	$VO(OC_2H_5)_3$	0.7	0.08
PE-gr-PAAc	9.1	$VO(OC_2H_5)_3$	1.21	0.10
	0.4	$TiCl_4$	0.15	0.30
PE-gr-PAN	10.0	$TiCl_4$	3.6	0.19
	3.4	VCl_4	0.2	0.03
PE-gr-PVA	11.0	$TiCl_4$	1.7	0.13
PE-gr-PDAS	1.2	VCl_4	0.52	0.25
PE-gr-P4-VP	2.0	$TiCl_4$	0.50	0.26

New types of metal-containing polymers can be effectively produced by construction of polymer supports in the form of gels which, in use, are capable of swelling, insoluble in the reaction medium, but permeable to the MX_n molecules, substrate and solvent (23-25). they are based on ethylene-propylene rubbers and also ternary copolymers of ethylene, propylene and nonconjugated diene, siloxane rubbers with the radically grafted vinylpyridine, acrylic acid (AAc), methylmethacrylate (MMA), etc. Further cross-linking of the rubber base allows the syntheses of three-dimensional networks to avoid the dispersion of these particles in the reaction media. MX_n is bound within these networks. Such polymers were termed "mosaic"; their structure is shown in Fig. 2. It is evident that the regions with metallopolymers are found inside the inert swelling matrix (fragment A).

As will be shown below, polymers with grafted ligands, which bind the MX_n, are used as highly efficient catalysts for various reactions.

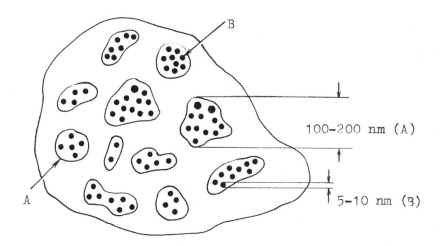

Fig. 2.   The morphological structure of gel-immobilized metal-containing polymers. A is elastomer (polymer dispersing medium), B are grafted copolymers inclusions with metal-containing polymer domains.

PRODUCTION OF A POLYMER COVER ON THE SURFACE OF MINERAL SUBSTANCES

A special type of metal-containing polymers can be produced by MX_n immobilization on oxides such as SiO_2, MgO, Al_2O_3, the surface of which is modified by grafting or absorption of polymers capable of binding the MX_n. These are mixed-type materials used in processes where the framework flexibility which cannot be eliminated, even by polymer cross-linking, should be essentially excluded. At the same time, the fixing of the macromolecule at one end on the inorganic material surface provides a noticeable advantage, i.e., preservation of a homogeneous microreactor polymer coil matched with the possibility of its easy separation from the reaction products, e.g., in catalysis. Among the most attractive techniques one can mention are, grafting vinyl group-containing compounds onto the oxide surface, followed by their further participation into poly- and copolymerization (26):

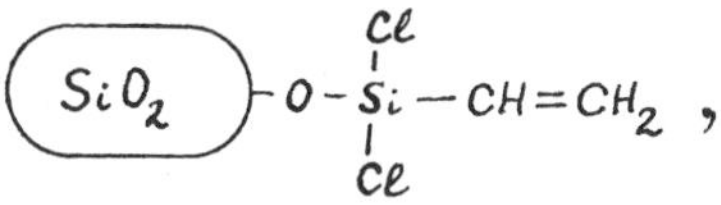

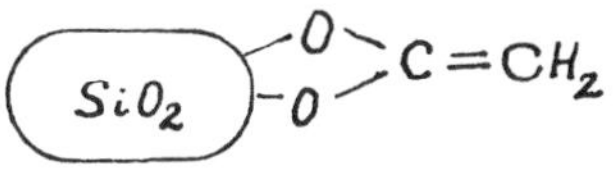

$$\boxed{SiO_2} - O - \underset{\underset{Cl}{|}}{\overset{\overset{Cl}{|}}{Si}} - CH=CH_2 \ , \qquad \boxed{SiO_2} \overset{-O}{\underset{-O}{>}} C=CH_2$$

The kinetics of radical polymerization of MMA absorbed on SiO_2 under γ-radiation by ^{60}Co has been studied in much detail (27). The termination-free post-radiation graft polymerization of vinyl monomers onto inorganic materials, e.g., SiO_2, occurs at the $\geqslant SiO^{\bullet}$ centers (28). The degrees of grafting can vary widely depending on the potential applications of such compositions. Their polymer cover can be subjected to different polymer-analogous conversions. The interaction of MX_n with them is similar to that with polymer-grafted macro-ligands.

PRODUCTION OF METAL-CONTAINING POLYMERS BY MCM POLYMERIZATION AND COPOLYMERIZATION

In the last few years, a new discipline dealing with the polymerization conversions of MCM, i.e., compounds incorporating a multiple bond capable of opening and an metal ion chemically bound to the organic part of the molecule, has emerged at the interface between the organometal and high-molecular chemistry. By the type of the transition metal in MCM, they can be classed into monomers with a covalent, an ionic, and a π- bond (29, 30):

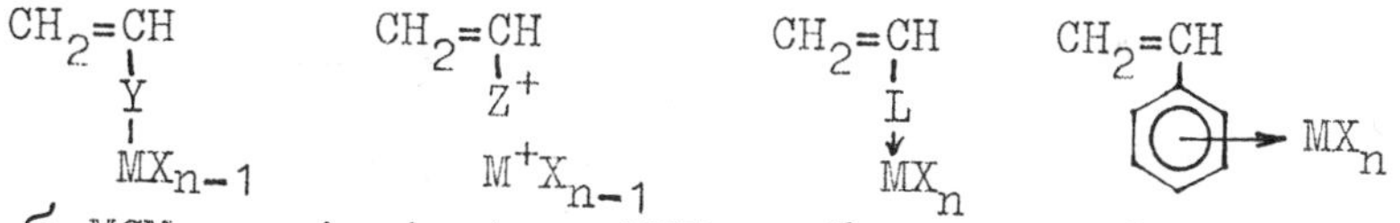

$$\underset{\underset{\underset{6-MCM}{}}{MX_{n-1}}}{\underset{\overset{|}{Y}}{\overset{|}{CH_2}=CH}} \qquad \underset{\underset{\text{ionic-type MCM}}{M^+X_{n-1}}}{\underset{\overset{|}{Z^+}}{\overset{}{CH_2}=CH}} \qquad \underset{\underset{n\delta-MCM}{MX_n}}{\underset{\overset{\downarrow}{L}}{\overset{}{CH_2}=CH}} \qquad \underset{\pi-MCM}{\overset{CH_2=CH}{\bigcirc}\!\!\!\to MX_n}$$

Multiple bonds can be of various types: vinyl (styryl, acrylate), allyl, ethynyl, vinylethynyl, diene, etc. In this case, the metal ion can be bound to one or several polymerizable groups.

Some aspects of MCM polymerization conversions have been generalized in a number of monographs dealing with polymerization of ionizable (31) and complex-bound monomers (32), Pb and Sn organic monomers (33) as well as manufacturing aspects of organoelement monomers and polymers (34).

The interval of MCM reactivity is extremely wide from those polymerizable as early as at the stage of formation to those incapable of polymerization even under relatively rigid conditions. Thus

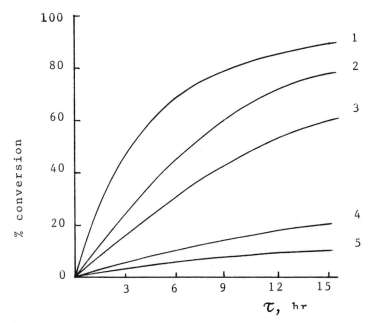

Fig. 3. The change of conversion degree of acrylic acid (1) and
of Co(2+) (2), Ni(2+) Fe(3+) (4), and Cu(2+)
(5) acrylates during the polymerization process
(C_m=0.9 mol/1; G_{AIBN}=2.5.10^{-2} Mol/1; ethanol, 78°C).

transition metal acrylates are readily polymerized via a radical mechanism, their polymerization rate diminishing (35) in the order AAc>Co(2+)>Ni(2+)>Fe(3+)>Cu(2+) (Fig. 3).

The kinetic analysis indicates (35) that polymerization of transition metal acrylates proceeds by the same elementary processes as that of their "metal-free" analogs. However, in many cases the polymerization may be complicated by an individual effect of the transition metal ion. In particular, the kinetic scheme for the Ti(4+)-containing MCM (36) can be described by the following system of equations (see Fig. 4):

$$\text{In} \xrightarrow{k_i} 2R^{\bullet}$$

$$R^{\bullet} + m \underset{}{\overset{K_{p1}}{\rightleftarrows}} (R^{\bullet}m)$$

$$(R^{\bullet}m) + m \xrightarrow{k_p} (R\,m_n)$$

$$R^{\bullet} + m \xrightarrow{k_{p'}} R_1^{\bullet}$$

$$(R^{\bullet}m) + (R^{\bullet}m) \xrightarrow{k_t} P$$

$$(R^{\bullet}m) + R^{\bullet} \xrightarrow{k_{t'}} P$$

$$R^{\bullet} + R^{\bullet} \xrightarrow{k_{t''}} P$$

where K_i is the rate constant for the initiator decomposition (In) into the free radicals ($R^{\bullet}$); K_{p1} is the equilibrium constant for the monomer-radical complexing (m); K_p and $K_{p'}$ are the rate constants for the chain propagation reaction involving the Ti(4+)-coordinated and ordinary radicals respectively; K_t, K_t', K_t'' are the reaction termination rate constants. The rate of the Ti(4+)-monomer polymerization is expressed by the equation:

$$w_p = k_{p'} \left(k_i [In]_0 / k_t k_{p1} \right)^{1/2} [M]_0^{1/2}$$

which correlates well with the experimental data. The unusual, in terms of concentration, kinetic principles of radical polymerization follow from the appearance of the Ti(4+)-coordinated radicals, involved mainly in chain termination:

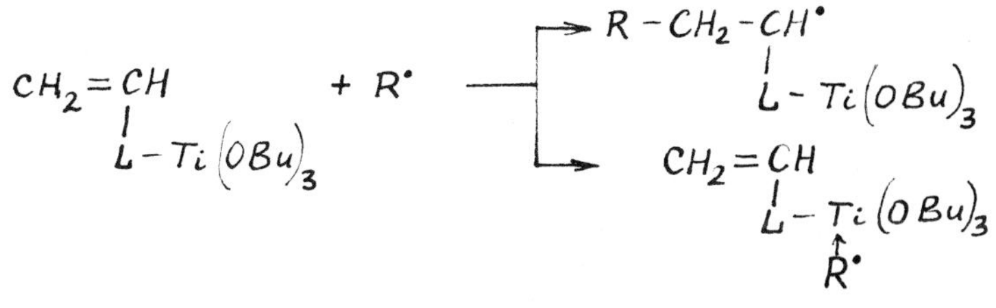

41

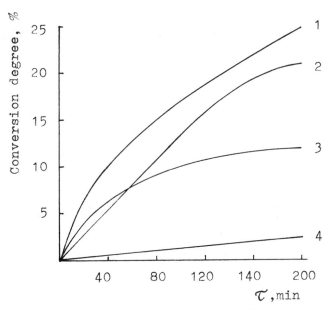

Fig. 4. Time dependence of conversion degree

for $(C_4H_9O)_3TiOR$ monomers, where R:

$CH_2=CH-C≡C$ (1), $-C(Me)_2C=C-CH=CH=CH_2$ (2)

$-CH_2CH_2OCO(Me)C=CH_2$ (3), $-CH_2-C=CH-CH=CH$ (4)

Another feature of MCM polymerization relates to the redox-processes (especially typical of MCM based on Fe(3+) and Cu(2+), to result in the intramolecular chain termination:

$$\sim CH_2-\overset{\cdot}{C}H \quad \longrightarrow \quad \sim CH_2-\overset{+}{\overset{\cdot}{C}H}$$
$$\overset{|}{L-M^{n+}} \qquad \qquad \overset{|}{L-M^{(n-1)+}}$$

Also of interest are stereoregulation in the radical MCM polymerization. The fact is that the metal containing group promotes the situation when in each chain propagation event the center changes its stereochemical configuration to the reverse one, which is expected to give rise to alternation of the units, i.e., a syndiotactic polymer:

$$R^{\cdot} + CH_2=CH \quad \longrightarrow \quad R-CH_2-\overset{\cdot}{C}H \quad \overset{m}{\longrightarrow} \quad R-CH_2-CH-CH_2-\overset{|}{C}H-CH_2-CH\sim$$

syndiotactic attachment

42

As this takes place, the orientation effect of the transition metal coordinated with the functional monomer groups and polymeric radical is shown to be important (32). Thus, even in radical MCM polymerization, we can monitor the elementary acts of metal insertion into the polymer chain and control the special configuration of the synthesized polymer. The major approaches to the solution of the stereoregulation problems consist in the search for finding ways to increase differences in free energies for activating the isotactic (ΔF_i) and syndiotactic (ΔF_s) additions in the chain propagation acts:

$$\delta \, (\Delta F) = \Delta F_i - \Delta F_s$$

Copolymerization of MCM with traditional monomers is the main technique of metal insertion into a polymer chain, and it is more widely used than their homopolymerization. However, copolymerization laws in such systems are difficult to analyze because of their multiparameter dependence of the kinetics and copolymerization characteristics on the process, parameters such as pH, solvent nature and even concentration ratio (30). The metal-containing group in MCM is, as a rule, an electron-donor substituent (scheme Q-e). The copolymerization yields complexes of different comonomers, effecting the polymer composition and structure. In our view, the most remarkable one is copolymerization of transition metal diacrylates with MMA, styrene, etc. (37), as well as vinylpyridine and vinylimidazole MX_n complexes and formation of ternary copolymers of the following composition (38):

$$\left(CH_2 - \underset{\underset{C \diagdown OCH_3}{\overset{\displaystyle O}{|}}}{\overset{\displaystyle CH_3}{\underset{|}{C}}} \right)_n \qquad \left(CH_2 - \underset{\underset{C \diagdown OH}{\overset{\displaystyle O}{|}}}{\overset{\displaystyle CH_3}{\underset{|}{C}}} \right)_m \qquad \left(CH_2 - \underset{\underset{C \diagdown O - MX_{n-1}}{\overset{\displaystyle O}{|}}}{\overset{\displaystyle CH_3}{\underset{|}{C}}} \right)_p$$

n=88.1–88.7%   m=11.12–11.25%   p=0.05–1.94%
                                 M=Co(2+), Cu(2+)

Fundamental studies into the copolymerization of tri-alkyl- or triaryltin methyl methacrylate with styrene, MMA, acrylonitrile (AN) and other compounds have been carried out (39, 41). The main principle is that the MCM are distributed randomly along the chain, the alternating tendency increasing with alkyl chain growth. Experiments have shown (40) the alternation of the units caused by a negative double bond polarization due to the localization of a significant charge inductively shifted to the multiple bond at the oxygen atom:

$$H_2C = C - C \underset{O \cdots}{\overset{O \cdots}{\diagup}} Sn R_3$$
$$\underset{CH_3}{|}$$

Chemical and thermal stability of metal-containing polymers can be
effectively improved by copolymerization of chelate-type MCM (see, e.g.,
(42)) based on N-(2-pyridyl)-methacrylamide or methacryloylacetophenone
(42):

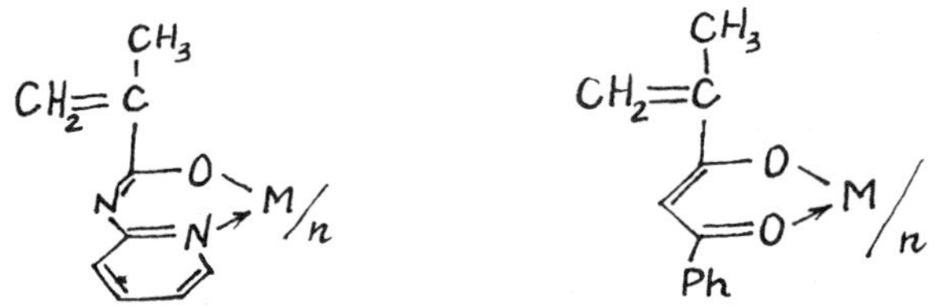

M = Cu(2+), Ni(2+), Co(2+), Mn(2+), Fe(3+), Cr(3+)

n = 2 or 3

Homopolymerization of these MCM is rather complicated but they can be
copolymerized with MMA, styrene and other tra-
ditional monomers. This is also true of metalloporphyrin monomers with
one or two side vinyl groups (43, 44).

Apart from metal-containing polymers with mononuclear metal
groups, the copolymerization yields metal-containing polymers with
polynuclear or heterometallic complexes in the side chain. In
particular, a triosmium vinylpyridine cluster copolymerizes with styrene
(45):

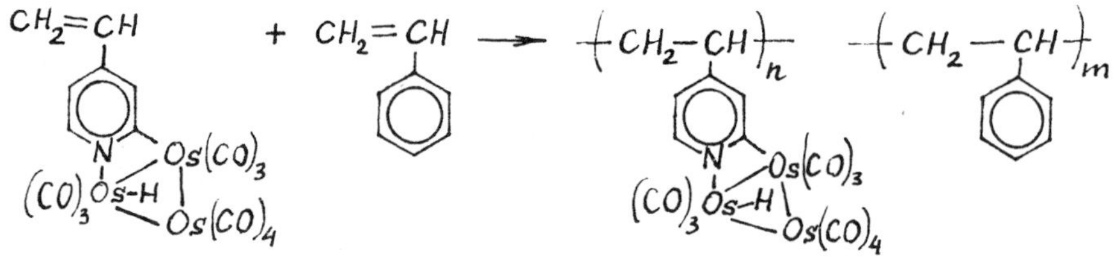

The copolymerization of such monomers is the only way to obtain polymer
metal clusters with a high purity of the product obtained and integrity
of the cluster groups as well as a structural homogeneity of the
resulting metallopolymers.

Copolymerization of the heterometallic MCM yields metallopolymers
with a variety of metals involved in their structure. Thus, using
dicyclopentadienyltitanium methacrylate and Ni or Cu diacrylate, it has
been possible to obtain (46) the following copolymers (Table 3):

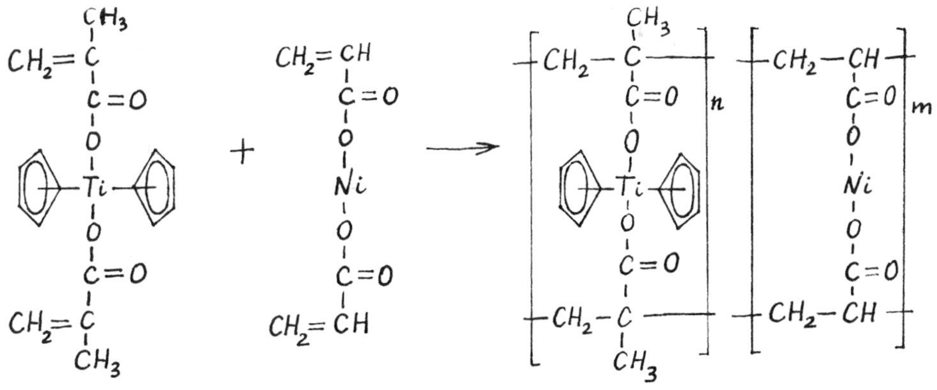

$$CH_2=\!\!\overset{\overset{\displaystyle CH_3}{|}}{C} \quad CH_2=CH \qquad\qquad \left[CH_2-\overset{\overset{\displaystyle CH_3}{|}}{C}\right]_n \left[CH_2-CH\right]_m$$

(Copolymerization scheme of the $Cp_2Ti(OCOC(CH_3)=CH_2)_2$ monomer with the Ni-containing acrylate monomer, yielding the copolymer.)

Table 3. Parameters of Copolymerization of Heterometallic Monomers

M_1	M_2		r_1	r_2	Q_1	Q_2	e_1	e_2
NiR'_2	R''_2	TiR_2	0.95	0.56	0.53	0.56	-2.65	-1.19
CuR'_2	R''_2	TiR_2	1.09	0.89	0.05	0.56	-2.13	-1.90

$R' = -OCOCH=CH_2$; $R'' = -OCOC(CH_3)=CH_2$; $R = -C_5H_5$

It is important that such processes can proceed not only in solution, but in the solid state, as well as, e.g., by mechano-chemical initiation or under high pressure in combination with shear strains.

As in the case of special-type macroligands, metal-containing polymers can be also synthesized by graft MCM polymerization onto the inert polymer support surface. Chemical and radiation chemical (direct or preirradiation) initiation methods are used for this purpose. Thus, triethyltin methacrylate was grafted onto various polymer supports by direct irradiation, the degree of grating being from 3 to 35% (47). Complex-bound monomers were graft-copolymerized similarly (48). Monomers containing Ti(4+) and V(5+) were grafted under vacuo by preirradiation of PE, PP and CEP. The metal content in the products varied from 0.10 to 0.12%.

The best results are obtained by preirradiation of polymer supports in the air with a subsequent decomposition of the hydroperoxides formed; the resulting immobilized radicals initiate graft polymerization of the transition metal acrylates (Fig. 5), MX_n acrylamide complexes and their complexes with allyl-type monomers (49, 50). MCM graft polymerization is essentially characterized by the same kinetic laws as those responsible for their homopolymerization; the synthesis of polymers with 1-3% of transition metal (e.g., acrylates) does not present any experimental difficulty. At the same time, graft polymerization of MCM is a sphere of polymer chemistry which just begins to shape as a science. In the not so distant future, one can expect considerable progress by applying this technique to the production of metal-containing polymers.

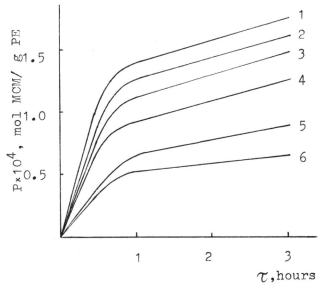

Fig. 5.   Kinetics of Co(2+) (2,3), Cu(2+) (2), Ni(2+) (4), Cr(3+) (5), and Fe(3+) (6) acrylates grafting (ethanol, C_m=0.04 mol/l, 79°C). Conditions: 1,2,5,6 - D=200 kJ/kg; 3,4 - D=100 kJ/kg; 1,3,4 - I=2.8 J/(kg.s); 2,5,6 - I=1.4 J/(kg.s).

Thus, the present review presents but a brief analysis of the current state of studies in the field of polymer chemistry in the USSR, as reviewed by the author.

Metal-containing polymers have already been widely recognized in many applications, sometimes in quite unexpected ones. This fact gives impetus to further MCM investigations in search for new syntheses, examination of their structure, molecular and supramolecular arrangement, etc. In this connection, we shall discuss the key areas of MCM applications.

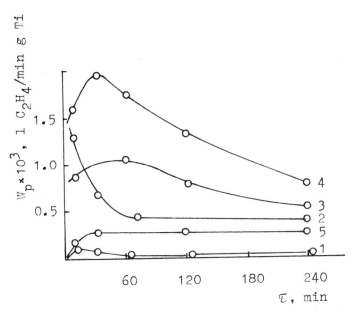

Fig. 6. Ethylene polymerization by heterogenized systems
on the base of $Ti(OC_4H_9)_4 - Al(C_2H_5)_2Cl$. Homogeneous system (1). Polymer carrier: polyethylene
grafted with polydiallylamine (2), polyethylene
grafted with polyallylamine (3), polyethylene
grafted with polyallyl alcohol (4), polyethylene
grafted with polyacrylic acid (5) (toluene, 70°C,
Al/Ti=100, $P_{C_2H_4}$=10 atm).

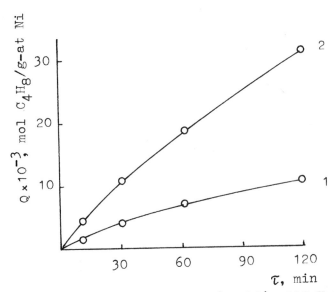

Fig.7. Ethylene dimerization by $Ni(napht)_2 - AlC_2H_5Cl_2$
homogeneous system (1) and heterogenized by polymer
carrier $Ni(OCOCH_3)_2 - PE-gr-PAAc$ one (2) (70°C,
Al/Ni=30, $P_{C_2H_4}$=10 atm).

The insertion of metal into a polymer gives rise to new properties in the product, caused by the metal itself, a kind of metallocomplex. One of them is catalysis of various reactions such as polymerization, hydration, oxidation, hydroformulation, etc. The results of these investigations have been reviewed in a number of monographs (51-54).

In optimal modifications such catalysts inherit the advantages of homogeneous and heterogeneous catalysts in that they are characterized by stability at elevated temperatures, higher activity due to a greater participation of the transition metal in the process, enhanced selectivity of the catalysed reactions and also facilitated separation of the substrate, reaction products and catalyst. The latter, in many cases, can be repeatedly regenerated. For example, $TiCl_3$ surface modification with MMA or its polymer doubles the activity of Ziegler-Natta-type systems in propylene polymerization and enhances their stereospecificity (55). Homo- and copolymerization of Ti(4+)-MCM projects are active components of ethylene polymerization catalysts (56). However, to produce effective heterogenized catalysts, the activity of which is two or three orders of magnitude that of the traditional ones, it is desirable to bring the reaction centers onto the polymer support surface. Metal-containing polymers synthesized by reaction MX_n with the grafted ligands or by graft polymerization of the MCM to the inert polymer support meet these requirements (57). Thus, under optimal conditions, the heterogenation of homogeneous $Ti(OC_4H_9)$-based systems increases the catalytic activity in ethylene polymerization by 100 to 150 times (PE yield being as high as 500-1000 kg/g Ti) (Fig. 6).

A remarkably high activity in ethylene polymerization is displayed by titanium-magnesium catalysts immobilized on polymer supports. Gel-immobilized catalysts for ethylene polymerization and dimerization (59) are characterized by a high stability. If a homogeneous $Ni(naphthenate)_2$ $EtAlCl_2$ system is practically completely deactivated after 2 hr of ethylene dimerization, its heterogenized Ni(2+) analog immobilized on a PE-grafted-poly(acrylic acid) is more active and stable in the dimerization conditions (60) for more than 40 hr (Fig. 7).

In general, the catalytic properties of metal-containing polymers based on Ti(4+), V(5+), V(4+), V(3+), and Ni(2+) in ethylene and propylene conversions depend on type of the functional groups which bind the MX_n to the polymer and on the transition metal distribution on the polymer i.e., isolated ions or cluster formations of a given size. However, polymer itself, as a specific macroligand, also takes part in the catalysis. The experiments have shown that the active centers are localized on the boundaries of the small clusters and are stabilized by their electron system.

Among other conversions of α-olefins, we shall mention their structural isomerization affected by the metal-containing polymers, i.e., oligomerization processes.

This process consists in the copolymerization of ethylene with -butene in situ to obtain linear polethylene of low density (LPELD). It can be essentially described as follows (61): the initial monomer m (ethylene, in this case) on the active center M_1, e.g., Ni(2+). by way of di-, oligomerization or some other reaction converts to an m_2 monomer (dimerization of ethylene into butene in this case) and on the active center M_2 such as Ti(4+), V(4+) etc., the m_1 - m_2 copolymerization proceeds as follows:

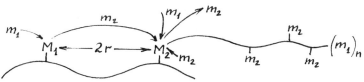

The dimensions of the microcell containing active centers of two types and immobilized on the polymer are about half the average distance between the centers of the bifunctional catalyst pairs. The M_1 - M_2 distance (r) can be controlled at the stage of synthesis of these heterometallic polymers. For example, in the immobilized azomethine heterometallic complexes, the average Ti(4+) - Ni(2+) distance is 2.2 A; they are synthesized by a consecutive lamination technique as shown below (62):

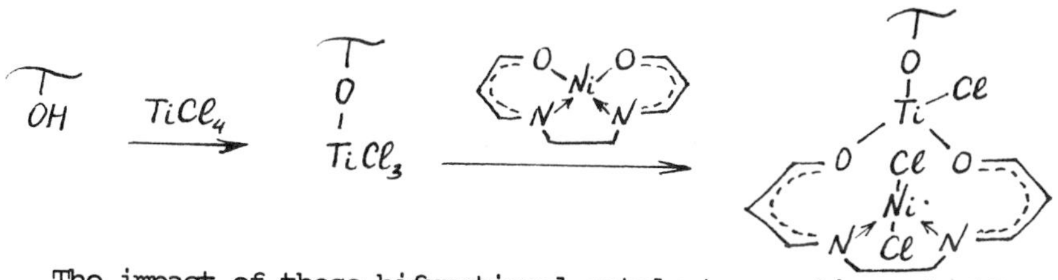

The impact of these bifunctional catalysts exerting a highly accurate synthesizing effect on the active centers can be compared with the operation of a well organized conveyer line where each worker performs his or her duty skill with a maximum speed (63). In our case, nickel produces a dimerizing and Ti a copolymerizing effect, and their mutual position provides for a rapid transfer of the stock from one active center to another. It should be noted that upon insertion of thirty C_2H_5 branchings per 1000 carbon atoms into PE, the properties of the resulting LPELD approximate those of PELD.

Metal-containing polymers are also applied to the catalysis of other processes such as polymerization and copolymerization of butadiene and isoprene (see, e.g., ref. (64)), copolymerization of diene and olefin monomers and polymerization conversions of acetylene-type monomers (65). Such investigations are likely to be oriented both theoretically and practically. Metallopolymers can be used as advantage in some other catalytic processes (54), among them hydrogenation of unsaturated compounds, oxidative conversions of hydrocarbons, in hydroformylation, polycondensation and other processes, etc. (Table 4). Catalysis of almost all reactions obeys the same or similar principles as in the case of polymerization. The position of metallopolymers in catalysis and their links with traditional catalysts can be illustrated as follows:

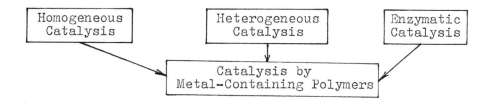

It may be concluded that catalysis by metal-containing polymers combines the main features of homogeneous, heterogeneous and enzyme catalysts.

BIOLOGICAL ACTIVITY OF METAL-CONTAINING POLYMERS

The biocidal properties of most metal-containing polymers have not been studied in detail. The biological activity of organotin polymers and copolymers to various fungi and microorganisms has been found only recently.

Polymerization of toxin-containing monomers, in particular, organotin species, is a promising way for the development of antifouling coatings for vessels and hydro-engineering structures against microorganisms, in which the pesticide groups are chemically bound to the polymer (66). They are attractive for the following reasons: a prolonged coating effect or ease with which polymer or lacquer solutions can be applied, an improved adhesion to a metal, the possibility of

Table 4. Metal-Containing Polymers in Catalysis.

Metal ion-containing polymers	Catalysed reaction
Ni(2+)	Dimerization of ethylene and isomerization of butylenes
Co(2+)	Butadiene polymerization; hydrocarbon oxidation; olefin hydroformylation
Pd(2+)	Hydrogenization of unsaturated and aromatic nitro compounds; oligomerization of phenylacetylene
Mo(6+), Mo(5+), Mo(4+)	Ethylene polymerization with Ziegler-Natta catalysts

using them as constructional materials, etc. Leaching a toxin out of a polymer under the action of sea water occurs at extremely low rates; viz., 0.004 mg/cm^2 a day so that it can be effective for 3-5 years. Biological protection can be also provided by grafting triethyltin methacrylate onto the surface. A satisfactory biological stability is attained by using as small as 0.5 - 1.0% grafted polymer (47).

Metal-containing polymers including those with Ti, Hg, Pb and
other heavy metals can possess a high biological activity. Thus, a
fungicide PMMA with improved thermostability has been obtained by MMA
copolymerization with 0.1 - 0.4% mercury methacrylate (67).

Polymers based on sodium sulfohexylmethacrylate are effective as
blood anticoagulants (68), iron-containing polyacrylic acid (its 1%
aqueous solution) is biologically active and capable of arresting
bleeding and is used in medicine as "feracryl" (69)). Potential uses of
MCM have not been fully explored yet.

COMPOSITE MATERIALS BASED ON METAL-CONTAINING POLYMERS

In principle, metal-containing polymers represent "ready-for-use"
composite materials. They possess a full spectrum of polymeric
properties enriched by the presence of transition or non-transition
metals in the macromolecules. The latter are potentially capable of
giving ionic and coordinative cross-links. Their metals provide for
electron transfers induced both by an electric field and high-energy
radiations. In addition, they display cohesion and adhesion
interactions, etc.

The improved features of metallopolymeric materials are mainly
caused by a cross-linking ability of a metal. It is realized through an
additional interchain-coordinative interaction of metal ions with
heteroatoms, as well as due to multiple bond attachments or condensation
processes (70-73).

One variant of cross-linking consists in a finalizing
polymerization of the residual double bonds (inner cross-linking agents)
in acrylate poly- and copolymers:

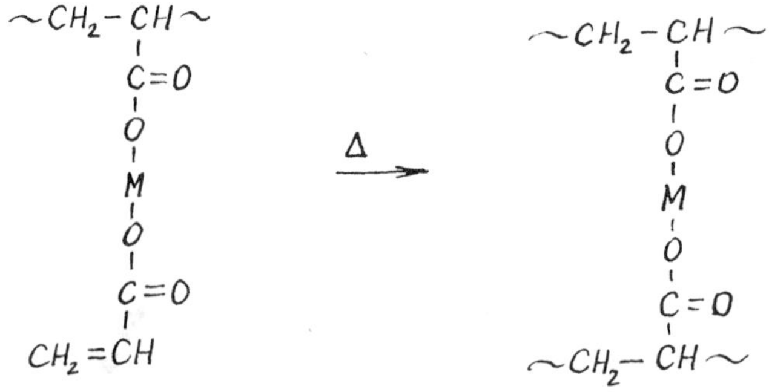

$$
\begin{array}{ccc}
\sim CH_2 - CH \sim & & \sim CH_2 - CH \sim \\
| & & | \\
C = O & & C = O \\
| & & | \\
O & \xrightarrow{\Delta} & O \\
| & & | \\
M & & M \\
| & & | \\
O & & O \\
| & & | \\
C = O & & C = O \\
| & & | \\
CH_2 = CH & & \sim CH_2 - CH \sim
\end{array}
$$

Thus, in a Co(2+) polyacrylate heated for 0.5 hr at 100, 150, 200 and 250°C, a considerable decrease in the intensity of the IR-spectra for the C=C (1640 cm^{-1}) bands is observed, and at 250°C this band practically disappears (Fig. 8).

Cross-linking leads to a markedly higher thermal stability of metal-containing (co)polymers. The destruction temperature (TD) of metal-containing macromolecules is often from 300 to 400°C and over. For example, polyoxyethyl butoxytitanium methacrylate withstands destruction up to 400°C, but Co(2+), Ni(2+) and Zn(2+) polyacrylates decompose around this temperature.

The thermo-mechanical curves have typical strain-temperature dependences (Fig. 9): transition from the glass-like to the viscous-flow state is characterized by the appearance of a high-elastic plateau; whereas its "metal-free" analog, i.e., polyacrylic acid, does not melt and is not subjected to high-elastic deformation. Thus, metal-containing polymers can combine the advantages of both thermoplastic and thermosetting materials. The former are ductile, can be readily processed but their process point is far higher than their operating temperature.

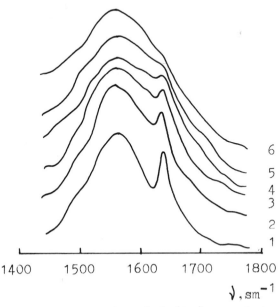

Fig. 8. IR-spectra fragments of Co(2+) acrylate (1), Co(2+) poly(acrylates) (2), and polymers heated in air for temperatures: 100 (3), 150 (4), 200 (5), and 250°C (6) (heating time - 0.5 hr).

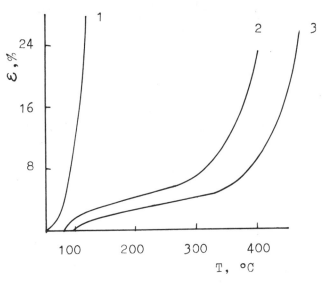

Fig. 9. Thermo-mechanical curves of polyacrylic acid (1),
Ni(2+) poly(acrylate) (2), and Co(2+) poly-
(acrylate) (3).

Table 5. Thermomechanical Properties of Ti(4+)-Containing Polymers and their "Metal-Free" Analogs

Polymer	T_g, °C	T_{fl}, °C	$T_{dec.}$, °C	E_t, %
Polypropargyl alcohol	63	63	–	–
Polypropargyloxytribu-toxytitanium	55/156	–	330	12
Polydimethylvinylethy-nylcarbinol	177	177	–	–
Polydimethylvinylethy-nylcarbinoloxytribu-toxytitanium	62/160	–	340	130
Polyoxyethyl methacry-late	54	157	–	5
Polyoxymethyl metha-cryloxytributoxyti-tanium	64/120	–	–	31

From this viewpoint, elastoplastics are preferred but they are fragile. MCM cross-linking will probably produce network polymers with improved durability and ductility at a high softening point, which can be exemplified by Ti(4+)-containing polymers (73) (Table 5). Thermomechanical properties of metal-containing copolymers depend on metal content therein, and the thermostabilization effect can be also attained by a simple mixing of metal-containing polymers with traditional thermoplastics.

The ability of MCM to be cross-linked by way of radiation makes it possible to use them as photosensitive materials (70, 74) as well as "protectors" against a variety of ionizations. In this case, the protective properties are coupled with optical transparency of the materials such as those based on Pb(2+) acrylates.

Another optical effect resulting from metal introduction into a polymer, shows up in the triple copolymers of MMA-MAA, methacrylates of Ba, Pb, Sr, and other metals (38, 73). In these copolymers, the total value of light transmission is greater than that in "metal-free" ones. The copolymers of alkylvinylsulfoxide (Cr(3+) complexes and MMA are distinguished by thermochromism (75) in that their colour changes from emerald green (at 20-30°) to violet (at 40°C). This can be attributed to the temperature-dependent changes in complex configuration: as the temperature decreases, the matrix rigidity increases to give a distorted configuration, and as it rises, the effect of the matrix diminishes and the complex acquires an octahedral form.

The presence of metal in a polymer may be responsible for its electrical conductivity, particularly on doping, piezoelectric activity and polyelectrolytic properties. Metal-containing polymers can be used to advantage as ingredients in various polymeric compositions for finishing mineral fillers, structural polymer dyeing and stabilization, etc. Thus, the addition of as little as 0.5 wt% and 2 wt% Ni polyacrylate to PE (72) increases the adhesion of such a composite to steel-3 by ca. 7 and 13 times, respectively (Fig. 10). This fact can be explained by a dual composition of metal-containing polymers containing metals and polymers. In such composition, the polymers act as an intermediate layer between a polymer and a metal.

Even a brief study illustrates a wide range of the potential uses of metallopolymers (see Fig. 11).

New applications of these unique polymers could be expected in the future. These can be facilitated by the synthesis of new metal-containing polymers, a detailed study of their structural arrangement at all levels, viz., molecular (with due regard to the chemical composition of the metal-containing polymer units, metal distribution within the chain and stereo-chemical arrangement of the chains themselves), supramolecular (intermolecular interactions, degree of the ordering and packing of the macromolecules) and also topological

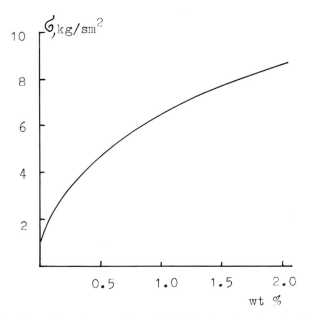

Fig. 10. The increasing of HDPE adhesion to steel-3 under action of Ni(2+) polyacrylates.

(in what way the polymer structural units are bound together). Special attention should be paid to the type of the structural arrangement of the metals themselves within the polymers, processes of cluster formation, antiferromagnetic exchange between the paramagnetic chain-bound ions, degree of cluster separation, the nature of interchain interactions, etc.

To elucidate all these problems, a close colloboration between the researchers with various professional interests at both the national and international levels would be required.

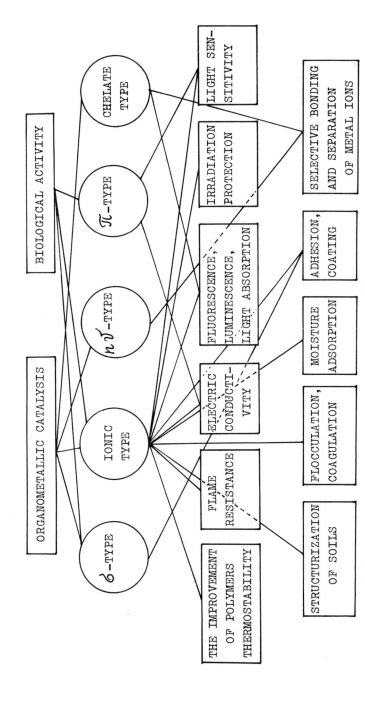

Fig. 11. The main ranges of diverse MCP applications.

REFERENCES

1. Catalysts Containing Supported Complexes, (Yu.I. Ermakov, Ed.), Novosibirsk, Institute of Catalysis, SO AN
   SSSR, 1977.
2. Catalysts Containing Supported Complexes, Proc. Symp.,
   Novosibirsk, Institute of Catalysis, SO AN SSSR,
   parts 1-3, 1980.
3. Yu.I. Ermakov and V.A.Likholobov (Eds.), Proc. 5th
   Intern. Symp. on Relations between Homogeneous and Heterogeneous Catalysts, Novosibirsk 1986, VNU, Utrecht,
   1986.
4. A.S.Polinskii, V.S.Pshezhetsky and V.A.Kabanov, Vysokomol. Soed., 25, 72 (1983); 27, 1014 (1985); Dokl. AN
   SSSR, 256, 129 (1981).
5. V.N.Tolmachev, V.Y.Gnidenko,Z.A.Lugovaya,XV Intern.
   Conf. on Coordination Chemistry, Abstracts, Nauka, Mos-
   kow, 119 (1973).
6. N.A.Plate, A.D.Litmanovich and O.V.Noa, Macromolecular
   Reactions, Khimiya, Moscow, 1977.
7. A.A.Popov, E.F.Vainstein and S.G.Entelis, J.Macromol.
   Sci., A, 11, 859 (1977).
8. S.L.Davydova and N.A.Plate, Coord. Chem. Rev., 16, 195
   (1985).
9. Yu.I.Kirsch, V.Ya.Kovner and A.I.Kokorin, Europ. Polym.
   J., 10, 671 (1974).
10. G.N.Arkhipovich, S.A.Dubrovskii, K.S.Kazanskii and A.N.
    Shupik, Vysokomol. Soed., 23, 1653 (1981).
11. G.N.Arkhipovich and A.N.Shupik, Z. Fiz. Khim., 59,
    1725 (1985).
12. A.D.Pomogailo and N.D.Golubeva, Kinet. Katal., 26, 947,
    (1985).
13. N.M.Bravaya, A.D.Pomogailo and E.F.Vainstein, Kinet.
    Katal., 25, 1140 (1984).
14. A.P.Filippov, Theoret. Experim. Khim., 19, 463 (1983).
15. A.P.Filippov, O.A.Polishchuk and M.A.Pechkovskaya,
    J. Neorg. Khim., 27, 353 (1982).
16. A.D.Pomogailo, Immobilized Polymer Metallocomplex Ca-
    talysts, Nauka, Moscow, 1988.
17. A.D.Pomogailo, D.A.Kritskaya and A.P.Lisitskaya, Dokl.
    AN SSSR, 232, 391 (1977).
18. D.A.Kritskaya, A.D.Pomogailo, A.N.Ponomarev and F.S.
    Dyachkovskii, Vysokomol. Soed., 21, 1107 (1979).
19. D.A.Kritskaya, A.D.Pomogailo, A.N.Ponomarev and F.S.
    Dyachkovskii, J. Appl. Polym. Sci., 25, 349 (1980);
    J. Polym. Sci., Polym. Symp., N68, 23 (1980).
20. B.L.Tsetlin, A.V.Vlasov and I.Yu.Babkin in: Radiation
    Chemistry, Nauka, Moscow, 1973.
21. F.S.Dyachkovskii, A.D.Pomogailo and N.M.Bravaya, J.
    Polym. Sci., Polym. Chem. Ed., 18, 2615 (1980).
22. N.M.Bravaya, A.D.Pomogailo and F.S.Dyachkovskii, Kinet.
    Katal., 24, 403 (1983).
23. V.A.Kabanov and V.I.Smetanyuk, Advances in Science and
    Technology, Kinet. Katal., VINITI, Moscow, 13, 213 (1984).
24. V.A.Kabanov, V.I.Smetanyuk and V.G.Popov, Dokl. Akad.
    Nauk SSSR, 225, 1377 (1975).
25. V.A.Kabanov and V.I.Smetanyuk, Makromol. Chem. Suppl.,
    172, 121 (1981).
26. V.V.Korshak, L.B.Zubakova, N.V.Kachurina and O.B.Bala-
    shova, Vysokomol. Soed., 21, 1132 (1979).
27. D.A.Kritskaya and A.N.Ponomarev, Vysokomol. Soed., 17B,
    67 (1975).

28. A.V.Olenin, A.L.Khristyuk and V.B.Golubev, Vysokomol. So-
    ed., 25, 423 (1983).
29. A.D.Pomogailo and V.S.Savostyanov, J. Macromol. Sci.-
    Rev. Macromol. Chem. Phys., 25, 375 (1985).
30. A.D.Pomogailo and V.S.Savostyanov, Metal-Containing Mo-
    nomers and the Associated Polymers, Khimiya, Moscow, 1988.
31. V.A.Kabanov and D.A.Topchiev, Polymerization of Ioniza-
    ble Monomers, Nauka, Moscow, 1975.
32. V.A.Kabanov, B.P.Zubov and Yu.D.Semchikov, Complex-Radi-
    cal Polymerization, Khimiya, Moscow, 1987.
33. D.A.Kochkin and I.N.Azerbaev, Organotin and -Lead Mono-
    mers and Polymers, Nauka, Alma-Ata, 1968.
34. L.M.Khananashvili and K.A.Andrianov, Organoelement Mono-
    mers and Polymers Processes, Khimiya, Moscow, 1983.
35. G.I.Dzardimaliyeva, A.D.Pomogailo, S.P.Davtyan and V.I.
    Ponomarev, Izv. Akad. Nauk SSSR, ser. Khim., N7, 1531
    (1988).
36. G.I.Dzardimalieva, A.O.Tonoyan, A.D.Pomogailo and S.P.
    Davtyan, Izv. Akad. Nauk SSSR, ser. Khim., N8, 1744
    (1987).
37. G.I.Dzardimaliyeva, Cand. Sci. Disser., Moscow, Institu-
    te of Chemical Physics USSR Academy of Sciences, 1987.
38. V.N.Serova, E.A.Gonyukh, L.Kh.Khazryatova and E.V.Kuz-
    netsov, in: Chemistry and Processes for Organoelement
    Compounds, Kazan University, 58 (1984).
39. Z.N.Rzayev and L.V.Bryksina, Vysokomol. Soed., 16A,
    1691 (1974).
40. K.P.Zabotin and L.V.Malysheva, Proceed. on Chemistry
    and Chemical Technology, Gorkii 1 (32) 47 (49) (1973).
41. Z.N.Rzayev, K.I.Gurbanov and S.G.Mamedov, Vysokomol.
    Soed., 26, 736 (1984).
42. I.E.Ufland, I.A.Ilchenko and A.D.Pomogailo, Izv. Akad.
    Nauk SSSR, ser. Khim., N2, 451 (1990).
43. G.P.Potapov and M.I.Aliyeva, Izv. VUZ'ov, Khim. Khim.
    Techn., 26, 1122 (1983).
44. A.B.Solovyova, A.I.Samokhvalov, T.S.Lebedeva, Dokl.
    Akad. Nauk SSSR, 290, 1383 (1986).
45. N.M.Bravaya, V.A.Maksakov, A.D.Pomogailo, Izv. Akad.
    Nauk SSSR, ser. Khim. (in press).
46. G.I.Dzardimaliyeva, V.A.Zhorin and A.D.Pomogailo, Dokl.
    Akad. Nauk SSSR, 287, 654 (1986).
47. V.Ya.Kabanov, I.A.Voronkov, D.A.Kochkin and V.I.Spit-
    sin, Dokl. Akad. Nauk SSSR, 200, 628 (1971).
48. Sh.A.Kurbanov, B.Sh.Khakimdzhanov, U.N.Musayev and R.S.
    Tillayev, Proc. Tashkent State University, N498, 26
    (1975).
49. V.S.Savostyanov, A.D.Pomogailo, D.A.Kritskaya and A.N.
    Ponomarev, Izv. Akad. Nauk SSSR, ser. Khim., N1, 42;
    N2, 353 (1985).
50. V.S.Savostyanov, A.D.Pomogailo, D.A.Kritskaya and A.N.
    Ponomarev, J. Polym. Sci., Polym. Chem. Ed., 27,
    1935 (1989).
51. G.V.Lisichkin, A.Ya.Yuffa, Heterogeneous Metallocomplex
    Catalysts, Khimia, Moscow, 1981.
52. E.A.Bekturov, L.A.Bimendina and E.A.Kudaybergenov, Poly-
    meric Complexes and Catalysts, Nauka, Alma-Ata, 1982.
53. V.D.Kopylova, A.N.Astanina, Ion-Exchange Complexes in
    Catalysis, Khimia, Moscow, 1987.
54. A.D.Pomogailo, Catalysis by Immobilized Complexes,
    Nauka, Moscow (in press).

55. L.M.Dubinina, V.V.Amerik, V.I.Kleiner and B.A.Krentsel. in: Homo- and Copolymerization of α-Olefins on Complex Catalysts, Nauka, Moscow, 1983.

56. G.I.Dzardimaliyeva, B.S.Selenova and A.D.Pomogailo, Proc. Vth Intern. Symp. on Relations between Homogeneous and Heterogeneous Catalysis, Novosibirsk, 1986.

57. A.D.Pomogailo and F.S.Dyachkovskii, Acta Polymerica, 35, 41 (1984).

58. A.M.Bochkin, A.D.Pomogailo and F.S.Dyachkovskii, Reactive Polymers, 9, 99 (1988).

59. V.A.Kabanov and G.A.Grishin, Kinet. Katal., 26, 1427 (1985).

60. A.D.Pomogailo, F.A.Khrisostomov and F.S.Dyachkovskii, Kinet. Katal., 26, 1104 (1985).

61. N.S.Enikolopyan, L.N.Raspopov, A.D.Pomogailo, N.D.Golubeva and A.G.Starikov, Vysokomol. Soed., 31A, 2624 (1989).

62. I.E.Ufland, A.D.Pomogailo, N.D.Golubeva and A.G.Starikov, Kinet. Katal., 29, 885 (1988).

63. V.A.Likholobov et al., J. D.I.Mendeleyev All-Union Chem., Soc., 34, 340 (1989).

64. B.A.Dolgoplosk, E.I.Tinyakova, Organometal Catalysis in Polymerization Processes, Nauka, Moscow, 1982.

65. Zh.S.Kiyashkina, A.D.Pomogailo and A.I.Kuzaev, J. Polym. Sci., Polym Symp., N68, 13 (1980).

66. V.F.Mishchenko et al., in: Growing and Biocorrosion Processes in Aqueous Media, Nauka, Moscow, 1981.

67. S.F.Zhiltsov and V.N.Kashaeva, in: PhisicoChemical Foundations of Polymer Synthesis and Processing, Gorkii, 1984.

68. V.F.Kurenkov, A.K.Vagarova and V.A.Myagchenkov, Europ. Polym. J., 18, 763 (1982).

69. V.Z.Annenkova, A.T.Platonova, G.M.Kononchuk et al., Khim., Farm. Zh., N3, 66 (1982).

70. A.V.Botsman and L.V.Marchenko, Ukrain. Khim. Zh., 52, 952 (1986).

71. Zh.O.Rustamova, M.G.Guseynov, G.A.Ibragimova and Z.N. Guseynova Plastmassy, N8, 52 (1983).

72. G.I.Dzardimaliyeva, B.S.Selenova, T.I.Ponomareva et al., Complex Organometal Catalysts of Olefin Polymerization, Chernogolovka, Institute of Chem. Phys, USSR Academy of Sciences, N10, 55 (1986).

73. B.I.Utey, B.M.Zuev, L.Kh.Akhmetgalyeva, E.B.Kuznetsov, Proc. Chem.-Technol. Institute, 56, 130 (1975).

74. S.G.Mamedova, Z.M.Rzaev and N.Sh.Rasulov, Vysokomol. Soed., 28B, 111 (1986).

75. V.A.Nikonov, G.V.Leplyanin, Yu.I.Mironov and A.R.Derzhinsky, Izv. Akad. Nauk SSSR, ser. Khim.,N11, 2647 (1984).

POLYMER-SUPPORTED CATALYSTS IN CONJUGATED DIENE

POLYMERIZATION

Guang-Quian Yu and Yu-Liang Li

Changchun Institute of Applied Chemistry
Academia Sinica
Changchun 130022, Jilin, China

INTRODUCTION

As is well known, the efficiencies of Ziegler-Natta catalysts are influenced by the nature and valence state of the transition metal, the types of ligands attached to the metal, and the catalyst morphology. Varying the ligands linked to the metal can change the distribution of the electron density of the metal (i.e., the bond polarity) in the active species, so that the catalyst activity is affected.

Polymer-supported catalysts have been developed in order to better utilize their potential catalytic activity. With this purpose in mind, our laboratory has paid a great deal of attention to polymer-supported metal complexes, especially to those with 4f electrons, such as La, Nd, Pr, and Eu. Some polymer-supported catalysts with high catalytic activity have been developed (1-9). In this paper we will discuss in detail our recent results on polymer-supported catalysts for conjugated diene polymerization; other work in this area will also be mentioned.

This article will be divided into two main sections. The first section will deal with the preparation and characterization of polymer-bound catalysts, focusing on the methods used to anchor metals to the polymer. We will also describe a classification that is based on the types of bonds used to link the metal to the carrier. Several spectroscopic techniques have been used to characterize the polymer-attached species. Infrared spectroscopy was the most useful method, especially when the attached species incorporated carboxylic group ligands. The second section is related to aspects of diene polymerization catalyzed by polymer-supported catalysts. Catalyst activities and factors that influence these activities will be discussed.

Inorganic and Metal-Containing Polymeric Materials
Edited by J. Sheats *et al.*, Plenum Press, New York, 1990

Since the ligands to be considered are often complicated organic macromolecules, it has become customary to employ appropriate abbreviations in order to avoid problems with nomenclature. The number of abbreviations used in this paper has been kept as small as possible; they are listed below.

Abbreviations

AA	acetate
Bd	butadiene
Bpy	bipyridyl, 2,2'-dipyridyl
DMSO	dimethylsulfoxide
EAAC	ethylene-acrylic acid copolymer
Et	ethyl
hep	heptylate
i-Bu	isobutyl
i-C_4H_9	isobutyl
naph	naphthenate
PAN	polyacrylonitrile
PE-COO	carboxylated polyethylene
PE-gr-PAA	polyethylene-graft-polyacrylic acid
PE-gr-PAAA	polyethylene-graft-N-acetylacrylamide
PE-gr-PAAM	polyethylene-graft-acrylamide
PE-gr-PAN	polyethylene-graft-acrylonitrile
PE-gr-PAVK	polyethylene-graft-poly(aminovinyl ketone)
PE-gr-PMVK	polyethylene-graft-methyl vinyl ketone
PE-gr-PSAA	polyethylene-graft-N-salicyl acrylamide
PE-gr-PVIA	polyethylene-graft-polyvinyl imidazole
PE-gr-4Vpy	polyethylene-graft-poly(4-vinylpyridine)
Phen	phenanthroline
PP-gr-PAA	polypropylene-graft-polyacrylic acid
PP-gr-4Vpy	polypropylene-graft-poly(4-vinylpyridine)
PS-COO	carboxylated DVB-crosslinked polystyrene
PS-gr-PAA	polystyrene-graft-polyacrylic acid
Py	pyridine
SAAC	styrene-acrylic acid copolymer
SMC	styrene-2-(methylsulfinyl) ethyl methacrylate copolymer
St	stearate

PREPARATION AND CHARACTERIZATION OF POLYMER-SUPPORTED CATALYSTS

Classification

As mentioned in the introduction, polymer-supported catalysts are classified here mainly according to the type of bond used to anchor the metallic moities to the polymer carriers. Only catalysts used in diene polymerization are summarized in Table 1.

Table 1. Polymer-Supported Complexes for Diene Polymerization

Complex	Type of metal-support bonding	Valence of metal	Reference
Carboxylates			
SAAC·Nd	metal-oxygen	III	4,6,8
EAAC·Nd	metal-oxygen	III	7
PP-gr-PAA·Nd	metal-oxygen	III	4,9
PS-gr-PAA·Nd	metal-oxygen	III	present work
$(PS-COO)_n LnCl_{3-n}$ (Ln=Nd,Ce)	metal-oxygen	III	10
$(PE-COO)_n NdCl_{3-n}$	metal-oxygen	III	10
SAAC·Fe	metal-oxygen	III	1, pres.work
PE-gr-PAA·Co	metal-oxygen	II	11
SAAC·Nd-Fe	metal-oxygen	III	present work
SAAC·Nd-Na	metal-oxygen	III	present work
Sulfur-Containing Complexes			
SMC·Ln (Ln=La,Pr,Nd, Eu,Ho,Er,Tm,Yb)	metal-oxygen	III	4-6
Oxygen-Containing Chelates			
PE-gr-PAAA·Co	metal-oxygen	II	12
PE-gr-PSAA·Co	metal-oxygen	II	12
Nitrogen-Containing Complexes			
PE-gr-4PVy·Co	metal-nitrogen	II	11
PE-gr-PVIA·Co	metal-nitrogen	II	11
PAN·Co	metal-nitrogen	II	13
Nitrogen-Containing Chelate			
PE-gr-PAVK·Co	metal-nitrogen	II	12

The Mode of Attachment of the Metal Complex

The common methods to attach a metal complex to a support are either by deposition or by an ionic, covalent, or coordinative chemical bond. Less common methods involve the polymerization of a metal complex possessing a polymerizable functional group.

Polymeric metal carriers may be obtained by grafting a suitable ligand onto a preformed polymer or by homopolymerization or copolymerization of monomers carrying the desired ligand groups. The second method may be generally more desirable, because the polymer ligand structure and purity can be controlled more easily in this way. The grafting method, on the contrary, is random, difficult to control, and is often accompanied by side and undesired reactions.

The immobilization of the metal complex can be achieved by adding the metallic derivative to the already formed polymer, via reactions at the ligand functional groups in the polymer chains. When the metal complexes to be anchored are soluble, reaction with the coordinating polymer (soluble or insoluble) may be carried out by conventional synthetic methods. It is highly desirable that the metallic precursor is soluble. Otherwise the metal-polymer reaction between two solids suspended in a solvent leads to conditions not favorable to a high yield of product.

The best method to synthesize a supported catalyst seems to involve the following sequence of reactions:

synthesis of a suitable monomer ligand;
copolymerization of this monomer ligand;
coordination of the metal complex.

Catalysts Anchored by Metal-Oxygen Bonds

Preparation of the coordinating polymers. Polymers containing the carboxyl group can be prepared by radical polymerization and copolymerization, primarily with styrene or divinylbenzene and unsaturated monomers, e.g., acrylic acid, acrylates and methacrylates.

As an example, the copolymer SAAC (1,4,6,8), used as a macromolecular ligand, was made by the radical polymerization of styrene with acrylic acid in the presence or absence of a solvent such as benzene, dioxane, or tetrahydrofuran (using AIBN as initiator). Recently, we have prepared SAAC with different compositions and different distributions of monomeric units by means of the control of the polymerization conditions.

It is also possible to carboxylate preformed polymers; an example is carboxylated DVB-crosslinked polystyrene, obtained by the procedures shown in the scheme below (16):

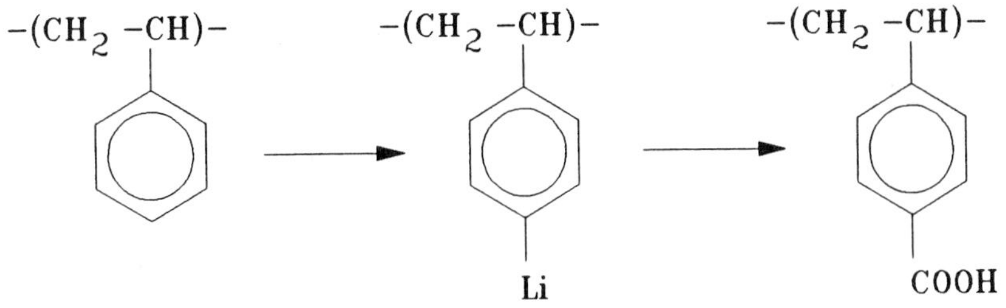

$$-(CH_2-CH)- \quad\longrightarrow\quad -(CH_2-CH)- \quad\longrightarrow\quad -(CH_2-CH)-$$

(benzene ring) → (benzene ring with Li) → (benzene ring with COOH)

Copolymer SMC can be obtained by the copolymerization of styrene with 2-(methylsulfinyl)ethyl methacrylate at 70°C in toluene with azobisisobutyronitrile as initiator (4,5).

Graft polymers containing the carboxyl ligand have also been prepared. For example, acrylic acid can be grafted to an inert support such as polypropylene or polyethylene by two main pathways. In one path the graft polymerization is initiated by radical initiators. For instance, polypropylene is suspended in H_2O and grafted with acrylic acid in the presence of water-insoluble radical initiators in water-miscible solvents (14). The other path involves the grafting of acrylic acid to polyethylene by gas phase polymerization (15).

PE-gr-PAAA was prepared by using PE-gr-PAN as starting material by way of the transformation shown below (12):

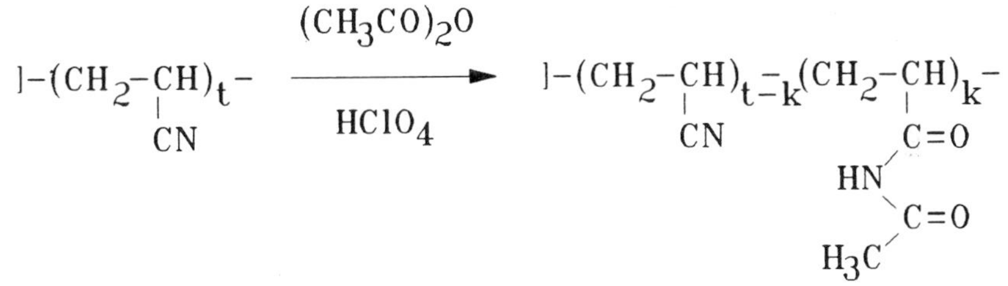

$$\text{]-(CH}_2\text{-CH)}_t\text{-} \quad\xrightarrow[\text{HClO}_4]{(CH_3CO)_2O}\quad \text{]-(CH}_2\text{-CH)}_{t-k}\text{(CH}_2\text{-CH)}_k\text{-}$$

with CN on left structure; right structure CN and C=O, HN, C=O, H_3C

Similarly, PE-gr-PSAA was obtained from PE-gr-PAAM:

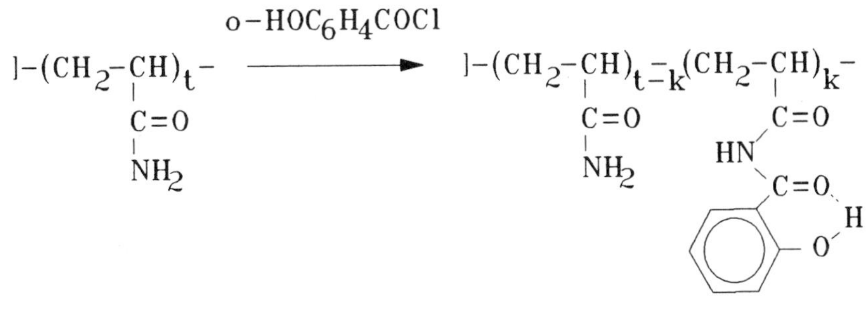

$$\text{]-(CH}_2\text{-CH)}_t\text{-} \quad\xrightarrow{o-HOC_6H_4COCl}\quad \text{]-(CH}_2\text{-CH)}_{t-k}\text{(CH}_2\text{-CH)}_k\text{-}$$

with C=O, NH_2 on left; right structure C=O, NH_2 and C=O, HN, C=O, benzene ring, H, O

PE-gr-PAN and PE-gr-PAAM were obtained by gas-phase or suspension polymerization methods (15). EAAC is available commercially.

Reaction of the coordinating polymers with metals. As far as the preparation of SAAC·M and EAAC·M (M=Nd, Fe or their mixture) is concerned, the metal chloride was reacted with a solution of SAAC, or EAAC (1,4,6-8).

Two methods were developed for the synthesis of SAAC·M. In the first method, the metal chloride was mixed directly with the polymer in methyl ethyl ketone; the reaction product was precipitated out immediately or by addition of a precipitant. In the second, the polymer suspension was reacted with a solution of the metal chloride in methanol. The reaction product was then washed with the solvent to remove the unreacted metal chloride.

A useful method to immobilize the metal on EAAC has also been developed (7,8). EAAC was dissolved in THF-chlorobenzene(3:1) by heating at 64°C. A solution of the metal chloride in methanol was added to this solution, and then a solution of ammonium hydroxide in THF was added dropwise with stirring. The reaction mixture was immediately poured into methanol, filtered, washed with methanol until the absence of Cl, and dried under vacuum at 40°C for 24 h. EAAC·M with different levels of metal loading was prepared by changing the molar ratio M/-COOH (see Table 2).

Elemental analysis of the reaction products of the support (SAAC or EAAC) with MCl_3 showed the absence of Cl. It is believed that the chlorine in MCl_3 has been exchanged by the carboxylic group in the carrier to form macromolecular carboxylate complexes with M as the crosslinking point.

It can be seen from Figure 1 that the bands at 1700 and 1235 cm^{-1} are attributed to $\nu(C=O)$ and $\nu(C-O)$ of the carboxylic group in SAAC. New bands appear at 1583 and 1417 cm^{-1} for SAAC·Fe, which are assigned to $\nu_{as}(COO)$ and $\nu_s(COO)$, respectively. The bands at 1583 and 1417 cm^{-1} become strengthened with the increase of iron loading in SAAC·Fe, whereas those at 1700 and 1235 cm^{-1} become weak.

Three modes of bonding with the carboxylic group are most probable in the case of SAAC·Fe formation:

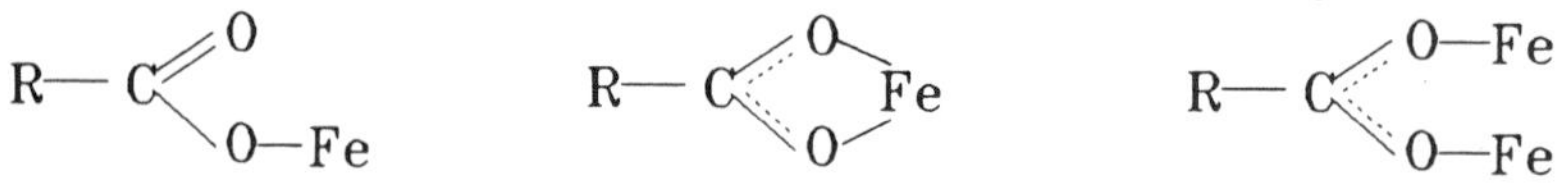

(A)   (B)   (C)

The difference between $\nu_{as}(COO)$ and $\nu_s(COO)$, $\Delta\nu$, reflects the symmetry of the carboxylic group on complex formation and the bond covalency. In the case of SAAC·Fe, a

Table 2. Composition Data for EAAC·Nd

Reactant		Product		
COOH content (mol/g)×10^3	Weight ratio NdCl$_3$·6H$_2$O/ EAAC	Element analysis (mol/g)×10^4		Mole ratio Nd/COOH
		Nd	Cl	
2.53	0.1	1.41	0.00	0.05
2.53	0.2	3.85	0.00	0.15
2.53	0.3	4.36	0.00	0.17
2.53	0.4	6.92	0.00	0.27
2.53	0.6	7.97	0.00	0.31

Table 3. Physical Properties of PP-gr-PAA and PP-gr-PAA·Ln

Complex	Color	T_s, (°C)	T_d, (°C)	Ln Content (mol/g)×10^4
PP-gr-PAA	white	158	359	—
PP-gr-PAA·La	white	156	331	1.1
PP-gr-PAA·Pr	light green	159	332	1.6
PP-gr-PAA·Nd	red violet	158	329	1.3
PP-gr-PAA·Eu	white	158	331	1.4
PP-gr-PAA·Er	light pink	158	331	1.6
PP-gr-PAA·Tm	white	158	331	2.1
PP-gr-PAA·Yb	white	158	330	1.2

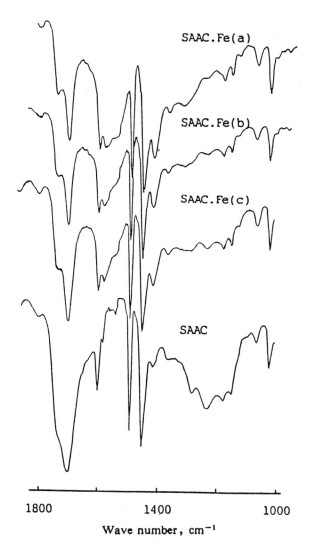

SAAC.Fe(a)

SAAC.Fe(b)

SAAC.Fe(c)

SAAC

1800   1400   1000

Wave number, cm⁻¹

Figure 1. IR absorption spectra of SAAC and
          SAAC·Fe complex: (a) Fe=0.68 mmol/g;
          (b) Fe=0.51 mmol/g; (c) Fe=0.35
          mmol/g.

$\Delta\nu$ of 166cm^{-1} was observed. Based on an investigation of IR spectra (17), it is possible that the complex SAAC·Fe possesses structure (C) and the Fe-O bond is higher in covalency.

Similarly, structures analogous to (A) and (B) are possible for rare earth complexes with carboxylic acids (18-22). It has been demonstrated by infrared spectra and XPS that SAAC·Nd and EAAC·Nd are of the bidentate carboxylate structure (B) and that the Nd-O bonds are rich in covalency (7,8).

In the same way, metals can bond to graft polymers. For example, rare earth metals were attached to PP-gr-PAA (4,9):

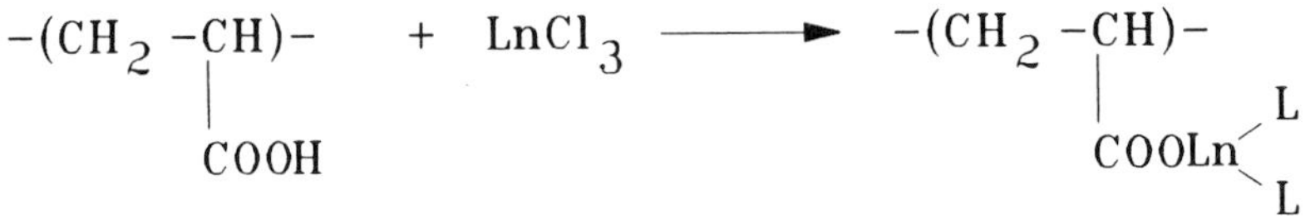

$$-(CH_2-CH)- \quad + \quad LnCl_3 \longrightarrow -(CH_2-CH)-$$
$$\underset{\text{COOH}}{|} \qquad\qquad\qquad\qquad \underset{\text{COOLn}\underset{L}{\overset{L}{\diagdown}}}{|}$$

where Ln = La, Pr, Nd, Eu, Er, Tm, and Yb; L = COO^-. The characteristic parameters of these complexes are listed in Table 3. Similarly, soluble Co(II) salts can be fixed to PE-gr-PAA (11):

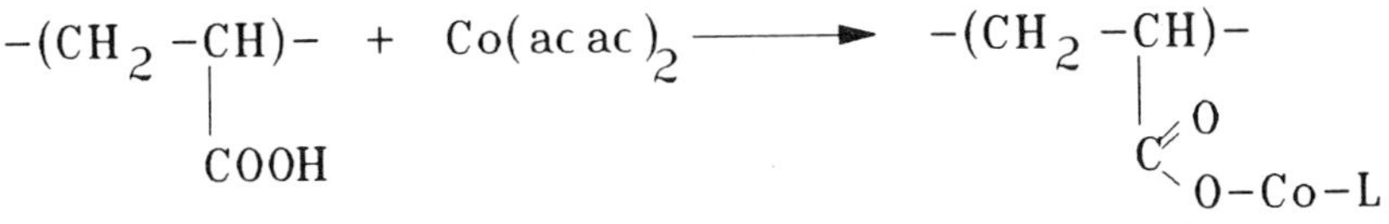

$$-(CH_2-CH)- \quad + \quad Co(ac\,ac)_2 \longrightarrow -(CH_2-CH)-$$
$$\underset{\text{COOH}}{|} \qquad\qquad\qquad\qquad \underset{C\overset{\diagup O}{\underset{\diagdown O-Co-L}{}}}{|}$$

where L is acac or -COO.

Carboxylated DVB-cross-linked polystyrene was exchanged with Ln by the following pathway (10):

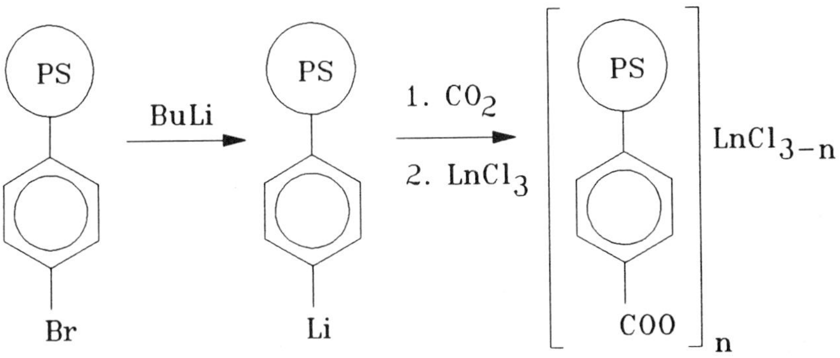

(Ln = Nd, Ce)

The resulting resins mainly contained $Nd(O_2C-PS)_3$ and $Ce(O_2C-PS)_3$ which had Nd and Ce loadings of 0.86 and 1.19 mmol of Nd and Ce per gram of polymer, respectively.

Neodymium derivatives of carboxylated polyethylene were obtained from oxidized polyethylene by the procedure:

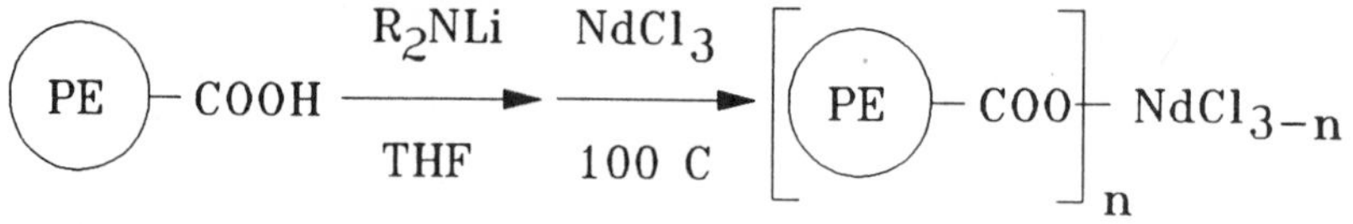

$$\left(\text{PE}\right)-\text{COOH} \xrightarrow[\text{THF}]{R_2NLi} \xrightarrow{\substack{NdCl_3 \\ 100\ C}} \left[\left(\text{PE}\right)-\text{COO}\right]_n NdCl_{3-n}$$

Analysis of the resulting product for Nd and Cl indicated that a mixture of neodymium chlorocarboxylates of average stoichiometry $(PE-CO_2)_{2.3}NdCl_{0.7}$ had formed.

The rare earth metals were supported on SMC by metal-O=S coordinative bonding (4, 5). It is clear from Table 4 that formation of the metal complex increases the thermal stability of the copolymer SMC. The color of $SMC \cdot LnCl_3$ closely resembles that of the respective parent chloride, changing shade from lighter to darker with an increasing amount of metal. The weight percentage of rare earth metal in the complex gradually increases from La to Yb, while the mole number remains nearly the same.

Chelate-type compounds are formed in the reaction of PE-gr-PAAA and PE-gr-PSAA with cobalt acetate (12):

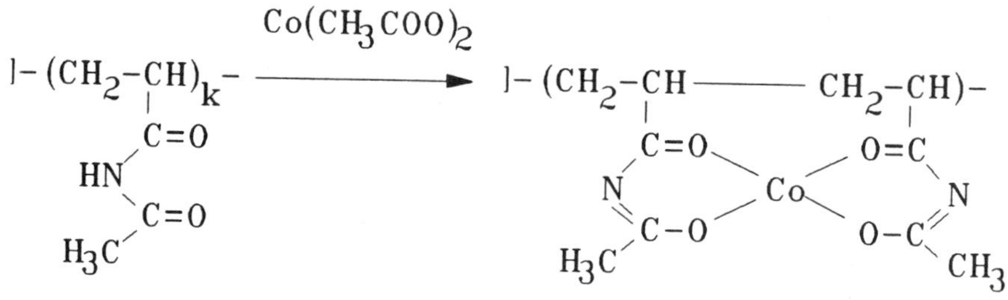

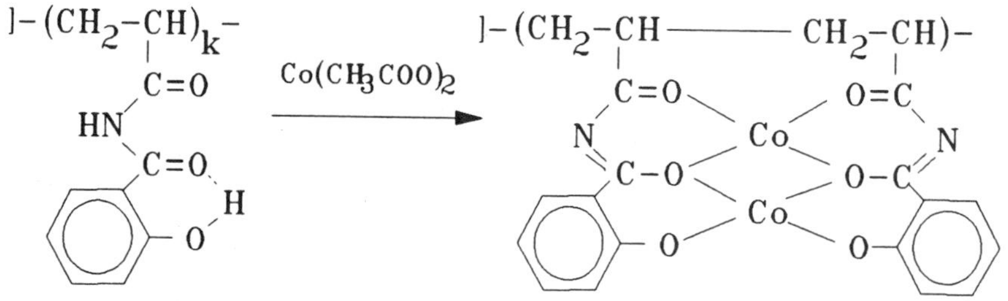

$$]-(CH_2-CH)_k- \xrightarrow{Co(CH_3COO)_2}]-(CH_2-CH \text{------} CH_2-CH)-$$

The existence of PE-gr-PAAA and PE-gr-PSAA in the form of trans-trans structures (I) and (II) creates conditions favorable for the formation of mono- and binuclear complexes of these ligands with Co.

Catalysts Anchored by Metal-Nitrogen Bonds

Preparation of the coordinating polymers. For preparation of PP-gr-P4VPy, monomers with the corresponding functional groups were incorporated into polypropylene or

Table 4. Physical Properties of SMC and SMC·LnCl$_3$

Complex	Color	T_s, ($^{\circ}$C)	T_d, ($^{\circ}$C)	Ln Content mol %
SMC	white	–	240	–
SMC·LaCl$_3$	white	239	252	8.39 (0.0604)
SMC·PrCl$_3$	light green	256	260	9.35 (0.0664)
SMC·NdCl$_3$	red violet	252	264	9.85 (0.0683)
SMC·EuCl$_3$	white	249	258	9.88 (0.0650)
SMC·HoCl$_3$	light yellow	252	264	10.34 (0.0627)
SMC·ErCl$_3$	light pink	248	250	10.62 (0.0635)
SMC·TmCl$_3$	white	248	252	11.30 (0.0669)
SMC·YbCl$_3$	white	243	256	11.16 (0.0645)

Table 5. Comparison of Activities of Iron Catalyst Systems

Catalyst System	Fe/Bd micromol/ gram	Al/Fe atom ratio	Temp. $^{\circ}$C	Time min.	Conversion %
SAAC·Fe-phen-Al(i-Bu)$_3$	1	70	20	10	92
Fe(St)$_3$-Bpy-Al(i-Bu)$_3$	4	30	room	240	9
Fe(hep)$_3$-Bpy-Al(i-Bu)$_3$	4	30	room	240	41
Fe(napth)$_3$-Bpy-Al(i-Bu)$_3$	4	30	room	240	44

polethylene by means of graft polymerization (11). The scheme shown below was used to prepare PE-gr-PAVK from PE-gr-PMVK (12):

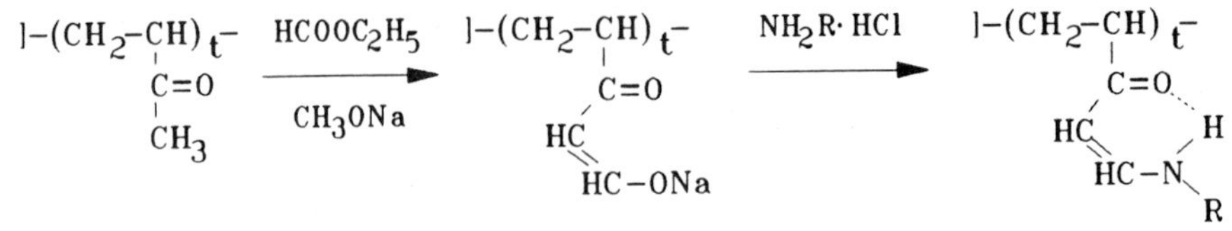

$$]-(CH_2-CH)_t^-\ \underset{CH_3ONa}{\xrightarrow{HCOOC_2H_5}}\ ]-(CH_2-CH)_t^-\ \xrightarrow{NH_2R\cdot HCl}\ ]-(CH_2-CH)_t^-$$

with substituents:
- First: C=O, CH₃
- Second: C=O, HC=, HC–ONa
- Third: C=O, HC, H, HC–N, R

The PE-gr-PMVK was obtained by gas-phase or suspension polymerization (15).

Reaction of the coordinating polymers with metals. The reaction of $CoCl_2$ with PP-gr-4-VPy resulted in the bonding of Co(II) to the polymer through a coordinative pyridine-cobalt interaction (11). $CoCl_2$ can be fixed onto PE-gr-PVIA in the same way. The reaction of PE-gr-PAVK with cobalt acetate in alcohol leads to chelate complexes of the type shown below (12):

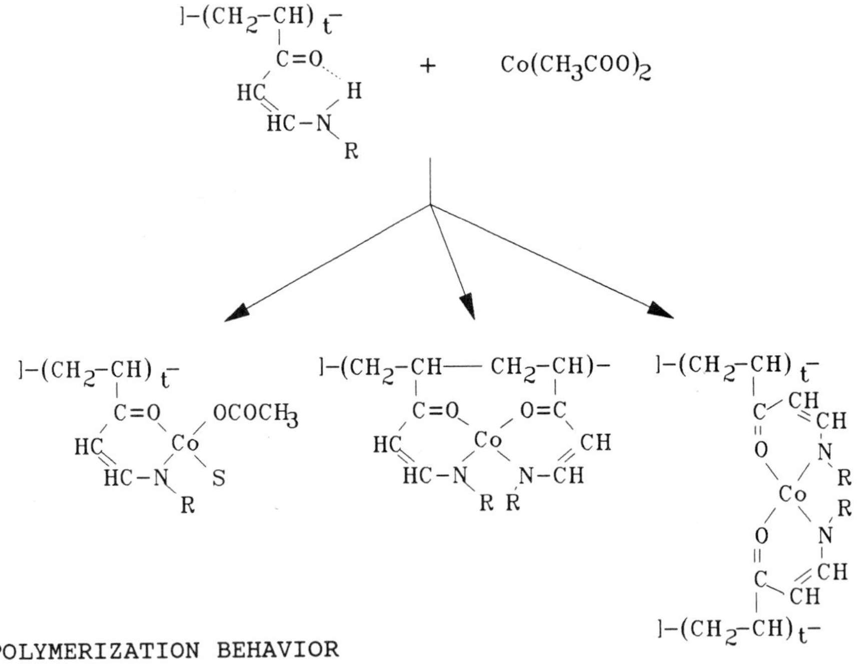

$$]-(CH_2-CH)_t^-\quad +\quad Co(CH_3COO)_2$$

with the ring substituent: C=O, HC, H, HC–N, R

leading to three products:

Product 1: $]-(CH_2-CH)_t^-$ with C=O, OCOCH₃, HC, Co, HC–N, S, R

Product 2: $]-(CH_2-CH\!-\!-\!-CH_2-CH)-$ with C=O, O=C, HC, Co, CH, HC–N, N–CH, R R

Product 3: $]-(CH_2-CH)_t^-$ with CH, CH, C, O, N, R, Co, R, O, N, C, CH, CH, $]-(CH_2-CH)_t^-$

POLYMERIZATION BEHAVIOR

Catalyst Activity

Improved polymerization activity probably represents the single most important advantage of polymer-supported catalysts. Table 5 illustrates the improved activity for butadiene polymerization of the polymer-supported SAAC·Fe system over corresponding low-molecular weight systems (1). The catalytic efficiency of SAAC·Fe was up to 99kg polybutadiene/(gFe·h), while in the case of Fe(naph)$_3$ the productivity was 0.49kg polybutadiene/(gFe·h). Thus, the catalytic activity of the SAAC·Fe system is 200 times that of the Fe(naph)$_3$ system.

Also for butadiene polymerization, the activity of the SMC·NdCl$_3$ polymer-supported system was found to be twice or three times that of the NdCl$_3$·4DMSO system (see Table 6) (4). The productivity of SAAC·Nd-Ph$_3$CCl-Al(i-Bu)$_3$ is so great that an efficiency of up to 170kg polybutadiene/(gNd·h) was achieved. Similar enhancements of activity for the EAAC·Nd polymer-supported systems can be seen in Table 7 (7).

For isoprene polymerization with Co(II) catalysts, homogeneous CoCl$_2$Py$_2$-Al(C$_2$H$_5$)$_2$Cl and its heterogeneous analog PP-gr-P4VPyCoCl$_2$-Al(C$_2$H$_5$)$_2$Cl were chosen for comparison of kinetic parameters (11). The results are shown in Table 8. The reaction orders with respect to monomer and with respect to Co for the homogeneous system were 0.4 and 1.4, respectively. The corresponding values for the heterogeneous system were 0.8 and 1.0, respectively. The rate constant for the chain propagation polymerization reaction k$_p$ was 6.5 or 10.7 for the homogeneous system and 3.5 or 2.0 for the polymer-supported system. The heterogeneous system was highly effective; each Co atom produced an active center for catalysis. In the homogeneous system, only 38% of the Co atoms participated in the formation of active centers.

One may ask what are the reasons for the high activities of the polymer-supported catalysts. It is generally agreed that homogeneous catalysis nearly always involves at least one intermediate in which the metal atom is coordinatively unsaturated. Such an unsaturated intermediate very often undergoes a tendency to dimerize, which is detrimental to catalytic activity. Anchoring the catalyst on a polymer support discourages the isolated, unsaturated metal atoms from combining, thus preserving the potential catalytic activity.

In addition, as mentioned above, the polymer-supported complexes possess the bidentate carboxylate group, and the metal-oxygen bonds in these complexes are rich in covalency. For Ziegler-Natta catalysts, one of the requirements for the formation of the active center is the alkylation of the metal ions by exchange between the alkyl group in alkyl aluminum and carboxylate. The covalency of the metal-oxygen bond in the support complex makes this exchange reaction easier.

Effect of Some Factors on Catalytic Activity

Effect of distribution of the functional monomer units in the polymer support. In our laboratory, investigation of the effect of the distribution of the functional monomer units in the polymer support on catalytic activity has been of interest.

Table 6. Activity of Various Catalyst Systems[a]

Catalyst System	Nd/Bd, micromol/g	Al/Nd, atom ratio	Time, min	Activity, kg PB/(gNd·h)	Viscosity, dl/g	Microstructure %		
						cis-1,4	trans-1,4	1,2
SAAC·Nd-Ph$_3$CCl-Al(i-C$_4$H$_9$)$_3$	0.1	2000	20	137	3.5	98.8	0.7	0.5
	0.2	1500	10	169	2.2	98.4	1.2	0.4
	0.2	1000	10	143	2.8	98.5	0.9	0.6
	0.2	500	15	83	3.3	98.0	1.3	0.7
	0.3	500	15	93	3.4	98.8	0.7	0.5
SMC·NdCl$_3$-Al(i-C$_4$H$_9$)$_3$	0.3	1000	15	56	2.7	98.1	1.1	0.8
	0.3	1000	30	33	2.8	98.0	1.2	0.8
	0.3	300	30	17	6.5	98.4	1.0	0.6
	0.2	200	30	19	8.5[b]	98.6	0.7	0.7
NdCl$_3$·4DMSO-Al(i-C$_4$H$_9$)$_3$	0.3	1000	30	16	2.7	98.7	0.8	0.5
	0.3	300	30	8	5.9	99.0	0.6	0.4

[a]Monomer butadiene: M = 16–17 g; M/hexane 1.6 g/mL; ternary system, Cl/Nd (atomic ratio) = 4.

[b]Block polymerization, M = 21 g.

Table 7. Polymerization of Butadiene Catalyzed by EAAC·Nd
and AA·Nd Catalytic Systems

Catalyst System	Nd/Bd, (mol/g)x10^7	Conversion, %
EAAC·Nd-i-Bu$_3$Al-Et$_2$AlCl	6	80
AA·Nd-i-Bu$_3$Al-Et$_2$AlCl	6	4
AA·Nd-i-Bu$_3$Al-Et$_2$AlCl	20	10

Polymerization conditions: Al/Nd(mol ratio) = 100; 6h; 50°C;
Cl/Nd(atom ratio) = 3.

Table 8. Comparison of Kinetic Parameters for Isoprene
Polymerization by Homogeneous and Heterogenized
Cobalt Systems

Parameter	System	
	homogeneous	heterogenized
E, kcal·mol^{-1}	16.5(0.5)	13(0.5)
Reaction order by monomer	0.4	0.8
Reaction order by cobalt	1.4	1.0
k, $\underline{M}^{-1}s^{-1}$ (kinetic method)	6.5	3.5
k, $\underline{M}^{-1}s^{-1}$ (radiochemical method)	10.7	2.0
n, mol/mol Co	0.38	1.0
polymer stucture, %		
1,4-cis	40 - 46	
1,4-trans	30 - 32	

The equation of Mayo-Lewis (23) relates the copolymer composition to the monomer composition, and is given by:

$$\frac{dM_1}{dM_2} = \frac{M_1}{M_2} \frac{r_1 M_1 + M_2}{r_2 M_2 + M_1} \quad (= \frac{m_1}{m_2} \text{ for low conversions})$$

where M_1 and M_2 refer to the monomer composition and m_1 and m_2 to the polymer composition. r_1 and r_2 are the reactivity ratios of M_1 and M_2, respectively. According to statistical theory, the probability that a sequence of m_1 units contains n members is:

$$P(m_1)_n = P_{11}^{(n-1)}(1-P_{11}) = \left[\frac{1}{(1+(1/r_1 F))}\right]^{n-1} \frac{1}{1+r_1 F}$$

Similarly:

$$P(m_2)_n = P_{22}^{(n-1)}(1-P_{22}) = \left[\frac{1}{(1+(F/r_2))}\right]^{n-1} \frac{1}{1+(r_2/F)}$$

These P values represent the fraction of all the m_1 (or m_2) sequences formed by n members. F is a function of the ratio f between the molar concentration of m_1 and m_2 in the copolymer and can be obtained from the copolymerization equation written by Fineman and Ross (24):

$$F = \frac{f-1+ \sqrt{(f-1)^2 + 4r_1 r_2 f}}{2r_1}$$

Table 9 shows the values of the distribution function P for sequences of different length (up to n=5) for styrene-acrylic acid copolymers of different composition. This table also shows the catalytic activities of the iron complexes supported on the copolymers having the corresponding composition. The probability of existence of long sequences of styrene, or of acrylic acid, increases with increasing content of the considered monomer in the copolymer. As the content of acrylic acid in the copolymer increases, the distribution of short sequences of styrene increases, while the corresponding distribution for acrylic acid decreases slightly. The catalytic activity of SAAC·Fe with the highest distribution of long sequences of acrylic acid is low. It is possible that the functional groups -COOH in the SAAC with the high distribution of long sequences of acrylic acid are not efficiently dispersed in the polymer chains. As a consequence, the iron ions in the complexes do not easily contact the alkyl aluminum to form active catalytic centers.

Table 9. Distribution Function of Sequences of Different Lengths in SAAC and Effect on Activity of SAAC·Fe

Mole % of Acrylic Acid	Conversion %	Microstructure, %			n	$P(Ph-CH=CH_2)_n$ %	$P(CH_2=CHCOOH)_n$ %
		cis-1,4	trans-1,4	1,2			
18	83	51.9	0.9	47.2	1	21.77	99.05
					2	17.03	0.94
					3	13.32	–
					4	10.42	–
					5	8.15	–
31	87	49.8	6.4	43.8	1	43.77	97.39
					2	24.61	2.55
					3	13.84	–
					4	7.79	–
					5	4.38	–
36	89	45.4	6.8	47.8	1	54.05	96.10
					2	24.84	3.75
					3	11.41	0.15
					4	5.24	–
					5	2.41	–
39	45	43.9	0.6	55.5	1	60.69	94.94
					2	23.86	4.80
					3	9.38	0.24
					4	3.69	–
					5	1.45	–

When the distribution of short sequences of acrylic acid in the copolymers is predominant, and the short sequence distribution is approximately equal to that of the long sequences for styrene, the complexes show optimum catalytic activity. This regularity was observed not only in the iron system, which produces polybutadiene with equibinary structure (cis-1,4 and 1,2 about 1:1), but also in the neodymium system producing high cis-1,4 polybutadiene.

Effect of the ratio of metal loading to the functional group. In order to examine the relationship between the constitution and the catalytic activity of polymer-supported neodymium complexes, three groups of SAAC were synthesized by different methods with different contents of -COOH. Table 10 clearly indicates that regardless of the method used (A or B) for the synthesis of SAAC, so long as the Nd/-COOH molar ratio in SAAC·Nd was about 0.20, the catalytic activities of the supported neodymium complexes SAAC·Nd were in the optimum state. This may indicate that the environment of the metal ions in SAAC·Nd is of great advantage in forming the active centers with the third component (alkyl chloride or aluminum alkyl chloride) and alkyl aluminum.

Among the three groups shown for each synthetic method in Table 10, the group for which the content of -COOH is intermediate (about 12 wt%) has a much better activity than that found for the corresponding higher -COOH content group (about 16%). At the higher content of functional group the catalytic activity was found to be poorer than for the lower content group, even if the Nd/-COOH molar ratio was 0.20. Obviously, in the case of the higher functional group content, we must increase the content of bound metal neodymium (see runs no. 93, 82, 74, 182, 272, 162, 214, 202, 17 in Table 10) in order to maintain the ratio of neodymium and -COOH at 0.20. It is evident that an increase in the content of the bound neodymium leads to an increase in the amount of Nd per unit support volume. A portion of Nd is embedded, and only the Nd metal on the surface of the formed catalyst will be activated, as usually occurs in heterogeneous catalysts. When SAAC was prepared by method B in the presence of solvent THF or dioxane, the neodymium complexes have higher activity.

SAAC·Fe and EAAC·Nd show some similarities in butadiene polymerization (Table 11). For example, as found for the Nd systems, catalytic activity was highest when the Fe/-COOH molar ratio was nearly 0.2.

Effect of the structure of the complex. The activities of the mononuclear chelates of Co(II) are quite different from that of the binuclear Co(II) chelate for butadiene polymerization.

In combination with organoaluminum compounds, the immobilized Co(II) chelates have a high and stable activity

Table 10. Relationship between SAAC·Nd Properties and Catalytic Activity[a]

Run #	Synthetic Method for SAAC	Functional Groups wt%	Functional Groups mmol/g	Quantity of Bound Metal mmol/g	Quantity of Bound Metal mol/mol of functional group	Conversion of butadiene, %
74	A	8.8	1.96	0.37	0.19	84
73				0.50	0.26	82
72				0.57	0.29	76
71				0.63	0.32	60
82		11.5	2.55	0.45	0.18	87
81				0.60	0.24	80
83				0.73	0.29	72
85				0.80	0.31	64
95		15.9	3.54	0.45	0.13	31
94				0.60	0.17	64
93				0.67	0.19	78
92				1.00	0.28	48
17	B	9.3	2.07	0.37	0.18	92
18	(solvent, dioxane)			0.49	0.24	90
19		11.9	2.65	0.64	0.31	82
202				0.52	0.20	94
203				0.61	0.23	92
204				0.69	0.26	85
205				0.85	0.32	62
214		15.2	3.88	0.57	0.17	84
213				0.76	0.23	80
212				0.83	0.25	76
161	B	8.6	1.92	0.31	0.16	74
162	(solvent, THF)			0.39	0.20	90
163				0.62	0.32	60
272		12.0	2.67	0.51	0.19	94
273				0.61	0.23	90
274				0.78	0.29	82
181		16.6	3.68	0.65	0.18	58
182				0.74	0.20	74
183				0.91	0.25	54

[a]polymerization conditions: monomer concentration, 10g/100mL hexane; Nd/Bd 0.3 micromol/g; third component, PhCH$_2$Cl; Cl/Nd(mole ratio), 3.5; cocatalyst, Al(i-C$_4$H$_9$)$_3$; Al/Nd(mole ratio), 200; catalyst aging concentration, 10 micromol/mL toluene; 50°C; 6h.

Table 11. Effect of Fe/-COOH Molar Ratio on Activity

Functional Group: 2.66 mmol/g

Quantity of Bound Metal:

mmol/g	mol/mol of functional group	Conversion, %	Viscosity, dl/g
0.35	0.13	71	13.2
0.42	0.16	76	11.3
0.51	0.19	78	11.1
0.53	0.20	81	10.6
0.87	0.33	76	12.1

polymerization conditions: Fe/Bd(micromol/g) = 1; 20°C;
Al/Fe(molar ratio) = 70; 30 min.

Table 12. Activities of Various Catalyst Systems

System	Nd/Bd, micromol/g	Conversion, %
SAAC·Fe-i-Bu$_3$Al-Et$_2$AlCl	0.05	0.0
SAAC·Nd-i-Bu$_3$Al-Et$_2$AlCl	0.05	48.5
SAAC·Nd-Fe-i-Bu$_3$Al-Et$_2$AlCl	0.05	55.0
SAAC·Nd-Na-i-Bu$_3$Al-Et$_2$AlCl	0.05	64.0

polymerization conditions: Al/Nd(molar ratio) = 200; 50 °C;
Cl/Nd(atom ratio) = 3.5; 6 h.

in the stereospecific polymerization of butadiene (12).
Figure 2 shows that the catalytic system PE-gr-PAAA·Co(II)-
Et$_2$AlCl is more active than its homogeneous analog based on
the chelate of Co(II) with N-acetylbenzamide. It is
noteworthy that the immobilized mononuclear chelates of
Co(II) with PE-gr-PAVK and PE-gr-PAAA have a comparable
activity. This indicates that the nature of the chelate unit
(N,O- and O,O-, respectively) has inappreciable influence on
the activity of the complexes in the polymerization of
butadiene. At the same time, the binuclear Co(II) chelate
with PE-gr-PSAA exhibits a lower (by a factor of 5-6)
activity than its mononuclear analogs.

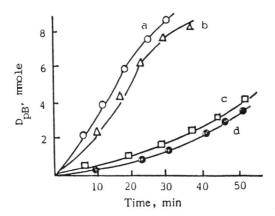

Figure 2. Kinetic curves for polymerization of
butadiene by catalytic systems based on
cobalt chelates with PE-gr-PAAA(a),
PE-gr-PAVK(b), N-acetylbenzamide(c), and
PE-gr-PSAA(d). 323°C; [Co]=1.2x10^{-4};
[butadiene]=2.2; Al/Cl=100 mole/liter;
cocatalyst-Et$_2$AlCl; solvent-toluene.

Effect of dispersion of the metal fixed on the polymer
chain. We have synthesized the mixed-metal complexes
SAAC·Nd-Fe and SAAC·Nd-Na in order to examine the effect of
the dispersion of the metal anchored on the polymer chain on
the catalytic activity. The purpose of the addition of the
second metal is to govern the dispersion of the metal acting
as the active species, and then to improve its activity. As
was expected, SAAC·Fe-Al(i-Bu)$_3$-Et$_2$AlCl (in contrast to
SAAC·Fe-Phen-Al(i-Bu)$_3$) and SAAC·Na-Al(i-Bu)$_3$-Et$_2$AlCl systems
showed no activity for butadiene polymerization under the
given experimental conditions, whereas SAAC·Nd-Al(i-Bu)$_3$-
Et$_2$AlCl system can catalyze butadiene to polymer (Table 12).
It may be considered that Fe and Na play only the role of
adjusting the dispersion of Nd immobilized on the polymer
chain; Nd serves as the active center. It is obvious from
Figure 3 that the catalytic activities of SAAC·Nd-Fe and
SAAC·Nd-Na systems were higher than that of the SAAC·Nd

system. It is reasonable that such high activities for
SAAC·Nd-Fe and SAAC·Nd-Na systems be attributed to the
efficient dispersion of Nd atoms supported on the SAAC chain
by the addition of Fe and Na. When the content of the
carboxylic group in SAAC was 2.1 mmole/g, the highest
activities of SAAC·Nd-Fe and SAAC·Nd-Na systems were at molar
ratios of 0.03 for Fe/Nd (Figure 4) and 0.70 for Na/Nd
(Figure 5).

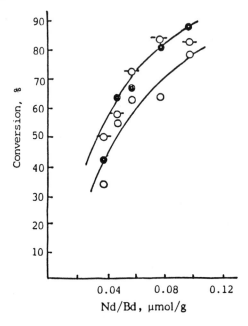

Figure 3. Comparison of the activity of various metal
complexes with SAAC. Polymerization conditions: Al/Nd (mol ratio)=200; Cl/Nd (atom
ratio)=3.5; 50°C; 6 h; (O) SAAC·Nd, (-O-)
SAAC·Nd-Fe, (●) SAAC·Nd-Na.

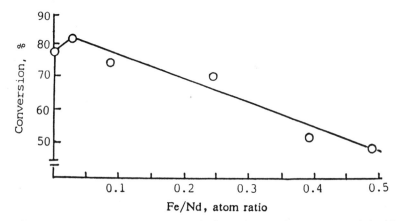

Figure 4. Effect of Fe/Nd in SAAC Nd-Fe on activity.
Polymerization conditions: Nd/Bd (micromol/g)
=0.1; Al/Nd (mol ratio)=200; Cl/Nd (atom
ratio)=3.5; 50°C; 6h.

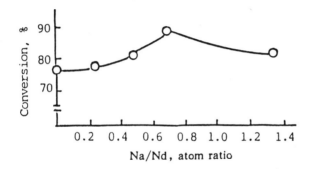

Figure 5. Effect of Na/Nd in SAAC·Nd-Na on activity.
Polymerization conditions: same as Figure 4.

Effect of various aluminum alkyls. In butadiene
polymerization with SAAC·Nd-AlR$_3$-Al(C$_2$H$_5$)$_2$Cl catalyst systems
(4), the order of AlR$_3$ activity was Al(i-C$_4$H$_9$)$_3$ > Al(i-
C$_4$H$_9$)$_2$H > Al(C$_2$H$_5$)$_3$. The activity of the system increased
with increasing Al/Nd ratio.

Effect of the third component. In the ternary system
formed using SAAC·Nd(4), the activity was very different for
different kinds of aluminum alkyl halides as the third
component. The following order of activity was found:
Al(C$_2$H$_5$)$_2$Cl > Al(C$_2$H$_5$)Cl$_2$ > Al(C$_2$H$_5$)$_2$Br. When we used alkyl
chloride in place of aluminum alkyl halide, the catalytic
activity depended on the type of alkyl chloride as well as
the solvent of catalyst preparation, as shown in Table 13.
Triphenylchloromethane is the best of the alkyl chlorides.
The activity of this catalyst is also dependent on the amount
of the third component used. The highest activity was found
when triphenylchloromethane was added at a Cl/Nd ratio of 3-
5.

Effect of the solvents used in the polymerization. In
the system consisting of EAAC·Nd (7), the butadiene
polymerization rate was obviously different in different
solvents. The polymerization rate in aromatic solvents was
much lower than that in hexane and decreased with increasing
basicity of the solvents. In the case of aromatic solvents,
it is possible that there is a competitive coordination of
the arene and monomer to Nd of the catalyst complex:

Table 13. Effect of Various Alkyl Chlorides on the Polymerization of Butadiene

RCl	Nd/Bd, micromol/g	RCl, mmol/mL		Conversion, %	Viscosity, dl/g	Microstructure, %		
		Toluene	Hexane			cis-1,4	trans-1,4	1,2
Ph_3CCl	0.2	0.1		92	9.0	98.4	1.0	0.6
	0.2	0.5		92	8.5	99.0	0.4	0.6
	0.2		0.1	72	8.8	98.7	0.9	0.4
$PhCH_2Cl$	0.2	0.5		80	10.8	98.4	1.0	0.6
	0.2		0.5	22	13.5	98.8	0.8	0.4
$CH_2=CHCH_2Cl$	0.2	0.5		76	9.8	99.0	0.7	0.3
	0.2		0.5	14	10.2	98.6	0.9	0.5
CCl_4	0.2	0.5		66	11.9	98.9	0.8	0.3
	0.2		0.5	42	12.4	98.4	1.0	0.6
$C_2H_5CH_2CH_2Cl$	0.3	0.5		trace	-	-	-	-
$(CH_3)_2CHCH_2Cl$	0.3	0.5		trace	-	-	-	-
$(CH_3)_2CHCH_2CH_2Cl$	0.3	0.5		trace	-	-	-	-

Long-chain saturated alkyl chlorides: Cl/Nd atomic ratio = 3, Al/Nd atomic ratio = 200, 50°C, 8h.
Other chlorides: Cl/Nd atomic ratio = 3.5; Al/Nd atomic ratio = 200, 50°C, 6h.

REFERENCES

1. Li Yuliang, Pang Shufen, Han Dong, and Ouyang Jun, Cuihua Xuebao, **5**(2), 174 (1984).

2. Li Yuliang, Xue Dawei, and Ougang Jun, Polym. Comm., **2**, 111 (1985).

3. Li Yuliang, Han Dong, and Ouyang Jun, Cuihua Xuebao, **7**(3), 276 (1986).

4. Li Yuliang and Ouyang Jun, J. Macromol. Sci. - Chem., **A24**(3&4), 227 (1987).

5. Li Yuliang, Pan Shufen, Li Guangquan, and Ougang Jun, Acta Chem. Sinica, **45**, 801 (1987).

6. Li Yuliang and Ouyang Jun, Acta Polymerica Sinica, **1**, 39 (1988).

7. Yu Guangqian and Li Yuliang, Cuihua Xuebao, **9**(2), 190 (1988).

8. Li Yuliang, Liu Guangdon, and Yu Guangqian, J. Macromol. Sci. - Chem., **A26**(2&3), 405 (1988).

9. Li Yuliang, Yu Guangqian, and Ouyang Jun, Rare Earth, **3**, 53 (1989).

10. D. E. Bergbreiter, Chen Liban, and Rama Cchandren, Macromolecules, **18**, 1055 (1985).

11. N. D. Golubeva, A. D. Pomogailo, A. L. Kuzaev, A. N. Ponomarev, and F. S. Dyachkovskii, J. Polym. Sci.: Polym. Symp., **68**, 33 (1980).

12. A. D. Pomogailo, I. E. Uflyand, and N. D. Golubeva, Kinet. Katal., **26**(6), 1404, (1985).

13. Toyo Rayon Co. Ltd., Japanese Patent No. 68-15626.

14. Japanese KoKai Tokkyo Koho, 79-56693; Chem. Abstr., **91**, 57855.

15. D. A. Kritskaya, A. D. Pomogailo, A. N. Ponomarev, and F. S. Dyachkovskii, Vysokomol. Soedin., **21A**(5), 1107 (1979).

16. M. J. Farrall and J. M. J. Frechet, J. Org. Chem., **41**, 3887 (1976).

17. N. W. Alcock and V. M. Tracy, J. Chem. Soc. Dalton, **21**, 2243 (1976).

18. S. S. Krishnamurthy and S. Soundararajan, J. Less-Common Metals, **16**, 1 (1968).

19. D. A. Edwards and R. H. Hayward, Can. J. Chem., **46**, 3443 (1968).

20. K. P. Patil, G. V. Chandrashekar, M. V. George, and C. N. R. Rao, Can. J. Chem., **46**, 257 (1968).

21. V. N. Krishnamurthy and S. Soundararajan, J. Less-Common Metals, **13**, 263 (1967).

22. S. N. Misra, T. N. Misra, and R. U. Mehrotra, J. Inorg. Nucl. Chem., **25**, 201 (1963).

23. F. R. Mayo and F. M. Lewis, J. Am. Chem. Soc., **66**, 1594 (1944).

24. M. Fineman and S. D. Ross, J. Polym. Sci., **5**, 259 (1950).

HIGHLY CHARGED DOPANT IONS FOR POLYACETYLENE

Shiou-Mei Huang and Richard B. Kaner

Department of Chemistry and Biochemistry and the Solid State Science
Center, University of California, Los Angeles, Los Angeles, California
90024-1569

ABSTRACT

The strong reducing power of solvated electrons in liquid ammonia has been used to chemically n-dope the conducting polymer polyacetylene with the alkaline earth divalent countercations Ca^{+2}, Sr^{+2} and Ba^{+2}. This same process also enables the divalent lanthanide ions, Eu^{+2} and Yb^{+2}, to act as n-dopants. Large increases in conductivity (up to ~ 30 Ω^{-1} cm^{-1}) are observed after n-doping of free-standing polyacetylene films. Although the smaller alkaline earth metals do not dissolve in liquid ammonia, two methods have been developed which enable Mg^{+2} ions to act as n-dopants for polyacetylene. Using magnesium salts dissolved in sodium-ammonia solutions leads to mixed Mg^{+2}/Na^{+} doped polyacetylene films, while an electrolysis method produces pure Mg^{+2} doped polyacetylenes. The maximum conductivity of polyacetylene n-doped with the small Mg^{+2} countercations (~ 0.5 $\Omega^{-1}cm^{-1}$) is less than that of the other divalent dopants. Magnetic measurements show decreases in the concentration of unpaired spins on n-doping polyacetylene with all of the divalent countercations. This indicates that spinless charge carriers are responsible for conductivity. The n-doping of polyacetylene with the divalent countercations causes two new peaks to appear in infrared spectra and creates midgap absorptions in visible spectra. These changes are consistent with a conduction model involving solitons and bipolarons.

INTRODUCTION

Conducting polymers have been the focus of a great deal of research (1,2) since the discovery in 1977 that polyacetylene can be made to conduct electricity like a metal after the introduction of charged dopant ions (3). All conducting organic polymers possess conjugated carbon backbones with delocalized π-orbitals. On addition or removal of electrons to or from their π-system (n- or p-doping, respectively) these materials become good conductors of electricity. Counterions must be added along with the charge to maintain electrical neutrality. Until recently all of the tens of counterions known such as the anions AsF_6^-, PF_6^-, ClO_4^-, I_3^-, etc. or the cations Li^+, Na^+, K^+, Bu_4N^+, etc. were monovalent. Here we discuss the n-doping of polyacetylene with the introduction of divalent ions. These divalent cation dopants lead to some interesting changes in the electrical, magnetic and optical properties of their host polymers.

EXPERIMENTAL

Polyacetylene films were grown on the walls of a glass reactor using a concentrated Ziegler-Natta catalyst. The catalyst, titanium tetrabutoxide and triethylaluminum, was dissolved in either toluene at -78°C as described by Shirakawa, et al. (4, 5) or in pure silicon oil at room temperature as outlined by Naarmann and Theophilou (6, 7). The catalyst in each method is cured for at least one hour and degassed to remove the ethylene given off. After shaking the catalyst onto the walls of the reactor, it is exposed to one liter-atmosphere of purified acetylene gas. Within five minutes using the Shirakawa method or 20 minutes using the Naarmann and Theophilou method, lustrous coppery films of polyacetylene, ~ 20 μm thick, form. The films are washed repeatedly in dry pentane, dried and cut into pieces in a helium filled drybox.

The acetylene gas is purified by passing it through two bubblers filled with concentrated sulfuric acid. This removes any residual acetone which comes out of the gas cylinder. Moisture is removed from the acetylene gas by further passing it through two 50 cm columns, the first filled with potassium hydroxide and the second filled with phosphorus pentoxide.

Conductivity measurements were carried out at room temperature using a four-probe technique in the helium filled drybox. Four platinum pressure contacts were made to the free-standing polyacetylene films. Electrical feedthroughs allowed equipment including a digital multimeter (Hewlett Packard 3468A) to be connected externally to the drybox. All measurements were made at least three times on several independent samples.

Analytic data were obtained using an atomic absorption spectrometer (Varian Spectra AA-30). High purity standards were used for calibration. All results were based on the average of at least four separate sample solutions aspirated into the spectrometer. Samples rerun at a later time gave essentially no change in the analysis. Magnetic measurements were taken using an EPR spectrometer (IBM 2000).

THEORY OF CONDUCTION IN POLYACETYLENE

In order to understand the physical properties of polyacetylene doped with divalent ions, it is important to consider the theory of conductivity of polyacetylene doped with monovalent ions. One of the most unusual characteristics of polyacetylene is that small amounts of dopant ions give rise to enormous increases in electrical conductivity without causing any increase in the number of unpaired electrons. In fact, the small level of paramagnetism observed in pristine polyacetylene actually decreases on doping (8). This is in contrast to what occurs in traditional semiconductors, such as silicon, where dopants increase both conductivity and paramagnetism. An explanation has been offered by the soliton theory of conductivity (9,10).

Solitons are neutral defects which arise as a consequence of the degenerate ground state structure of *trans*-polyacetylene. *Trans*-polyacetylene has a conjugated carbon backbone as shown in Figure 1. Each carbon is sp^2 hybridized forming bonds with two adjacent carbons and one hydrogen. The p_z orbitals lie perpendicular to the polymer chains and overlap to form a band of π-bonding molecular orbitals (M.O.'s) and a band of π^*-antibonding M.O.'s. Due to a Peierls distortion the one-dimensional *trans*-polyacetylene chains "dimerize", forming alternating single and double bonds. This introduces a gap of ~1.4 eV between the valence band (π-M.O.'s) and conduction band (π^*-M.O.'s). Since the two forms of alternating single and double bonds are equal in energy, *trans*-polyacetylene has a degenerate ground state, as shown in Figure 1.

When the two degenerate forms of *trans*-polyacetylene meet in a linear chain, a defect called a soliton forms, as shown in Figure 2a. This neutral soliton possesses a spin of 1/2 and no charge. It can be considered a non-bonding π electron which has an energy half-way

Figure 1.   The structure of *trans*-polyacetylene showing its two degenerate ground
state configurations.

between the valence and conduction bands, as indicated in Figure 2b. Doping will occur
preferentially at this soliton site. N-doping creates a negative soliton, while p-doping forms a
positive soliton. A negative soliton and its corresponding filled midgap level is shown in
Figure 3a and b. A positive soliton would have an empty level at midgap. Both negative and
positive solitons have charge but no spin. Increased doping creates a band of charged soliton
midgap states. These states are not localized as may be implied by the simplified drawings in
Figures 2 and 3. In fact ~85% of the charge is spread over ~15 CH units (11). Solitons are
thus delocalized and highly mobile. The ability of a mobile soliton band to transport charge
without introducing free spins is a theory which has been cited extensively to explain the high
conductivities observed in doped polyacetylenes (12, 13).

       To maintain electrical neutrality on doping polyacetylene, counterions must associate
themselves with the charged solitons. The movement of these counterions is very slow
(relative to the movement of the solitons) and thus they do not contribute directly to the
electrical conductivity in the doped polymer. However, their size and charge could effect the
nature of the solitons. Neither of these possibilities has been extensively studied, although
increasing the size of the counterion is known to limit the maximum amount of doping (14).
Since the dopant counterions used so far have always been monovalent, we will consider
here the possibility of varying the charge on the counterion.

       Divalent or multivalent counterions can be used to test some interesting properties of the
soliton theory. There are two main possibilities for states which could be created by a
divalent dopant ion depending on whether the ion associates itself with a single chain or
forms a bridge between chains. If a divalent ion associates itself with two chains, a soliton
would be created on each chain, as shown in Figure 4a. This could in principle facilitate
interchain charge transport through the midgap states. Alternatively, a divalent dopant could
associate itself with a single chain creating a bipolaron state, as shown in Figure 4b. The
bipolaron consists of two charges on one chain held together by attraction to the same
divalent dopant ion. The bipolaron state is comparable but slightly different than that formed

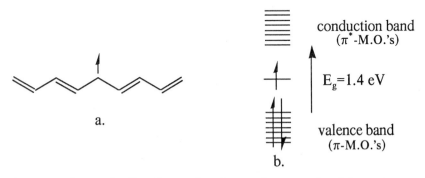

a.

conduction band
(π^*-M.O.'s)

E_g=1.4 eV

valence band
(π-M.O.'s)

b.

Figure 2.   a. A neutral soliton forms when the two degenerate bond alternating states
of *trans*-polyacetylene meet. b. The electron resides in a non-bonding

molecular orbital half-way between the π and π^* molecular orbital levels. A
neutral soliton has a spin of 1/2 and no charge.

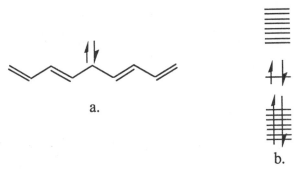

Figure 3.   a. A negative soliton in *trans*-polyacetylene. b. The filled midgap state has no spin and carries a negative charge.

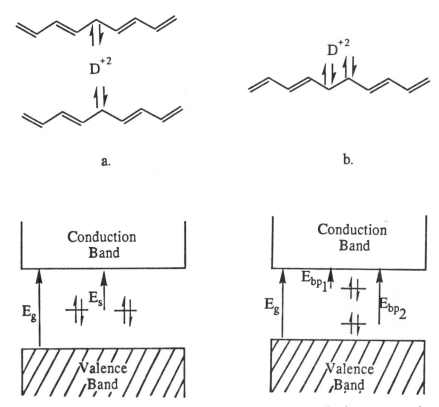

Figure 4.   When a *trans*-polyacetylene chain is n-doped with a divalent countercation the charge could either be a. distributed between two or more chains leading to the formation of two soliton states with their corresponding midgap soliton absorptions (E_s) or b. localized on a single chain leading to the formation of a bipolaron state with its two interband absorptions (E_{bp_1} and E_{bp_2}).

in polyparaphenylene, polypyrrole or polythiophene, where the quinoid structure confines two like monovalent charges (15). The close association of two like charges splits the midgap level into two bipolaron levels, as shown in Figure 4b. This localization of charge could inhibit intrachain charge transport in polyacetylene. Therefore the synthesis of divalent ion doped polyacetylenes is important to elucidate the effects on conductivity and determine which states are responsible for charge transport.

CHEMICAL SYNTHESIS OF DIVALENT DOPED POLYACETYLENES

A very strong reducing agent and the presence of suitable dopant counterions are needed to introduce divalent cations into polyacetylene. Such a strong reducing solution can be made by dissolving the alkaline earth metals, calcium, strontium or barium in liquid ammonia (equation 1).

$$M \xrightarrow{NH_3} M^{+2}(NH_3) + 2e^-(NH_3) \qquad M = Ca, Sr \ or \ Ba \qquad (1)$$

These solutions are blue at moderate alkaline earth metal concentrations and bronze at high concentrations (> 0.1M) due to the presence of free electrons solvated by ammonia molecules. When a film of polyacetylene is immersed in such a solution, the solvated electrons are capable of partially reducing the polyacetylene chains to polycarbanions (equation 2).

$$(CH)_x + xy \ e^- \longrightarrow [(CH)^{-y}]_x \qquad (2)$$

The alkaline earth cations associate themselves with the partially reduced polyacetylene chains and act as countercations to maintain electrical neutrality, resulting in the net reaction given in equation 3.

$$(CH)_x + xyM \xrightarrow{NH_3} [M_y^{+2}(CH)^{-2y}]_x \qquad M = Ca, Sr \ or \ Ba \qquad (3)$$

The partially reduced polyacetylene films formed are gold and have become good electrical conductors (16). It should be noted that in addition to the heavier alkaline earth metals, all the alkali metals (Li, Na, K, Rb and Cs) dissolve in liquid ammonia and they can be incorporated into polyacetylene as n-dopants using this method. An earlier preliminary report to our work had suggested the use of sodium-ammonia solutions for n-doping polyacetylene (17).

The apparatus used for synthesizing divalent ion doped polyacetylenes is shown in Figure 5. It consists of three chambers capped by 12 mm Teflon stopcocks. After cleaning and drying the apparatus to remove any traces of moisture, it is brought into a drybox containing purified helium. The appropriate alkaline earth metal is then scraped to remove any surface contamination and a tared amount is placed in the left-hand chamber. Free-standing polyacetylene films are cut, tared and placed in the right-hand chamber. The apparatus is next returned to the vacuum line. Ammonia, which has already been purified by distillation from sodium metal, is cryogenically transferred to the chamber containing the metal. The metal dissolves forming a deep blue solution.

The highly colored metal-ammonia solution containing solvated metal ions and solvated electrons is now used for doping. Most of the solution is poured into the right-hand chamber containing the polyacetylene at -70°C. A spontaneous n-doping reaction with the polyacetylene takes place and it is allowed to proceed for minutes or hours depending on the desired level of doping. Maximum conductivities are obtained after about two hours of doping time.

The procedure to remove any unreacted metal is extremely important. Great care is taken to prevent the formation of metal amides which are especially difficult to remove from the surface of the polyacetylene films. All metal-ammonia solutions are metastable and will

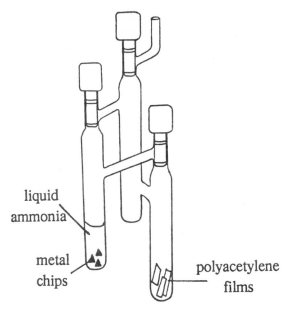

liquid
ammonia

metal
chips

polyacetylene
films

Figure 5.   The apparatus used in the chemical n-doping of polyacetylene from metal-
ammonia solutions.

eventually decompose to metal amides (18). Temperature fluctuations and impurities which
can catalyze such decompositions must be avoided.

The washing procedure is carried out by pouring the metal-ammonia solution into the
unused middle chamber. Fresh ammonia is then cryogenically distilled from this solution
back into the right-hand chamber containing the polyacetylene.   After stirring the
polyacetylene in the ammonia, the solution is poured back into the middle chamber. The
middle chamber has a short shunt to the polyacetylene chamber. This facilitates moving the
wash solution and avoids bumping during distillation. Residue from the wash solution can
be easily poured back into the left-hand chamber when needed. Repeated distillation of the
ammonia into the middle chamber, followed by washing in the right-hand chamber is carried
out. After no more color is observed in the fresh wash solution after pouring it over the
doped polyacetylene, the procedure is repeated once more. Very clean n-doped polyacetylene
is thus produced. Evacuation then removes any uncoordinated solvent.

This chemical n-doping procedure can readily be extended to the lanthanide ions Eu^{+2}
and Yb^{+2}. Europium and ytterbium metals are known to dissolve in liquid ammonia (19).
They form solvated divalent cations and solvated electrons in ammonia with the characteristic
blue color. Upon immersion of polyacetylene the solvated electrons spontaneously reduce
the polyacetylene chains to polycarbanions and the divalent lanthanide ions become
countercations to maintain charge neutrality.

CONDUCTIVITY AND ANALYTICAL RESULTS

Maximum room temperature conductivities measured for polyacetylene films n-doped
in liquid ammonia solutions are given in Table 1. The conductivity of pristine, undoped *cis*-
rich polyacetylene is $^-10^{-9}\ \Omega^{-1} cm^{-1}$, while that of pristine *trans*-polyacetylene is $^-10^{-5}\ \Omega^{-1}$
cm^{-1} (3). Therefore in all cases large increases in conductivities are observed after n-doping
in metal-ammonia solutions. The conductivity values, which range from $1\text{-}100\ \Omega^{-1}\ cm^{-1}$, are
on the low end of conductivities for metals.

92

Table 1. Conductivity and gravimetry data for n-doped polyacetylenes chemically synthesized in metal-ammonia solutions.

Dopant	Formula (by weight uptake)	Conductivity $(\Omega^{-1}\,cm^{-1})$
Na^+	$[Na_{0.30}(CH)]_x$	100
Ca^{+2}	$[Ca_{0.23}(CH)]_x$	20
Sr^{+2}	$[Sr_{0.06}(CH)]_x$	30
Ba^{+2}	$[Ba_{0.07}(CH)]_x$	20
Eu^{+2}	$[Eu_{0.06}(CH)]_x$	20
Yb^{+2}	$[Yb_{0.08}(CH)]_x$	1

For comparison to the divalent dopant ions, the conductivity of a monovalent ion (Na^+) is given in Table 1. The value of 100 Ω^{-1} cm^{-1} observed after n-doping using the sodium-ammonia solution is comparable to values obtained for sodium doped film using other methods such as sodium naphthalide/tetrahydrofuran solutions (20,21). It is clear that divalent dopant ions decrease the maximum conductivities observed when compared to monovalent ions. It is therefore likely that some localization of charge is occurring in the divalent doped polyacetylenes.

Conductivities of all the n-doped polyacetylene are stable over time if the films are kept in an inert environment. Great care needs to be taken since exposure to moisture or oxygen will rapidly diminish their conductivity.

The formulas given in Table 1 are based solely on weight uptake data after exposure of the polyacetylene films to metal-ammonia solutions followed by washing and vacuum drying. To determine the actual metal content of the doped polymers atomic absorption measurements were carried out. These show conclusively that ammonia molecules are coinserted in the polyacetylene films along with the metal cations. This can be compared to the coinsertion of the solvent tetrahydrofuran which has been found to coordinate with the alkali metal ions Li^+ and Na^+ (22,23). The sodium doped samples coinsert slightly more than about one ammonia molecule for each sodium ion. The real formula for the sodium sample is then $[Na^+_{0.17}(CH)^{-0.17}(NH_3)_{0.18}]_x$. This is consistent with the maximum amount of sodium n-doping found in electrochemically n-doped polyacetylenes (22). The more highly charged calcium ions have a solvent sphere consisting of a bit more than three ammonia molecules per divalent calcium ion. The real formula for the calcium doped sample is then $[Ca^+_{0.087}(CH)^{+0.174}(NH_3)_{0.30}]_x$. This again gives a more reasonable value for the actual charge spread over each (CH) chain.

MAGNETIC SUSCEPTIBILITY

The magnetic data for divalent ion doped polyacetylenes are compiled in Table 2. The electron paramagnetic resonance (EPR) signal of undoped *trans-* polyacetylene with a spin concentration of 3.8 x 10^{-8} emu/g was used as a standard to calculate the spin concentrations of the divalent doped polyacetylenes (20). It is clear from Table 2 that in all cases the number of unpaired spins decreases on doping (24). The unpaired spins in pristine *trans*-polyacetylene have been attributed to neutral solitons with a spin of 1/2 (25). Thus the decrease in magnetic susceptibility is likely due to conversion of neutral solitons to negative solitons or bipolarons which have no spin.

The peak to peak line width (ΔH_{pp}) of pristine *trans*-polyacetylene, 1.1G, increases on n-doping with the divalent cations. This broadening of the EPR signal has been observed in polyacetylene n-doped with monovalent cations such as Na^+ (26,27). Sodium-ammonia samples also exhibit line broadening. This line broadening for the divalent doped polyacetylenes is indicative of localization of charge within the soliton and/or bipolaron states. The apparent increase in ΔH_{pp} observed for Ca^{+2}, Sr^{+2}, Eu^{+2} and Ba^{+2} parallels their radii: Ca^{+2} (~1.1Å), Sr^{+2} (~1.3Å), Eu^{+2} (~1.3Å) and Ba^{+2} (~1.5Å). This is consistent with increased spin/orbit coupling as observed with increasing size for monovalent cations (27). All the EPR signals from the divalent doped polyacetylenes were symmetric which is indicative of homogeneous doping (28).

The temperature dependence of the EPR line width for a ~12.1 mol% calcium doped polyacetylene sample is given in Figure 6. As the sample is cooled to 113 K, the linewidth ΔH_{pp} decreases from ~2.5 G to ~0.7 G. This is consistent with increased mobility for the charge carriers as the lattice vibrations are slowed down. Similar temperature dependent spectra have been observed for monovalent doped polyacetylene (28).

INFRARED AND OPTICAL PROPERTIES

Dopant induced infrared absorptions are a characteristic of charged soliton states in polyacetylene (29,30). Both n-doping with incorporation of monovalent countercations such as Na^+ and p-doping with addition of monovalent counteranions such as I_3^- or AsF_6^- gives rise to the dopant induced infrared-active modes u_1 and u_2 (31,32). The u_1 mode, which involves a contraction of the weaker C-C "single" bond on one side of the defect center and expansion of these bonds on the other side, occurs at ~980 cm^{-1}. The u_2 mode involves analogous displacements of the stronger C-C "double" bonds and is found at ~1360 cm^{-1}. The oscillation of charge back and forth across the charged soliton in response to lattice vibration gives rise to these strong infrared absorption bands.

Infrared spectra of polyacetylene films lightly doped with the alkaline earth countercations Ca^{+2}, Sr^{+2} or Ba^{+2} show two new absorptions at ~900 cm^{-1} and at ~1400 cm^{-1} (33). These absorptions increase with heavier doping. However, increasing the doping

Table 2. EPR data for divalent doped polyacetylene.

Composition (by weight uptake)	ΔH_{pp} (G)	χ (emu/g)
trans-$(CH)_x$	1.1	3.8×10^{-8}
$[Ca_{0.08}(CH)]_x$	1.4	7.1×10^{-11}
$[Sr_{0.06}(CH)]_x$	4.1	6.5×10^{-10}
$[Ba_{0.04}(CH)]_x$	6.5	7.1×10^{-11}
$[Eu_{0.04}(CH)]_x$	4.4	$3.3 \times 10-9$

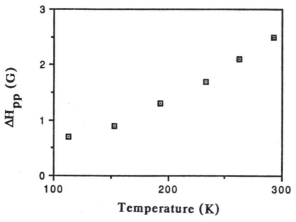

Figure 6.   The EPR line width vs. temperature for a calcium doped polyacetylene film.

level makes the polymer films more conductive and reflective (less absorbing), which rapidly extinguishes their IR signal. The two new infrared peaks at ~900 and ~1400 cm^{-1} have also been observed after incorporation of the lanthanide countercation Eu^{+2} (17). These dopant induced infrared absorptions can be attributed to charge oscillation across the defect center analogous to the monovalent ions. However, these peaks could be consistent with either soliton or bipolaron states.

To explore more fully the nature of the defects induced by the n-doping of polyacetylene with divalent cations, optical absorption experiments have been carried out (17). Spectra were obtained on thin semitransparent films of polyacetylene grown on flat glass. The ~2000Å thick films were synthesized using one-third the normal catalyst concentration and a five second exposure to 60 torr of acetylene gas (10). After washing and drying the thin *cis*-rich polyacetylene films, they were thermally isomerized to *trans*-polyacetylene at 150°C for two hours *in vacuo*(34).

Optical absorption experiments on the thin semitransparent *trans*-polyacetylene revealed essentially no absorption until the band gap of ~1.4 eV was reached. The strong absorption beyond ~1.4 eV in the interband region, is consistent with previous studies on undoped polyacetylene(10). After n-doping with either Ca^{+2} or Eu^{+2} countercations, increased absorption is found in the midgap region (0.7 - 0.9 eV).

A film comparably n-doped with Na$^+$ countercations also shows a midgap absorption which can be attributed to charged soliton states (10). A direct comparison of the Ca^{+2} and Eu^{+2} doped samples with the Na$^+$ doped films reveals that the divalent cation doped polyacetylenes exhibit increased oscillator strength both above and below the midgap transition of the monovalent doped polyacetylene (35). These two levels are the characteristic signature of charged bipolaron states. Bipolarons could be created by confining pairs of solitons. This would occur when a divalent cation associates itself with a single (CH)$_x$ chain, as illustrated in Figure 4b. Some disorder such as a non-random distribution of counterions could also play a part in spreading the midgap state. As the midgap absorption increases a corresponding decrease in oscillator strength is found above the band gap. This bleaching at the band edge is also an expected consequence of confining soliton pairs. Bipolaron formation is expected to preferentially take oscillator strength from the band edge, while soliton formation takes oscillator strength uniformly from the interband region (36).

N-DOPING OF POLYACETYLENE WITH MAGNESIUM COUNTERCATIONS

Since the lighter alkaline earth metals do not dissolve in liquid ammonia, they are more difficult to use as n-dopants for polyacetylene. However, two methods have been developed which can incorporate Mg^{+2} ions into polyacetylene. These methods are the addition of magnesium salts to sodium-ammonia solutions and electrolysis. We will treat each in turn.

Addition of Magnesium Salts to Sodium-Ammonia Solutions

A modified ion-exchange method has been developed which uses solvated electrons created by dissolving sodium metal in liquid ammonia to partially reduce the polyacetylene films. When magnesium ions are present in this reducing solution some will be incorporated into the polyacetylene as n-dopants in place of sodium ions.

Tared polyacetylene films were put in a blue sodium-ammonia solution containing dissolved magnesium perchlorate. The length of immersion time for the films was varied from 30 min to 2 hours. This led to different levels of Na^+ and Mg^{+2} incorporation in the polyacetylene films, as can be seen from Table 3. After washing, drying and weighing the polyacetylene, four-probe conductivity measurements were made. Each film was then decomposed in acid solution for at least two days and atomic absorption spectroscopy was used to determine their sodium and magnesium content. All samples were found to contain coinserted ammonia based on their weight uptake compared to the analytical results.

From Table 3 it is clear that this modified ion-exchange method is successful in substituting some Mg^{+2} ions for Na^+ ions. However, this process does not entirely replace the Na^+ ions with Mg^{+2} ions, and thus results in only mixed ion conductors. All conductivities observed are quite high and should be compared to the value of $100 \ \Omega^{-1} \ cm^{-1}$ reported in Table 1 for pure sodium doped polyacetylene. In every case the conductivity is diminished by the exchange of Na^+ with Mg^{+2} ions. From Table 3 it appears that there is a correlation between an increasing ratio of Mg^{+2} to Na^+ and a decreasing conductivity. This suggests that the Mg^{+2} ions decrease the conductivity, likely by localizing charge.

Several sodium/magnesium doped polyacetylene films were prepared for magnetic measurements. The polyacetylene films were all thermally isomerized to the *trans*-isomer before doping. The EPR data for three representative samples are given in Table 4. In all cases, a decrease in the unpaired spin concentration was observed. This again indicates spinless charge carriers. A broadening of the EPR peak width was also observed in each case. Although most of the samples had peak widths of 2-3 G a sample with $\Delta H_{pp} = 12.5 G$ was found. The broadening is indicative of some charge localization. It's also clear from Table 4 that incorporating Mg^{+2} ions into thermally isomerized *trans*-polyacetylene is a difficult process.

Table 3. Composition and conductivity for Mg^{+2}/Na^+ doped polyacetylene films.

Composition	Conductivity ($\Omega^{-1} \ cm^{-1}$)	Mg^{+2}/Na^+ Ratio
$[Mg_{0.018}Na_{0.049}(CH)_x]$	62	0.37
$[Mg_{0.030}Na_{0.059}(CH)_x]$	27	0.51
$[Mg_{0.032}Na_{0.037}(CH)_x]$	11	0.86
$[Mg_{0.047}Na_{0.046}(CH)_x]$	3	1.02

Table 4. Magnetic data for Mg^{+2}/Na^+ doped polyacetylenes.

Composition	χ (emu/g)	ΔH_{pp} (G)
$[Mg_{0.007}Na_{0.015}(CH)_x]$	1.3×10^{-8}	1.9
$[Mg_{0.012}Na_{0.014}(CH)_x]$	8.7×10^{-9}	2.5
$[Mg_{0.002}Na_{0.022}(CH)_x]$	2.7×10^{-9}	2.8

Electrolysis

Another way of creating solvated electrons in liquid ammonia is by use of an electrolysis method (37). A platinum mesh cathode and a magnesium anode were immersed in liquid ammonia containing magnesium perchlorate. When an external potential of ~3 volts or greater was applied across the electrodes, a blue color was produced at the cathode. After several minutes of electrolysis the entire solution turned blue. This solution, containing solvated electrons and Mg^{+2} ions, was then used to chemically n-dope polyacetylene. After wahing and drying the doped films, four-probe conductivity measurements were carried out. The highest conductivity observed so far for Mg^{+2} doped polyacetylene is 0.6 Ω^{-1} cm^{-1}. The small Mg^{+2} ions thus appear to limit the maximum conductivity for these divalent dopant ions. This electrolysis method could likely be extended to Be^{+2} and Al^{+3} ions and such experiments are currently in progress.

CONCLUSIONS

The reducing power of solvated electrons in metal-ammonia solutions has been shown to be a very effective way of introducing many highly charged dopant cations into polyacetylene. The characteristic blue colored solutions containing solvated electrons and solvated metal ions can readily be produced by dissolving any of the alkali metals, the heavier alkaline earths (Ca, Sr or Ba) or the lanthanides Eu or Yb in liquid ammonia. The lighter alkaline earth metal magnesium can be forced to dissolve in liquid ammonia by electrochemical means. Immersion of polyacetylene films into any of the metal-ammonia solutions results in spontaneous n-doping. The maximum conductivities achieved with the divalent dopant ions are ~30 Ω^{-1} cm^{-1}. These can be compared to the 100 Ω^{-1} cm^{-1} conductivity value for comparable doping with the monovalent cation Na^+. The lower values for the divalent ions indicate some localization of charge carriers. The smallest of the divalent dopant ions considered, Mg^{+2}, has an even lower maximum conductivity of 0.6 Ω^{-1} cm^{-1}. This is likely due to further localization of charge. When the conductivities of mixed Mg^{+2}/Na^+ doped polyacetylenes were examined a correlation between an increasing Mg^{+2}/Na^+ ratio and a decreasing conductivity was observed.

Examination of the magnetic properties of all of the divalently doped polyacetylenes showed that on doping the number of unpaired electrons decreased. This indicates that spinless charge carriers are responsible for conductivity. A model involving charged solitons and bipolarons as the spinless carriers has been presented. The two new absorption peaks found in the infrared spectra of lightly Ca^{+2}, Sr^{+2}, Ba^{+2} and Eu^{+2} doped polyacetylene are consistent with new absorptions found in monovalent doped polyacetylenes which have been attributed to charged soliton states. Visible spectra of the divalent ion doped polyacetylenes show midgap absorptions indicative of charged soliton states with some charged bipolaron states. It is thus reasonable that most of the divalent ions associate themselves with two or more polyacetylene chains creating soliton type states, while some of the divalent dopants associate themselves with a single chain creating bipolaron states. This localization of charge would explain the lowering of conductivity when divalent ions are compared to monovalent ions.

ACKNOWLEDGEMENTS

The authors would like to thank Dr. James Kaufman for help and discussions on optical spectra. This work was supported by the National Science Foundation Grant No. CHE 8657822 through the Presidential Young Investigator Award Program along with matching funds provided by Ford Motor Company and the Petroleum Research Fund Grant No. PRF 21622-AC7-C administered by the American Chemical Society.

REFERENCES

1. T.J. Skotheim, ed., "Handbook of Conducting Polymers", Vols. 1 and 2, (Marcel Dekker, NY, 1986).
2. Proccedings of the International Conference on Science and Technology of Synthetic Metals, *Synth. Met.*, **17-19** (1987); *ibid*, **17-29** (1989).
3. H. Shirakawa, E.J. Louis, A.G. MacDiarmid, C.K. Chiang and A.J. Heeger, *J Chem. Soc., Chem. Comm.,* 578 (1977).
4. H. Shirakawa and S. Ikeda, *Polym. J.*, **2**, 231 (1971).
5. H. Shirakawa, T. Ito and S. Ikeda, *Polym. J.*, **4**, 460 (1973).
6. N. Theophilou, R. Aznar, A. Munarki, J. Sledz, F. Schué and H. Naarmann, *Synth. Met.*, **16**, 337 (1986).
7. H. Naarmann and N. Theophilou, *Synth. Met.,* **22**, 1 (1987).
8. S. Ikehata, J. Kaufer, T. Woerner, A. Pron, M.A. Druy, A. Sivak, A.J. Heeger and A.G. MacDiarmid, *Phys. Rev. Lett,* **45**, 423 (1980).
9. W.P. Su, J.R. Schrieffer and A.J. Heeger, *Phys. Rev. Lett.*, **42**, 1698 (1979).
10. A. Feldblum, J.H. Kaufmann, S. Etemad, A.J. Heeger, T.-C. Chung and A.G. MacDiarmid, *Phys. Rev. B.*, **26**, 815 (1982).
11. W.P.Su, J.R. Schrieffer and A.J. Heeger, *Phys. Rev. B.*, **22**, 2099 (1980).
12. M. Peo, S. Roth, K. Dransfeld, B. Tieke, J. Hocker, H. Gross, A. Grupp and H. Sixl, *Solid State Comm.,* **35**, 119 (1980).
13. J.C. Scott, M. Krounbi, P. Pfluger and G.B. Street, *Phys. Rev. B.*, **28**, 2140 (1983).
14. F. Ignatious, B. François and C. Mathis, *Synth. Met.*, **28**, D109 (1989).
15. A.J. Heeger, *Phil. Trans. R,. Soc. Lond. A*, **314**, 17 (1985).
16. R.B. Kaner, S.-M. Huang, C.-H. Lin and J.H. Kaufman, **28**, D115 (1989).
17. S.C. Gau, Ph.D. Thesis, University of Pennsylvania, 1982.
18. J.J. Lagowski and G.A. Moczygemba, in "The Chemistry of Non-Aqueous Solvents." Vol. II, J.J. Lagowski, ed., Academic Press, New York (1967); W.L. Jolly and C.J. Hallada, in "Non-Aqueous Solvent Systems", T.C. Waddington, ed., Academic Press, New York (1965).
19. J.C. Worf and W.L. Kerst, *J. Phys. Chem.*, **60**, 1590 (1956).
20. T.- C. Chung, A. Feldblum, A.J. Heeger and A.G. MacDiarmid, *J. Phys. Chem.*, **74**, 5504 (1981).
21. J.J. André, M. Bernard, B. François and C. Mathis, *J. Phys.*, **44**, C3-199, (1983).

22. B. François and C. Mathis, *J. Phys.*, **44**, C3-21, (1983).

23. L.W. Shacklette, J.E. Toth, N.S. Murthy and R.E. Baughman, *J. Electrochem. Soc.*, **132**, 1529 (1985).

24. J.H. Kaufman, S.M. Huang, R.K. Shibao and R.B. Kaner, *Sol. State Comm.* (in press).

25. J.C.W. Chien, *J. Polym. Sci., Polym. Lett. Ed*, **19**, 249 (1981).

26. B. François, M. Bernard and J.J. André, *J. Chem. Phys.*, **75**, 4142 (1981).

27. P. Bernier, in "Handbook of Conducting Polymers", Vol. I, pp. 1099-1125, T.J. Skotheim, ed., Marcel Dekker, New York (1986).

28. F. Moraes, J. Chen, T.-C. Chung and A.J. Heeger, *Synth. Met.*, **11**, 271 (1985).

29. E.J. Mele and M.J. Rice, *Phys. Rev. Lett.*, **45**, 926 (1980).

30. S. Etemad, A. Pron, A.J. Heeger, A.G. MacDiarmid, E.J. Mele and M.J. Rice, *Phys. Rev. B.*, **23**, 5137 (1981).

31. B. François, M. Bernard and J.J. André, *J. Chem. Phys.*, **75**, 4142 (1981).

32. C.R. Fincher, Jr., M. Ozaki, A.J. Heeger and A.G. MacDiarmid, *Phys. Rev. B.*, **19**, 4140 (1979).

33. S.-M. Huang and R.B. Kaner, *Sol. St. Ionics*, **32/33**, 575 (1989).

34. A. Feldblum, A.J. Heeger, T.-C. Chung and A.G. MacDiarmid, *J. Chem. Phys.*, **77**, 5114 (1982).

35. J.H. Kaufman, S.-M. Huang, R.K. Shibao and R.B. Kaner, *Sol. State Comm.* (in press).

36. K. Fesser, A.R. Bishop and D.K. Campbell, *Phys. Rev. B.*, **27**, 4804 (1983).

37. A.D. McElroy, J. Kleinberg and A.W. Davidson, *J. Am. Chem. Soc.*, **72**, 5178 (1950).

NEW POLY(IMIDAZOLEAMIDES) SYNTHESIZED FROM HCN TETRAMER

D.S. Allan*, A.H. Francis†, P.G. Rasmussen*†

*Macromolecular Science and Engineering Program
†Department of Chemistry
The University of Michigan
Ann Arbor, MI 48109-1055

Introduction

The search for new polymeric materials having useful properties has been made by developing new mixtures of old materials, as polymer blends or copolymers, or by the development of new polymer backbones. In this paper we describe the synthesis and properties of an unusual heteroaromatic polyamide derived from the cyanoimidazole nucleus.

Aromatic polyamides are of interest as fiber-forming materials of high strength and thermal stability. The homoaromatic polyamides poly(p-phenylene terephthalamide) and poly(m-phenylene isophthalamide) have been very successful as fire-resistant high strength fibers. Poly(p-benzamide) also has good fiber properties, but has not had the same commercial success. All three of these polymers, and indeed aromatic polyamides in general, suffer from low solubility and difficult processability.

We sought to incorporate cyanoimidazoles into heteroaromatic polyamides. The presence of heteroatoms in the aromatic ring provides a chemical "handle" by which the polyamide might be modified to improve processability or physical properties. Furthermore, the nitrile group confers acidity to the nitrogen atom at the 1-position of the imidazole ring. This acidity might be exploited to improve solubility of the polymer in basic solvents for improved processing.

Cyanoimidazoles as a class are synthesized conveniently from diaminomaleonitrile (DAMN, **1**)—a tetramer of hydrogen cyanide—and electrophilic carbon reagents. Reaction of DAMN with cyanogen chloride affords 2-amino-4,5-dicyanoimidazole (**2**) in moderate yield [1].

The amine **2** is somewhat atypical of aromatic amines. The pKa for deprotonation of the 1-position is a relatively acidic 6.2. The 2-amino group is very non-nucleophilic. Both of these effects are due to the strong perturbation of the electronic structure of the imidazole ring caused by the presence of two strongly electron-withdrawing nitrile groups. The anion produced by deprotonation of the 1-position is stabilized by delocalization of the charge into the nitrile groups; similarly, the lone pair of the nitrogen of the amino group

Inorganic and Metal-Containing Polymeric Materials
Edited by J. Sheats *et al.*, Plenum Press, New York, 1990

is delocalized through the ring into the nitrile groups and is therefore less available for nucleophilic attack on electrophilic species.

The nitrile groups in 4,5-dicyanoimidazoles that are unsubstituted at the 1-position are equivalent due to tautomeric equilibration of the active proton between the two nitrogen atoms of the ring. Under controlled conditions, it is possible to hydrolyze a single nitrile group to the amide or completely to the carboxylic acid with the other nitrile remaining intact. Such controlled hydrolysis has been reported for 4,5-dicyanoimidazole [2]. 2 can be hydrolyzed under varying conditions to give the monoamide, diamide, or dicarboxylic acid. The mononitrile-monocarboxylic acid 3 is available from the monoamide by subsequent hydrolysis [3].

Basic hydrolysis of the nitriles requires an excess of alkali, since one equivalent of base is consumed by the buffering capacity of the acidic proton at the 1-position of the imidazole ring. Likewise, acid hydrolysis of the monoamide requires excess HCl since one equivalent of acid is consumed by the ammonia by-product.

The amino acid 3 has low solubility in water, although the compound itself binds water tightly. Elemental analysis of the amino acid fits the composition of the monohydrate, $C_5H_4N_4O_2 \cdot H_2O$. If the acid hydrolysis reaction is neutralized with sodium bicarbonate, the sodium salt of the amino acid precipitates as a white solid. The analysis for this product fits that calculated for the hemihydrate.

The highly functionalized imidazole 3 can be polymerized by active phosphite ester methods [4] to afford the A-B polyamide. The monomer is dissolved in a mixed solvent system of N-methylpyrrolidinone and pyridine containing dissolved lithium chloride. Upon addition of triphenyl phosphite, the reaction immediately takes on a yellow color. During the course of the polymerization, the color increases in intensity until the polymerization mixture appears black. The products are isolated as fine yellow powders by precipitation in methanol followed by a methanol wash.

The intense yellow color of the polymeric products is attributed to light absorption by an extended π-electronic system. Upon polymerization, the polymerizing functional groups are converted to amide linkages which are capable of participating in extended conjugation along the polymer backbone. To the extent that such conjugation contributes to the ground state electronic structure of the polymer, the π-system is delocalized and the electronic absorptions move from the UV into the visible regime.

Compound **3** → (reagents: PhO)$_3$P, NMP, LiCl, pyridine) → Compound **4**

In the course of the active ester polymerization, some nitrile hydrolysis apparently occurs. While clearly present, the nitrile stretching band in the infrared spectrum of the polymer is diminished in intensity. The diminution of the nitrile band could be due to a change in the environment of the nitrile upon polymerization or a decrease in the concentration of the group because of some side reaction such as hydrolysis. A reasonable fit to the combustion analysis data for the polymer can be made if a water molecule is included in the calculation. The presence of water in the polymer could be the result of hydration of the hydrophilic functional groups in the molecule or the result of hydrolysis of the some of the nitrile groups to amide groups or a mixture of both.

Chemical and Physical Properties

The polymer is soluble in polar aprotic solvents such as N-methyl pyrrolidinone and in aqueous base. The solubility in base is due to deprotonation of the acidic hydrogen at the 1-position of the imidazole ring. In imidazoles, the pKa of the 1-nitrogen is approximately 14, but in cyanoimidazoles the pKa of this position ranges from 2 (in 4,4',5,5'-tetracyano-2,2'-biimidazole) to approximately 9-10 in monocyano substituted imidazoles [5]. The acidity of this position provides a handle for solution processing or for chemical modification of the polymer.

The intrinsic viscosity of the polymer prepared using the phosphorylation reaction is 0.2 dl/g in NMP at 30°C. The physical properties of the materials prepared thus far are those expected for rigid chains of moderate molecular weight. Brittle films can be cast from aqueous ammonium hydroxide solution or from polar aprotic solvents. The films which form are brittle.

At cryogenic temperatures, the polymer exhibits photoluminescent behavior characteristic of semiconductors. When irradiated at 77K with the 450 nm line of an argon ion laser the polymer exhibited a luminescence maximum of 550 nm (Figure 1). Similar measurements on a dimer model compound (*vide infra*) showed a luminescence maximum at higher energies. The luminescence occurred with high quantum yield from all samples, an observation which is atypical of noncrystalline materials. The emissions covered a wide frequency range; however, the peak maximum with the highest energy was from the model dimer. The correlation of red shift of the luminesence maximum with other estimates of molecular weight suggests that the conjugation length increases on average as the chain length increases. The observation of these emissions is indicative of the presence of a band gap relatively free of trapping states and impurities, and suggests that negative carrier doping by chemical or electrochemical means may lead to stable materials.

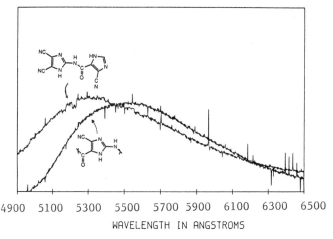

WAVELENGTH IN ANGSTROMS

Figure 1. Laser Photoluminescence Spectrum

Conformational Analysis

Since conjugation of the backbone requires some extended planarity of the polymer chain, we have investigated the backbone conformation both by model studies and computational means. We synthesized the dimer model **5** and solved its crystal structure.

The dimer is planar within experimental error and contains two waters of hydration per dimer molecule (Figure 2). Guinier X-ray powder diffraction on the polymer shows it to be partially crystalline as well.

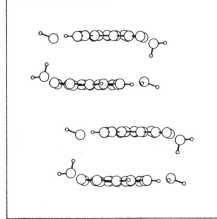

Figure 2. X-Ray Crystal Structure of the Dimer Model

This result suggests that the polymer will prefer either of the planar conformations shown below of the four possible planar dyads (Figure 3). In the notation of polypeptide chemistry, these conformations correspond to φ=0 or 180° and ψ=180°. Conformer (a) yields a linear polymer backbone, while conformer (b) results in a helical chain.

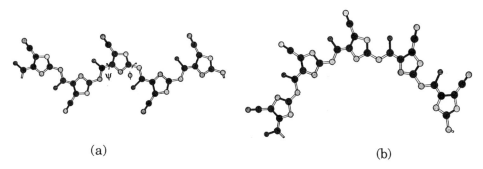

(a)   (b)

Figure 3. Planar Conformations of the Polyamide

We have investigated electrochemical reductions of monomeric cyanoimidazoles by cyclic voltammetry experiments [6]. These measurements show that cyanoimidazoles are good candidates for negatively doped conducting polymers. The native polymer is an insulator ($\rho > 10^8$ cm^{-1}ohm^{-1}) by pressed pellet measurements at room temperature. Further experiments, including doping and photoconductivity measurements, are underway.

Thermal Stability

The polymer contains very little hydrogen and has a very high nitrogen content. The lack of hydrogen limits kinetic pathways for thermal decomposition of the polymer, conferring thermal stability. The differential scanning calorigram shows a broad peak centered at 100°C attributed to loss of adsorbed water and a narrower peak at 220°C attributed to loss of bound water. The calorigram is relatively featureless between 220 and 350°. Above 350°C, strongly endothermic decomposition is evident.

Experimental

2-Amino-4(5)-cyano-5(4)-imidazolecarboxamide

2-Amino-4,5-dicyanoimidazole (3.83 g, 28.8 mmol) was dissolved in sodium hydroxide solution (0.58 $\underline{N}$, 60 ml, 1.2 eq). The solution was heated at room temperature for 22 h, after which time a small precipitate of diamide side-product had formed. The solution was filtered and returned to the flask for reflux for 3 h. The solution then was acidified to pH=2 with hydrochloric acid. The precipitate that formed was isolated by filtration to give 3.75 g (86%) of crude product. The product was purified by recrystallization from 410 ml of water to yield 3.33 g (77%) of the desired product. IR(KBr) 3478, 3440, 3348, 3206, 2232, 1693, 1646, 1611, 1567, 1294, 1096. MS(EI) 151 (M+), 134, 106, 80, 53, 43.

Anal. calcd. for $C_5H_5N_5O$, C, 39.74, H, 3.31, N, 46.36; obs. C, 39.38, H, 3.31, N, 46.22.

The diamide by-product was present to the extent of 1.4%. IR(KBr) 3478, 3385, 3346, 3258, 3197, 3020, 1695, 1680, 1656, 1637, 1583, 1491, 1413, 1271, 758, 718, 627, 485. MS(EI) 151(M^+), 134(100%), 106, 80, 53, 43. Anal. calcd. for $C_5H_7N_5O_2$, C, 35.51, H, 4.17, N, 41.41; obs. C, 35.80, H, 4.13, N, 41.53.

2-Amino-4(5)-cyano-5(4)-imidazolecarboxylic acid

2-Amino-4(5)-cyano-5(4)-imidazolecarboxamide (6.36 g, 42.1 mmol) was combined with hydrochloric acid (5 ml conc. HCl, diluted to 100 ml) and heated to reflux. The amide dissolved after about 1 h of heating. After 18 h of heating, a white precipitate had formed. The flask was cooled and the precipitate was isolated by filtration and dried to give 3.35 g (52%) of the desired product. The filtrate volume was reduced and a second crop of product weighing 0.36 g (5.6%) was obtained. IR(KBr) 3407, 3160, 2253, 1687, 1604, 1397, 1122 cm^{-1}. MS(EI) 151(M^+), 134, 108, 81, 65, 54, 44(100%), 38. Anal. calcd. for $C_5H_4N_4O_2 \cdot 1H_2O$, C, 35.30, H, 3.55, N, 32.93; obs. C, 34.97, H, 3.46, N, 31.19.

Direct Polycondensations with Triphenyl Phosphite and Lithium Chloride

Representative procedure: 2-Amino-4(5)-cyano-5(4)-imidazolecarboxylic acid (0.510 g, 3 mmol), anhydrous lithium chloride (0.5 g), pyridine (2 ml, 1.95 g, 24.7 mmol), and N-methylpyrrolidinone (7 ml) were combined in a dry 25-ml three-necked roundbottom flask equipped with magnetic stirrer, reflux condenser, and oil bath. The mixture was warmed to 85°C to dissolve the solid materials. Triphenylphosphite (1 ml, 1.18 g, 3.82 mmol) was added. The solution turned yellow in minutes. After heating at 85°C for 2 h, the mixture was cloudy. Heating was continued for a total of 18 h, after which time the mixture was dark red-yellow.

The flask was cooled and the mixture poured into 250 ml of methanol. A bright yellow solid precipitated immediately. The solid was filtered and washed with methanol to give 0.270 g (66%) of product. Reaction in the absence of pyridine did not cause significant changes; reaction in the absence of lithium chloride lead to greatly diminished yield.

Characterization: IR(KBr) 3386, 3138, 2247, 1704, 1641, 1579, 1419, 1347, 1312, 1248 cm^{-1}. Mass spectroscopy under a variety of ionization conditions (electron impact, chemical ionization, fast atom bombardment) failed to give any spectrum. A portion of the solid was dissolved in ammonium hydroxide solution (diluted 1:5) and cast on a watch glass. A brittle yellow film was deposited on evaporation of the water. Microscopic examination of the film did not reveal any discernible crystals. X-ray powder diffraction (Guinier) further demonstrated that the material is noncrystalline. Anal. calcd for $C_5H_2N_4O$ (repeat unit), C, 44.79, H 1.5, N 41.78, O 11.93 ; calcd. for $C_5H_4N_4O_2$ (hydrate), C 39.48, H 2.65, N 36.83, O 21.04; obs. C, 40.54, H, 2.71, N, 31.01. UV-Vis(NMP) λ_{max} 424 nm(ε 47.2 $\dfrac{1}{g \cdot cm}$ ~ 6.32 $\dfrac{1}{mol \cdot cm}$). [η] = 0.19 dl/g (NMP, 30.00°C).

Acknowledgments

The authors gratefully acknowledge support for this work from NSF-DMR-8820628. DSA acknowledges predoctoral fellowship support from the NSF. We are grateful to Y.K. Kim for preparation of one of the compounds used in this study.

References

1. Donald, D.S.; Webster, O.W. *Adv. Het. Chem*, **1987**, *41*, 1-40 (review).

2. Yamada, Y.; Kumashiro, I.; Takenishi, T. *Bull. Chem. Soc. Jpn.*, **1968**, *41*, 1237-1240.

3. Rasmussen, P.G.; Allan, D.S.; Apen, P.G.; Thurber, E.L. *Polym. Prep.*, **1988**, *29(1)*, 325-6.

4. Higashi, F.; Goto, M.; Kakinoki, H. *J. Polym. Sci., Polym. Chem. Ed.*, **1980**, *18*, 1711-17.

5. Rasmussen, P.G.; Hough, R.L.; Anderson, J.B.; Bailey, O.H.; Bayón, J.C. *J. Am. Chem. Soc.*, **1982**, *104*, 6155-6.

6. Allan, D.S.; Bergstrom, D.F.; Rasmussen, P.G. *Synthetic Metals*, **1988**, *25*, 139-155.

SYNTHESIS AND STRUCTURE OF POLYIMIDE/METAL MICROCOMPOSITES VIA

IN SITU CONVERSION OF INORGANIC ADDITIVES

L. T. Taylor* and J. D. Rancourt

Virginia Polytechnic Institute and State University
Department of Chemistry, Blacksburg, VA   24061-0212

ABSTRACT

In contrast to the preparation of composites via the heterogeneous dispersion of fillers within the matrix resin, materials have been prepared by the homogeneous incorporation of metal or metal oxide precursors. Processing the intitially mutually soluble blend results in a composite material. In light of the uniqueness and general applicability of composites, a review of the literature as it relates to this homogeneous approach is presented. Most of the published studies concern high temperature stable polymers such as polyimides and a variety of relatively stable metal additives. The following review will therefore deal only with polyimide/metal-based composite films; wherein, the polymer functions essentially as an inert matrix for the additive conversion.

INTRODUCTION

The continuing development and application of high technologies has intensified the quest for new materials which exhibit special properties. Composites have emerged as a new generation material with tremendous potential in this regard. Composites are formed from a matrix (polymer) and a filler which may be metallic (e.g. metal, metal oxide or alloy) or nonmetallic (e.g., silicon nitride, silicon carbide, carbon fibers). The great potential of composites derives in part from the substantial modification of the thermal, mechanical and electrical characteristics of polymers which results when the polymer is combined with particulate, flake or fiberous fillers. But, the composite retains, to a large extent, the advantages of plastics such as cost, processability, density and aesthetics. The first modern attempt to combine polymers and metals occurred in 1951 when a glass fiber-reinforced polystyrene mine case was designed for the Army[1]. Since that time a number of composites have become prevalent in, for example, automotive, aerospace, electronics and construction industries.

Conventional polymer-metal composites involve the dispersion of filler in a polymer matrix. Two methods of packing filler particles in a polymer matrix have evolved during the past 30 years: "random" distributions and "segregated" distributions. Continuity in the "random" packing method relies on the formation of a network strictly by the chance contact between filler particles as governed by percolation theory. The composite may be described as a three dimensional lattice. Filler and polymer species are of similar effective size and shape and, therefore, each may occupy any lattice site. In the "segregated" distribution, small filler particles preferentially occupy the interstices of the larger polymer particles, thereby giving rise to a reticulated structure. "Random" conductive composites are amenable to injection molding or extrusion techniques; whereas, "segregated" conductive composites are normally processed by compression molding. The critical filler volume loading for "random" composites is considerably higher than for "segregated" composites.

More heat-resistant thermoplastic resins are usually preferred in polymer blends which contain metallic fillers such as iron, copper, aluminum and silver powder or flakes. At least six factors must be considered when preparing these types of polymer-metal composites: (1) particle size ratio, (2) particle shape, (3) prepolymer viscosity, (4) electrostatic attraction, (5) oxide layer thickness, and (6) polymer bead shear. Another factor to be considered in blending is the difference in density between the organic polymer and metallic filler. In other words, because the viscosity of the mixture is usually lower during processing, the denser metallic particles may tend to separate. Given the large number of variables in preparing a composite, these materials may suffer from unpredictable and uncontrollable heterogenieties and from lack of reproducibility due to the preparative "art" involving formulation of both the metal powder dispersion and the polymer-metal composite.

In contrast to the preparation of composites via the heterogeneous dispersion of fillers within the matrix resin, materials have also been prepared by the homogeneous incorporation of metal or metal oxide precursors. Processing the intitially mutually soluble blend results in a composite material. In light of the uniqueness and general applicability of composites, we wish to review the literature as it relates to this homogeneous approach. Most of the published studies concern high temperature stable polymers such as polyimides and a variety of relatively stable metal additives. Transition metal carbonyls, however, have been used with less thermally stable, more reactive olefin-based polymers[2-8]. The following review will deal only with polyimide/metal-based composite films; wherein, the polymer functions essentially as an inert matrix for the additive conversion.

RESULTS AND DISCUSSION

The preparation of composites prepared by high temperature decomposition reactions have mainly concentrated on polyimides and reactive inorganic additives (e.g., simple salts and complexes). In many cases polymer-metal composites have been unintentionally produced by workers whose primary goal was the modification of polyimide properties by incorporation of metal ions. For example, approximately 30 years ago Angelo[9] briefly reported in a patent the addition of metal ions to several types of polyimides for the purpose of forming particle-containing (<1μ)[10] transparent polyimide shaped structures. Ten years later a report[10] appeared concerning silver acetate incorporation into the poly(amide acid) derived from pyromellitic dianhydride (PMDA) and 4,4'-oxydianiline (ODA). Prior to this report a patent[11] was filed covering very similar work wherein a poly(amide acid) containing silver acetate was converted to a polyimide containing silver metal by heating the material at 300°C **in vacuo** for 30 minutes. The film was stated to be tough, flexible, opaque and metallic. Further heating in air at 275°C for 5-7 hours, however, was necessary to render the film electrically conductive. Most of these doped films exhibited a unique feature in that one side contained a tightly adhering, readily observable deposit which originated as a result of the thermal imidization process. In each case the resulting surface deposit was derived from the chemical dopant.

During the 1980's there has been a revival of interest in the **in situ** thermal decomposition of inorganic additives in polyimide precursors. Specifically, different metal systems and polyimide precursors have been explored, a better understanding of the structure of polyimide/metal composites has been achieved, and numerous contributing synthetic parameters have been deduced. The following dianhydride-diamine pairs which exhibit glass transition temperatures between 200 and 400°C have been employed: BTDA-ODA, BTDA-APB, BDSDA-ODA and PMDA-ODA, (Structures 1-4). The synthetic procedure involved

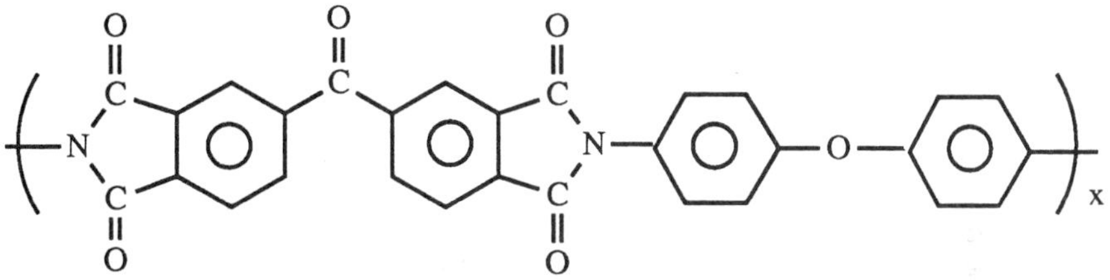

BTDA-ODA

1

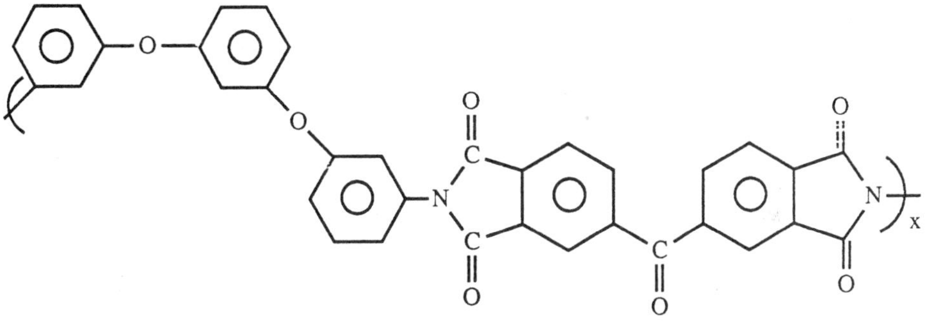

BTDA-APB

2

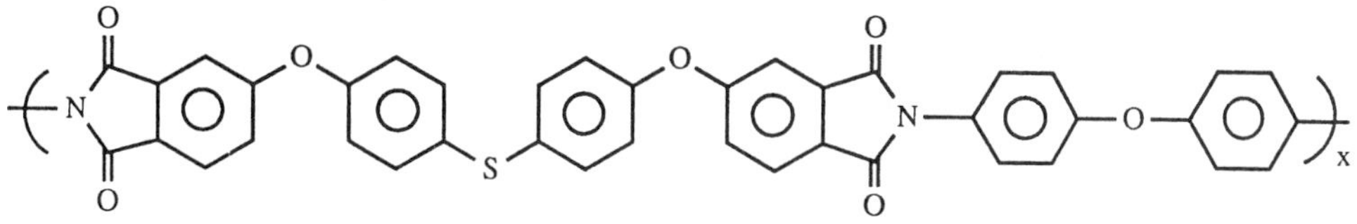

BDSDA-ODA

3

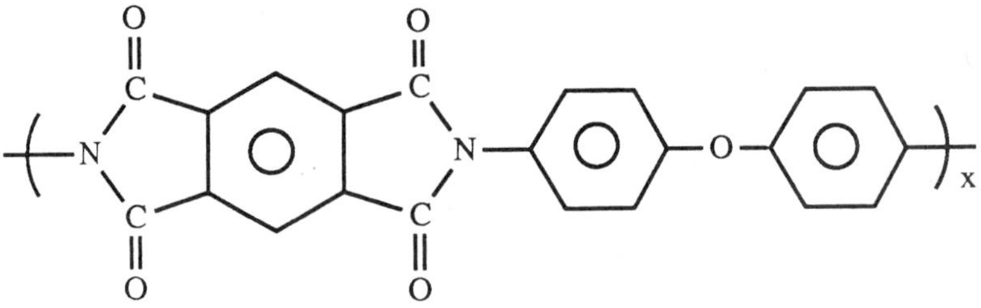

PMDA-ODA

4

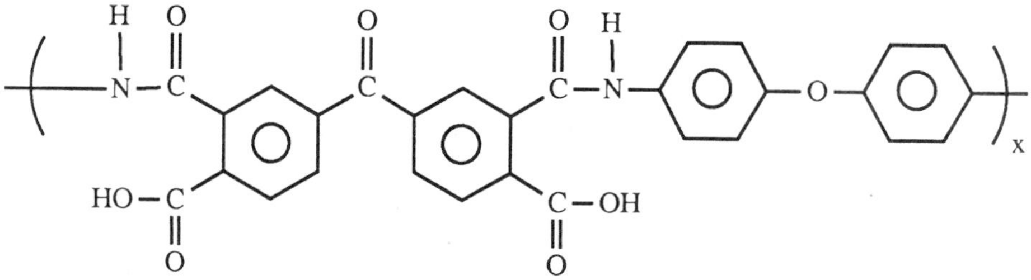

Poly(amide Acid) of BTDA-ODA

5

formation of a poly(amide acid) intermediate (**5**) in N,N-
dimethylacetamide, intimate mixing (dissolution) of the additive and
poly(amide acid) and subsequent thermal imidization of the precursor and
decomposition of the additive in controlled atmosphere with formation of
a polyimide/metal composite film. An alternate **in situ** method whereby
polymerization to the poly(amide acid) was performed in the presence of
the inorganic additive also proved satisfactory. Two patterns of dopant
reactivity have generally been observed[12-18]: (a) conversion of the
additive to the metallic state (Pd, Pt, Ag, Au) or (b) conversion of the
additive to the metal oxide (Li_2O, CuO, SnO_2, Co_3O_4). Table I lists the
optimized additives in each case. Free-standing films of approximately
1 μm thickness have been prepared.

The primary motivation for preparing metal ion modified polyimide
films has been to obtain materials having mechanical, electrical,
optical, adhesive and surface chemical properties different from
nonmodified polyimide films. For example, the tensile modulus of metal
ion modified polyimide films was increased (both at room temperature and
200°C) and elongation was reduced compared with the nonmodified
polyimide[19]. Although certain polyimides are known to be excellent
adhesives, lap shear strength (between titanium adherends) at elevated
temperature (275°C) was increased further by incorporation of
tris(acetylacetonato)aluminum(III)[20]. In another example, highly
conductive, reflective polyimide films containing a palladium metal

Table I. Additives for **In Situ** Thermal Decomposition

Conversion to Metal	Conversion to Metal Oxide
$Pd(S(CH_3)_2)_2 Cl_2$	$LiCl$
$Pt(S(CH_3)_2)_2 Cl_2$	$Cu(TFA)_2$*
$AgNO_3$	$SnCl_4$
$HAuCl_4 \cdot 3H_2O$	$CoCl_2$

*TFA = Trifluoroacetyacetonate anion

surface were prepared and characterized. The thermal stability of these films was reduced about 200°C, but they were shown to be useful as novel metal-filled electrodes[21].

Structure of Composite Films

Most of these systems have been characterized by X-ray photoelectron spectroscopy (chemical state of near-surface layer), Auger electron spectroscopy with depth profiling via argon ion etching (position and thickness of near-surface layer), transmission electron microscopy of ultramicrotomed cross-sections (physical internal structure), elemental analysis (extent of metal salt conversion), and surface electrical resistivity versus temperature profiles (continuity of near-surface layer). The data from these techniques were used cooperatively to develop a model for these microcomposite polyimide films. The model represents the sample as three distinct regions, Fig. 1. The bulk of the film contains either converted (e.g. Ag) or nonconverted (e.g. $CoCl_2$) additive in a predominately polyimide environment. An oxide-rich (e.g. Co_3O_4) or metal-rich layer (e.g. Ag, Au) interspersed with polyimide accounts for the second region. Finally, the surface region termed the polymer overlayer is composed of a gradient of both metal (or oxide) and predominately polyimide.

Direct current electrical resistivity determinations (bulk and each surface) of polyimide films either containing **in situ** generated Co_3O_4 or SnO_2 have proven valuable toward gaining information concerning the microstructure of these systems. The field and time dependence of the electrical current as well as the time dependence of charge accumulation have been evaluated[22]. In general, for both Co_3O_4 and SnO_2 containing films the bulk materials and cast-side surfaces were characterized as dielectric in nature with loss and conduction. The air-side charging characteristics, on the other hand, were markedly different from the volume mode charging characteristics. No short term polarization occurred in the samples implying little or no direct contribution from the polymer matrix to the air-side electrical properties. Also, the air-side electrical resistivity of BTDA-ODA polyimide films at room temperature was reduced substantially for both cobalt chloride modified and tin chloride dihydrate modified samples (i.e. six orders of magnitude and eleven orders of magnitude, respectively).

The qualitative nature of the air-side resistivity-temperature profiles for the ion-modified polyimides categorized them as either semiconductors or insulator[5]. The magnitude (10^{12}-10^{15} ohm) of the surface resistivity categorized the modified films as insulators. In addition, no instantaneous polarization and no time-dependent polarization current was observed as expected for an insulator. On the other hand, the lack of dielectric response in the air-side charge-time profiles for the majority of these films suggested that the air-side surface was actually a semiconductor. The equation $\rho_v = \rho_s xd$ can be used to estimate the bulk resistivity (ρ_v) of the air-side surface oxide layer provided the surface resistivity (ρ_s) and the oxide layer thickness (d) are known. Such a calculation is valid because the Auger depth profile and electrical resistivity measurements (air-side compared with volume) indicate that metal ion modified polyimide films were

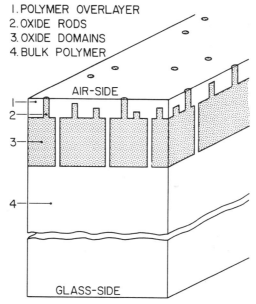

1. POLYMER OVERLAYER
2. OXIDE RODS
3. OXIDE DOMAINS
4. BULK POLYMER

AIR-SIDE

GLASS-SIDE

Fig. 1.   Structural Model of Polyimide/Metal Composite Films.

highly anisotropic if cured in air. For example, the measured surface resistivity of a cobalt chloride modified polyimide film is $1.1 \times 10^{12} \Omega$ at room temperature, which results in a calculated bulk resistivity of $1.5 \times 10^7 \Omega$ cm based on a cobalt oxide thickness of 1350 Å. Kuomoto and Yanagida[23] have measured the bulk electrical resistivity of Co_3O_4 (99.99% purity) in the temperature range between 200 and 900°C. Extrapolation of their data in the region between 200 and 320°C (Figure 1 of Ref. 23) to 25°C results in a determined bulk resistivity value of $1.2 \times 10^8 \Omega$-cm, which compares favorably with the calculated value for several cobalt chloride modified films.

Preparation technique, purity, method of analysis, and oxide structure may each influence the apparent bulk film resistivity. For example, details regarding lithium chloride-cobalt chloride codoped polyimide films have been published[24]. In the co-doped films, mixed oxides of only about 200 Å thickness gave surface resistivities at 150°C in vacuum of approximately 10^{10} Ω whereas, in order to obtain a comparable surface resistivity with only Co_3O_4, a thickness of about 1800 Å was required. Furthermore, the mixed oxide film has a 10%-15% lower activation energy for surface conduction than the film containing the single oxide.

Apparent activation energies for bulk and air-side electrical conduction of Co_3O_4 and SnO_2 containing films (37-81 Kcal/mole) were obtained from the slope of Arrhenius plots[25]. The volume mode

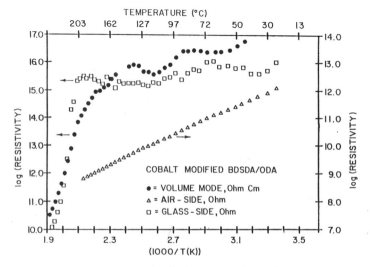

Fig. 2.   Surface and volume resistivity as a function of temperature for cobalt chloride modified BDSDA-ODA polyimide.

electrical resistivity determinations indicated several linear regions, Fig. 2. The bulk activation energy calculated below the glass transition temperature depends on the polyimide as would be expected based on free volume considerations. The lower than expected activation energy (13-15 Kcal/mole) for the air-side and its nondependence on polyimide structure reinforced the notion that the conduction in this mode was electronic regardless of whether the films were cured in dry or moist air.

Variable temperature electrical resistivity determinations indicate
however that the near-surface region (air side surface) of the polymer
contained discrete metal oxide domains as opposed to a continuous metal
oxide layer. That is, up to the polymer glass transition temperature
surface resistivity decreased with increasing temperature in keeping
with the expected semiconductor properties of the metal oxide. Heating
the polymer through its T_g, however, suddenly causes the surface
resistivity to thermoreversibly increase. These data suggested that
some regions of the surface metal oxide domains may be separated by a
very small amount of polyimide. In other words, as the polyimide film
is heated through the glass transition region the polymer may expand

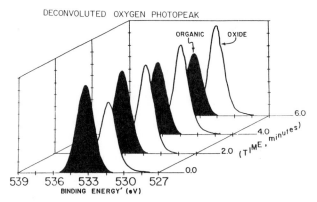

Fig. 3. XPS oxygen photopeak ion milling depth profile of cobalt
chloride modified BTDA-ODA polyimide.

thereby increasing the distance between the metal oxide domains and
consequently increasing the resistivity of the polymer in the glass
transition region by more than an order of magnitude. Air-side X-ray
photoelectron spectroscopy (XPS) photopeaks for carbon, oxygen
(carbonyl) and nitrogen also confirmed the presence of polyimide at the
surface. Controlled ion milling coupled with atomic concentration
analysis of the milled area via both XPS and AES further supported the
presence of a polyimide/Co_3O_4 gradient overlayer. Signals due to
polyimide (C,N) initially decreased with increasing sampling depth
while the signal due to metal increased. Overall, the signal due to
the oxygen photopeak, having a contribution from both the organic
(polyimide) oxygens and oxide oxygen (cobalt oxide), increased with
increasing sampling depth. However the curve-resolved oxygen photopeak
indicated that the organic oxygen (higher binding energy) peak area
actually decreased but the oxide oxygen (lower binding energy) peak
area increased with increasing sampling depth Fig. 3.

In another study, the near-surface distribution of gold was examined by nondestructive variable take-off angle XPS[26]. While the amount of gold detected was small at all angles, it increased with increasing sampling depth for both sides of the BTDA-ODA film. This suggested that both outermost film surfaces (glass-side and air-side) were virtually all polyimide which decreases in concentration as sampling depth increases.

The extent of oxide and/or polymer overlayer formation is very sensitive to process conditions. For example, in the case of cobalt chloride doped BTDA-ODA polyimide films, a moist air cure atmosphere produced the lowest air-side surface electrical resistivity; while, curing in dry nitrogen or a dry air atmosphere produced the highest resistivity films. In contrast to cobalt chloride modified polyimides, the moist air cured tin modified polyimide had the higher resistivity compared with those cured in dry air, Fig. 4. The fact that a dry air or a moist air atmosphere produced films having very different electrical characteristics is interesting in view of X-ray photoelectron spectroscopic data. The binding energy values for the metal and oxide oxygen photopeaks in the cobalt and tin cases were nearly identical with dry and moist air curing, as are the peak shapes.

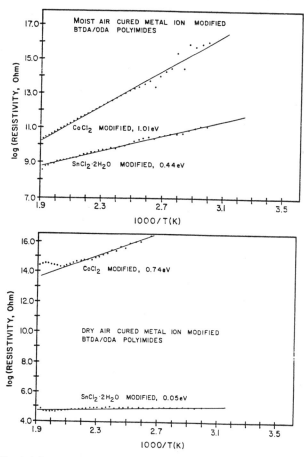

Fig. 4. Variable temperature surface electrical resistivity of modified polyimide films cured in dry air and moist air.

117

Employing atomic composition data, obtained by XPS, one is, however, led to believe that the atmosphere may affect the type of oxide that forms. Co/O ratios for dry and moist air cured films are 1.7 and 0.9, respectively; whereas, Sn/O ratios are 1.6 and 1.7. Thus, the electrical properties are apparently influenced by both the type of oxide formed during the thermal cure and by the spatial distribution and amount of metal oxide within the surface of host polymer. The thickness of both the oxide layer and polymer overlayer also appear to be very important here in achieving improved surface electrical resistivity.

An additional feature of these gradient composites is that, although some of the dopant is converted to metal oxide, the bulk resistivity is still many orders of magnitude greater than it should be

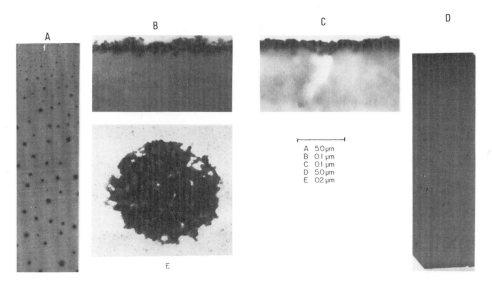

Fig. 5. Transmission electron micrographs of ultramicrotomed cross-section of cobalt chloride modified BTDA-ODA polyimide film before (A, B and E) and after (C and D) extraction.

for the amount of incorporated metal salt (e.g. $CoCl_2$ and $SnCl_2$). This leads to the speculation that bulk precipitation of the metal salt has occurred because the dielectric constant of the polymer, upon removal of the solvent, is not high enough to support dissociation of an appreciable amount of the dopant. The bulk structural features envisioned here, were verified using transmission electron microscopic evaluations of ultramicrotomed cross sections (Fig. 5, parts A, B and E)[27] of cobalt chloride modified BTDA-ODA polyimide films cured in a moist air atmosphere. Energy dispersive analysis of X-rays (EDAX) verified that the particles dispersed throughout the films were

composed primarily of cobalt and chlorine. It was attempted, on the
basis of these data, to use extraction techniques to modify the polymer
structure and properties via removal of any nonconverted dopant.
However, the polyimide film, at the level of ion incorporation
utilized, was an efficient barrier against extraction. Proof that if
the polyimide were more permeable then greater extraction efficiency
would have been attained was deduced by the following experiment. The
same TEM grid, used to obtain the micrograph shown in Figure 5, part A,
was soaked in distilled water. Prior to soaking the particles were
primarily composed of cobalt and chlorine and should therefore be
soluble in water. After soaking the grid for 24 hours the micrographs
shown in Figure 5, parts C and D, were obtained. All the spherical
particles were absent but most of the near-surface deposit was present.
The spherical particles were apparently removed because the
ultramicrotomed cross-section thickness, comparable to the diameter of
these particles, makes the particles ($CoCl_2$) accessible to the water
which can dissolve them. Co_3O_4, although accessible to the water, is
insoluble and was retained by the polyimide.

Gold/Polyimide Composites

The metallization process involving gold-doped polyimides is unique
in that both sides of the film contained gold. Hot stage optical
microscopy was employed to study this phenomenon in detail[28]. The

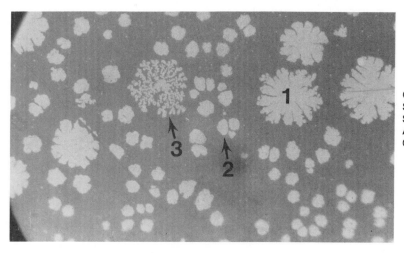

GLASS
SIDE
SURFACE
AT 125X.
OPTICAL.

Fig. 6.   Reflected-light micrograph (125X) of the glass-side surface of
a $HAuCl_4 \cdot 3H_2O$ modified BTDA-ODA polyimide film.

glass side of gold-modified BTDA-ODA film was much different than the
air-side. In general, the air side surface was comprised of metallic
gold which exhibited shapes suggestive of colloidal gold. The glass-
side had several types of gold aggregates which could be classified
according to their shape, Fig. 6 . A Class I aggregate is the large
structure (75-110 μm diameter) which is slightly branched at the
particle periphery. Class II particles are smaller (10-30 μm

diameter), more compact and are the most abundant shapes observed on the glass-side. Class III particles are highly ramified and have a compact center with dendritic arms emanating from a central core. The distribution of particle size and particle shape were thought to be influenced by the imidization reaction. A triocular microscope fitted with a 35 mm camera and hot stage was useful in monitoring the events which occurred during the polymer metallization process. The nucleation rate was measured by placing a gold modified BTDA-ODA film (previously cured to 100°C) in the hot stage chamber of the optical microscope, ramping the temperature to a given isotherm, and photographing the changes which occurred as a function of time. The graph shown in Fig. 7 summarizes the nucleation rate at 160°, 175° and 200°C by indicating the total number of both the large, dendritic particles (Class I) and the smaller spherical particles (Class II) observed as a function of time at each isotherm. The shape of the nucleation rate curves could be correlated to the type of aggregates formed. In Region I, after an initial induction time, small particles nucleated slowly and eventually grew into the large structures labeled as Class I aggregates in Figure 6. After the growth rate of these large particles slowed, many new small particles (Class II) quickly formed, as is indicated by the large increase in nucleation rate in Region II. In Region III no new particles formed at the indicated isothermal temperature.

Since nucleation and growth were studied between 160° and 200° it was reasonable to assume that the metallization process may be related to the imidization reaction. The change in the mechanical properties of a poly(amide acid) gold modified film which occur during imidization was monitored using a dynamic mechanical thermal analyzer (DMTA). A maximum in tan δ at 93°C corresponded to the glass transition temperature of the poly(amide acid). As the sample temperature increased, there was a subsequent increase in the storage modulus (E') between 160° and 225°C. This increase in modulus caused by stiffening of the polymer matrix was found to have a substantial effect on diffusion controlled processes (i.e. surface metallization) in the modified polymer. For example a "critical" degree of imidization was speculated to be necessary for the onset of the Class II particle nucleation. These observations were explained as follows. Heating the poly(amide acid) film at the isothermal temperature apparently situates the colloid particles in a polymer matrix 60° to 100°C above its T_g. Presumably, colloid particles migrate easily in this "rubbery" matrix, diffuse to the growth centers and form the Class I structures as the sample is further heated. Eventually the T_g of the polymer is equal to

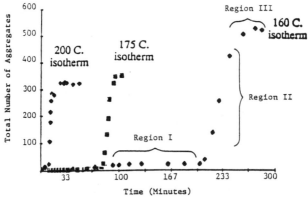

Fig. 7. Number of gold particles as a function of time at three temperature isotherms.

or surpasses the isotherm temperature. The corresponding increase in modulus retards the motion of the colloidal particles. The colloidal particles no longer migrate the relatively long distance necessary to populate the large Class I aggregates. Rather, short distance local migration is therefore preferred and gives rise to a higher population of the small Class II aggregates.

Copper/Polyimide Composites

Highly anisotropic copper-containing polyimide films have been produced through the homogeneous incorporation of bis(trifluoroacetyl-acetonato)copper(II), $Cu(TFA)_2$, within the polyimide precursor followed by thermal cyclodehydration[29]. By judious choice of polyimide precursor, doping level, and curing atmosphere two specific types of surface structure have been obtained. Bi- and tri-layered, Fig. 8, films have resulted which contain copper oxide or metallic copper surface strata, respectively[30]. The surface resistivities of these films were decreased up to seven order of magnitude relative to nondoped polyimide films. While this procedure yielded a significant increase in conductivity, the measured resistivities fall significantly short of those theoretically expected for the copper and copper oxide domains encountered. This was believed to be due to the poor inherent continuity that evolves during the production of these multilayered systems. Migration efficiency was also rather poor in these systems, because the surface metal-containing deposit accounted for only 20% of the dopant used.

Fig. 8.   Transmission electron micrograph of ultramicrotomed cross section of the surface region of a wet air-cured $Cu(TFA)_2$ modified BTDA-ODA polyimide film.

To provide a better understanding of the process-property relationships that are operative in the production of metal-modified polyimides and thereby improve connectivity, maximize surface migration and attain near-theoretical surface resistivities a statistically designed experiment (Plackett-Burman) was undertaken. Potential advantages of this approach include (1) higher efficiency than one-factor-at-a-time experimentation, and (2) the minimization of the effects of error through hidden replication and randomization inherent in these designs. Table II displays the process variables chosen for this design that had been shown to (or were thought to) influence the final structure of copper doped films. Reference 31 provides greater detail of the experimental approach used here.

Table II. Processing Conditions Used in the Experimental Strategy

Processing Variable	Variable Range	
	Minima	Maxima
Dopant Level	0.5x	2.0x
Cure Atmosphere	Moist Air	Dry Air
Casting Substrate	Glass	Polyimide
B-Stage Time	20 min.	120 min.
Conversion Temperature	200°C	300°C
Conversion Time	30 min.	120 min.

Table III describes six properties which were used in characterizing certain aspects of the metal containing deposit on the film surface. The processing variables that proved relevant in controlling the magnitude of the property (relative to a 90% confidence interval) are given in decreasing order of significance.

From the information given in Table III, the processing variables may be split into two groups. Group I, which consists of conversion temperature and dopant level, are processing variables that affect dopant migration/phase separation and are responsible for the amount of additive that reaches the surface of the film. For example at 200°C, the amount of solvent in the film along with the presence of a large

Table III. Film Properties and the Influence of Processing Variables

Property Monitored	Critical Variable
A. Surface Copper Concentration	1. Conversion Temperature 2. Dopant Level
B. Surface Deposit Thickness	1. Conversion Temperature
C. Surface Electrical Resistivity	1. Conversion Temperature 2. Conversion Time
D. Surface Fluorine Concentration	1. Dopant Level 2. Conversion Time 3. Curing Atmosphere
E. Surface CuO Concentration	1. Conversion Temperature 2. Curing Atmosphere

amount of poly(amide acid) keeps the dopant solubilized in the polymer matrix. At this point migration is minimal and surface enrichment will not occur. But, at 300°C, the rate of imidization rises significantly, depleting the bulk of poly(amide acid) functionality along with loss of almost all solvent. When this occurs, a large portion of the dopant phase separates within the matrix, allowing the dopant that does not become trapped locally to migrate to the surface. Dopant level also affects surface enrichment by increasing the localized dopant concentration within the film.

A second group of variables were defined that affect the chemical state of the copper enriched surface. Figure 9 exhibits a proposed reaction scheme for the decomposition of $Cu(TFA)_2$. In the presence of high temperature and atmospheric oxygen and/or water, $Cu(TFA)_2$ may undergo decomposition directly to CuO. Alternatively, high temperature may first induce the decomposition of $Cu(TFA)_2$ to CuF_2 which may then undergo hydrolysis/dehydration to CuO. Inorganic fluorine actually was detected on the surface of a number of films, albeit in small amounts, and only in the presence of significant quantities of organic fluorine which these authors used as an elemental tag for the unconverted dopant. Curing atmosphere and conversion time were both theorized to affect this reaction scheme. Specifically, the use of a moist air cure atmosphere should favor the reaction toward a higher CuO concentration independent of which reaction route is followed. While, longer conversion times in either curing atmosphere were predicted to result in near-complete formation of CuO from $Cu(TFA)_2$.

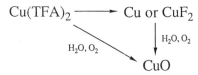

Fig. 9.   Possible reaction schemes for the **in situ** decomposition of $Cu(TFA)_2$. By-products are not shown.

The factorial design indicated that one of the optimal conditions needed to produce the most continuous copper oxide and thereby most electrically conductive film was conversion temperature. Thermal processing to even higher temperatures than those defined in the factorial design was thereforeundertaken. In this regard several comparable copper doped films were initially produced in a wet air cure atmosphere (300°C for 0.5 hr.). At this point, the films were "post-processed" at 350°C, in wet air for an additional hour. Table IV compares the surface resistivities of a post-processed film with a film produced during the factorial design experiment (i.e. film with the lowest resistivity). Upon inspection, post-processing produced a film with resistivity as low as the best film produced (cured for 2.0 hours at 300°C) in the factorial experiment. Comparing the experimental and calculated resistivities suggested that the continuity of the surface layer in the post-processed film approached that of a wholly continuous medium. These properties were obtained with the added benefit of much improved mechanical properties. The post-processed films were more flexible and did not exhibit unusual surface phase separation. Scanning electron microscopy (SEM) Fig. 10, revealed an increase in the surface concentration of copper domains following additional cure. An increase in the thickness of the copper oxide layer from approximately 700 Å to 1300 Å was also noted by transmission electron microscopy. Post-processing was considered by the authors to be analogous to sintering; wherein, small particles fuse to become a large solid mass.

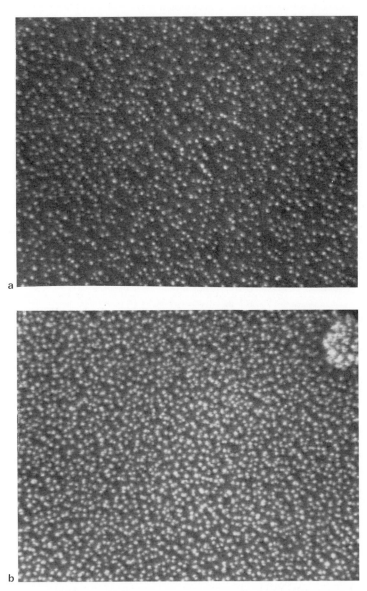

Fig. 10. Scanning electron micrograph of a copper doped polyimide film (a) prior to, and (b) following post-processing.

Silver/Polyimide Composites

The production of metallic surfaces by employing silver salts has been reinvestigated; however, an understanding of these materials is not as extensive as previously described for cobalt, tin and gold modified films. Auerbach, in 1984, reported[32] that doping PyraML-5057[33] and PyraLN-2530[33] with silver nitrate, followed by thermal curing under a carbon blanket at 360°C in an oxygen-containing atmosphere produced polyimide films with surface resistivities of 1-100 ohms/square. More recently[34], he reported the use of a low-power laser (10-30 mW) to effect silver reduction in similar spin-coated 0.75 μm thick polyimide films. An important aspect of this latter work was that small, well-defined patterns on the film could be etched and thus reduced to achieve conductivities on the order of pure silver of comparable dimensions. A micro structure for these materials was not proposed. An evaluation of the continuity or discreteness of the silver and polyimide phases would be of great interest as would a comparison with the better understood **in situ** generated gold system.

Auerbach later extended his study to copper dopants[35]. A variety of copper-containing materials were evaluated but CuCl and $Cu(CH_3CN)_4(BF_4)$ worked best. Samples were prepared by dissolving poly(N-vinylpyrro-lidone) and a copper dopant into a solvent such as N-methyl pyrrolidinone. The resulting solution was cast at 3000 rpm onto a glass or polyimide substrate and dried at 140°C for 30 minutes. The thickness of the copper doped polymer film was between 1 and 5 μm depending on the viscosity of the casting solution. Laser radiation was focused onto the thin film and scanned across the polymer surface to produce a highly conductive copper line. The measured resistance was only a factor of five above the calculated value for a copper conductor of comparable dimensions. Scanning electron microscopy of the conductive region showed granular copper spheres approximately 0.1 μm in diameter.

Summary

In situ generation of metal, metal salt and metal oxide produces materials with distributions of particles that would be difficult to achieve via heterogeneous doping. Surface metal or metal oxide coatings can also be obtained that would not result from conventional heterogeneously modified polymers. The chemistry and response of these microcomposite polymers to process conditions need to be explored further. As the structure and properties of the systems that have been herein described are better understood and new metal/polymer systems are explored, their application in electronic, adhesive and composite systems should emerge.

ACKNOWLEDGMENT

The authors gratefully acknowlege the financial support of the National Aeronautics and Space Administration.

REFERENCES

1. R. P. Kusy, "Metal-Filled Polymers", Ed. S. K. Bhattacharya, Marcel Dekker, Inc., NY, p. 2.
2. C. H. Griffiths, M. P. O'Hare and T. W. Smith, J. Appl. Phys., 50, 7107 (1979).
3. T. W. Smith and D. Wychlick, J. Phys. Chem., 84, 1521 (1980).
4. P. H. Hess and P. H. Parker, J. Appl. Polym. Sci., 10, 1915 (1966).

5. M. Berger and T. A. Manuel, J. Polym. Sci., Part A-1, 4, 1509 (1966).

6. S. Reich and E. P. Goldberg, J. Polym. Sci. Polym. Phys. Ed., 21, 869 (1983).

7. R. Tannenbaum, E. P. Goldberg and C. L. Flenniken, "Metal-Containing Polymeric Systems", Eds. J. E. Sheats, C. E. Carraher and C. V. Pittman, Plenum Press, NY, 1985, p. 303.

8. F. Galembeck, C. C. Ghizoni, C. N. Inpe, C. A. Ribeiro, H. Vargas and L. C. M. Miranda, J. Appl. Polym. Sci., 25, 1427 (1980).

9. R. J. Angelo and E. I. DuPont DeNemours and Co., "Electrically Conductive Polyimides", U. S. Patent 3,073,785 (1959).

10. N. S. Lidorenko, L. G. Gindin, B. N. Yegorov, V. L. Kondratenkov, I. Y. Ravich and T. N. Toroptseva, AN SSR, Doklady, 187, 581 (1969).

11. A. L. Endrey and E. I. DuPont DeNemours and Co., U. S. Patent, No. 3,073,784 (1963).

12. T. L. Wohlford, J. Schaaf, L. T. Taylor, T. A. Furtsch, E. Khor and A. K. St. Clair, in "Conductive Polymers", R. B. Seymour, Editor, P. 7, Plenum Publishing Corp., New York, 1981.

13. A. K. St. Clair and L. T. Taylor, J. Macromol. Sci. Chem., A16, 95 (1981).

14. L. T. Taylor and A. K. St. Clair, "Polyimides", Vol. 2, K. L. Mittal, ed., Plenum Publishing Corp., New York (1984), p. 507.

15. R. K. Boggess and L. T. Taylor, J. Polym. Sci. Polym. Chem., 25, 685 (1987).

16. S. A. Ezzell and L. T. Taylor, Macromolecules, 17, 1627 (1984).

17. E. Khor and L. T. Taylor, Macromolecules, 15, 379 (1982).

18. D. G. Madeleine and L. T. Taylor, "Recent Advances in Polyimide Science and Technology", W. D. Weber and M. R. Gupta, eds., Mid-Hudson Chapter of the Society of Plastic Engineers, Inc., Poughkeepsie, NY, (1987).

19. L. T. Taylor and A. K. St. Clair, J. Appl. Polym. Sci., 28, 2393 (1983).

20. L. T. Taylor, A. K. St. Clair and NASA Langley Research Center, "Aluminum Ion Containing Adhesives", U. S. Patent 284,461 (1981).

21. T. A. Furtsch, H. O. Finklea and L. T. Taylor, "Polyimides", Vol. 2, K. L. Mittal, ed., Plenum Publishing Corp., New York (1984), p. 1157.

22. J. D. Rancourt, G. M. Porta and L. T. Taylor, Solid Thin Films, 158, 189 (1988).

23. K. Komoto and H. Yanagida, Commun. Am. Ceram. Soc., 64, 156 (1981).

24. J. D. Rancourt and L. T. Taylor, Macromolecules, 20, 790 (1987).

25. J. D. Rancourt, R. K. Boggess, L. S. Horning and L. T. Taylor, J. Electrochem. Soc., 134, 85 (1987).

26. D. G. Madeleine, S. A. Spillane and L. T. Taylor, J. Vac. Technol., A5, 347 (1987).

27. J. D. Rancourt and L. T. Taylor, ACS Symp. Ser., 367, 395 (1988).

28. D. G. Madeleine, T. C. Ward and L. T. Taylor, J. Polym. Sci. Polym. Phys., 26, 1641 (1988).

29. G. M. Porta, J. D. Rancourt and L. T. Taylor, Materials, 1, 269 (1989).

30. G. M. Porta and L. T. Taylor, J. Mater. Res., 3, 211 (1988).

31. G. M. Porta, J. D. Rancourt and L. T. Taylor, "Polyimides: Materials, Chemistry and Characterization", C. Feger, M. M. Khojasteh and J. E. McGrath (Eds.), Elsevier Sciences Publishers B. V., Amsterdam (1989) p. 251.

32. A. Auerbach, J. Electrochem. Soc., 131, 937 (1984).

33. DuPont registered trademark.

34. A. Auerbach, J. Electrochem. Soc., 132, 1437 (1985).

35. A. Auerbach, Appl. Phys. Lett. 47, 669 (1985).

FUNCTIONALLY-SUBSTITUTED TETRAMETHYLCYCLOPENTADIENES: SYNTHESIS AND USE IN THE CONSTRUCTION OF ORGANOTRANSITION METAL MONOMERS AND POLYMERS

Charles P. Gibson*, David S. Bem, Stephen B. Falloon, and Jeffrey E. Cortopassi

Department of Chemistry, West Virginia University, Morgantown, WV 26506

Introduction

Molecules which contain the cyclopentadienyl (η^5-C_5H_5) ligand comprise one of the most important classes of organotransition metal (OTM) compounds. The η^5-C_5H_5 ligand is especially useful as a spectator ligand since it binds quite strongly to the metal, occupies several coordination sites, and is fairly unreactive. In recent years, numerous reports have indicated that compounds constructed from the pentamethylcyclopentadienyl (η^5-C_5Me_5) ligand often exhibit markedly different reactivities than their cyclopentadienyl-containing congeners due to the much greater steric bulk and superior electron-donating ability of the η^5-C_5Me_5 ligand.[1]

While a variety of polymers have been constructed from functionally-substituted cyclopentadienyl metal complexes,[2-5] polymers constructed from analagous functionally-substituted tetramethylcyclopentadienyl metal compounds have not been studied. With this in mind, we decided to develope a synthetic program which would enable us to create and study such materials. In the remainder of this chapter, we will discuss a particularly useful method for synthesizing functionally-substituted tetramethylcyclopentadienes (HC_5Me_4R),[6] and the subsequent use of these molecules in the syntheses of functionally-substituted organometallics. In addition, we will describe some of our recent results in which show that these organometallics may be used in the syntheses of OTM-containing polymers.

Results and Discussion

The general approach that we have adopted for the preparation of HC_5Me_4R molecules is based on a pentamethylcyclopentadiene (HC_5Me_5) synthesis which was originally reported by Burger *et al.*,[7] and later (in an improved form) by Kohl and Jutzi.[8] In

Inorganic and Metal-Containing Polymeric Materials
Edited by J. Sheats *et al.*, Plenum Press, New York, 1990

this preparation (Scheme I), 2,3,4,5-tetramethylcyclopent-2-enone (**1**) was first synthesized

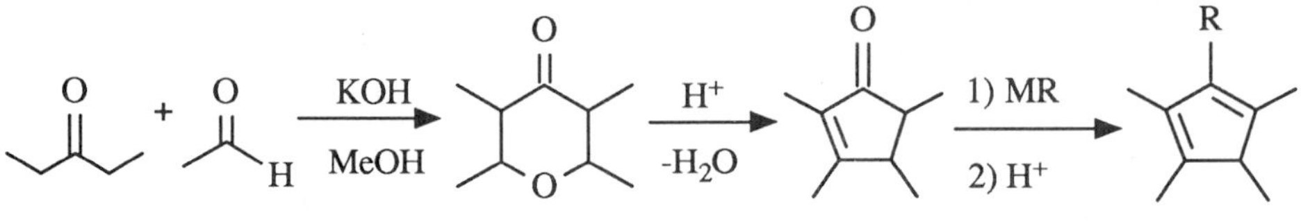

Scheme I

in two steps from 3-pentanone and acetaldehyde. Treatment of **1** with methyl lithium (MR = CH_3Li) or methylmagnesium bromide (MR = $BrMgCH_3$), followed acid-catalyzed dehydration, gave the desired HC_5Me_5. By substituting appropriate organolithium or Grignard reagents, this strategy may be used as a general route to HC_5Me_4R molecules, which may be suitable for use in the creation of functionally-substituted organometallics and OTM-containing polymers.

One of the first uses of this strategy was reported by Buzankai and Schrock.[9] In this case, the alkynyl-substituted $HC_5Me_4(CH_2CH_2C{\equiv}CEt)$ (**2**) was synthesized upon the addition of 1-lithio-3-hexyne to **1**, followed by acidic workup (Scheme II). The crude

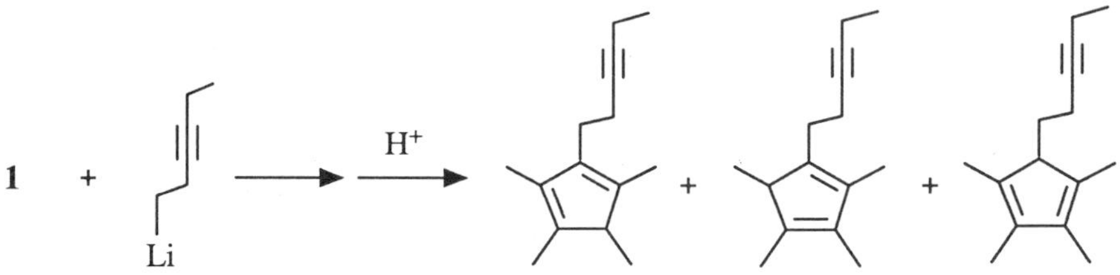

Scheme II

mixture of products was then distilled at reduced pressure (1 torr), and the fraction which distilled from 82-85 °C was collected. Characterization of this fraction indicated that **2** was present as a mixture of three double-bond isomers (**2a-c**).

Since three closely related isomers of HC_5Me_4R were isolated in this synthesis, unequivocal NMR spectroscopic characterization of the products was difficult due to overlapping peaks. This characterization problem, which is quite common in HC_5Me_4R syntheses, may be solved by converting the product mixture to a single MC_5Me_4R product (M = Li or Na) via reaction with n-butyl lithium, diisopropyl amide (LDA), or sodium amide. Other advantages of converting HC_5Me_4R mixtures to the corresponding MC_5Me_4R is that the products are generally solids which are more easily purified and handled (in a inert atmosphere), and are in a form which is generally the most useful for subsequent syntheses of functionally-substituted organometallics. In the case of **2a-c**, treatment of the mixture with n-butyl lithium led to the isolation of a single product,

$(Li^+)[C_5Me_4(CH_2CH_2C\equiv CEt)^-]$ (3) which gave unambiguous ^1H and ^{13}C NMR spectra, and was appropriate for use in subsequent syntheses of organometallics.

The motives underlying the original synthesis of 3 were based on the observation that certain tungstacyclobutadienes react with internal acetylenes to give (substituted) cyclopentadienyl complexes.[10] Consequently, alkynyl-substituted organometallics were deemed to be appropriate for use in the syntheses of heterobimetallic complexes. To this end, the alkynyl-substituted $(\eta^5-C_5Me_4CH_2CH_2C\equiv CEt)Rh(CO)_2$ (4), was synthesized via reaction of $[RhCl(CO)_2]_2$ with 3 (Scheme III). Subsequent treatment with

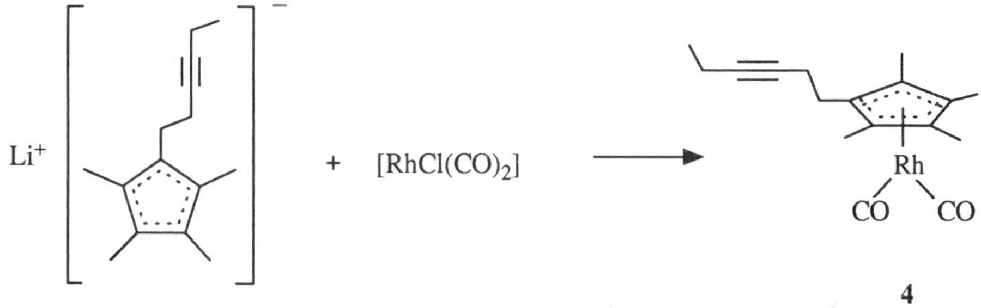

3	**Scheme III**	4

$W(C_3Et_3)(OCMe_2CMe_2O)(OCMe_3)$ gave the desired tungsten-rhodium bimetallic complex. In addition, several closely related bimetallics were also reported. While alkynyl-substituted tetramethylcyclopentadienyl metal compounds might be useful in syntheses of OTM-containing polymers, the use of 2-4 in the construction of polymers was not discussed.

Recently, the preparation of the ω-alkenyl-substituted $HC_5Me_4(CH_2CH_2CH=CH_2)$ (5) was briefly described by Okuda and Zimmerman (Scheme IV).[11] In this synthesis, 1 was

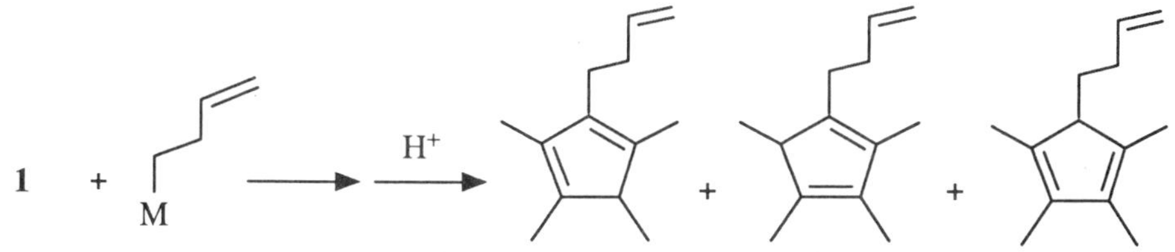

Scheme IV

first treated with $MCH_2CH_2CH=CH_2$ (M = Li or MgBr), and then with aqueous acid. Separation of the organic layer, followed by vacuum distillation (0.001 torr) of the resulting crude mixture and collection of the 100-105 °C fraction afforded an oil which contained 5. While characterization of this fraction was not reported, it may be assumed that the three isomeric forms of the compound (5a-c) were present, along with some contaminants.

Analytically pure samples of mixtures of isomers **5a-c** were prepared by first treating the impure mixture of **5** with *n*-butyl lithium to give the corresponding $(Li^+)[C_5Me_4(CH_2CH_2CH=CH_2)^-]$ (**6**), then washing the impurities from **6** with pentane and ether, and finally converting **6** back to a mixture of **5a-c** via methanolysis.

Several new functionally-substituted organometallics were synthesized from **5a-c**, and from **6**. The synthesis of $(\eta^5\text{-}C_5Me_4CH_2CH_2CH=CH_2)Co(CO)_2$ was accomplished by reaction of **5a-c** with $Co_2(CO)_8$ in refluxing 3,3-dimethylbutene (Scheme V). In this case,

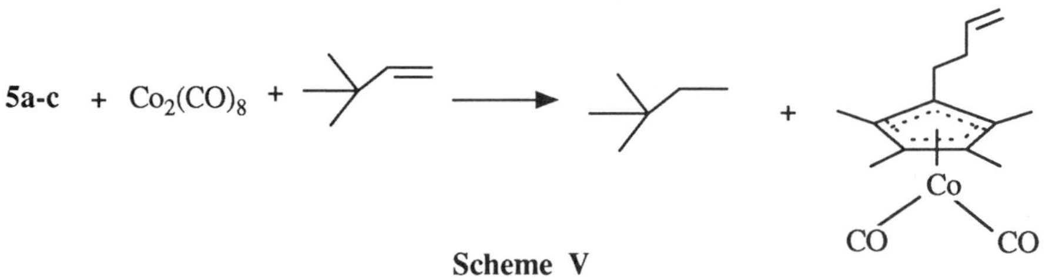

$$5a\text{-}c \quad + \quad Co_2(CO)_8 \quad + \quad \text{[structure]} \quad \longrightarrow \quad \text{[structure]} \quad + \quad \text{[structure]}$$

Scheme V

the solvent acted as a hydrogen acceptor, thereby preventing significant hydrogenation of the ω-alkenyl substitutant. The functionally-substituted metallocene $(\eta^5\text{-}C_5Me_4CH_2CH_2CH=CH_2)_2Fe$ (**7**) was synthesized via the reaction of **6** with $FeCl_2$, while reaction of **6** with $FeBr_2$ or $Fe(acac)_2$ in the presence of CO afforded the piano stool compounds $(\eta^5\text{-}C_5Me_4CH_2CH_2CH=CH_2)Fe(CO)_2Br$ (**8**) and $(\eta^5\text{-}C_5Me_4CH_2CH_2CH=CH_2)Fe(acac)CO$ (**9**), respectively (Scheme VI). Although the

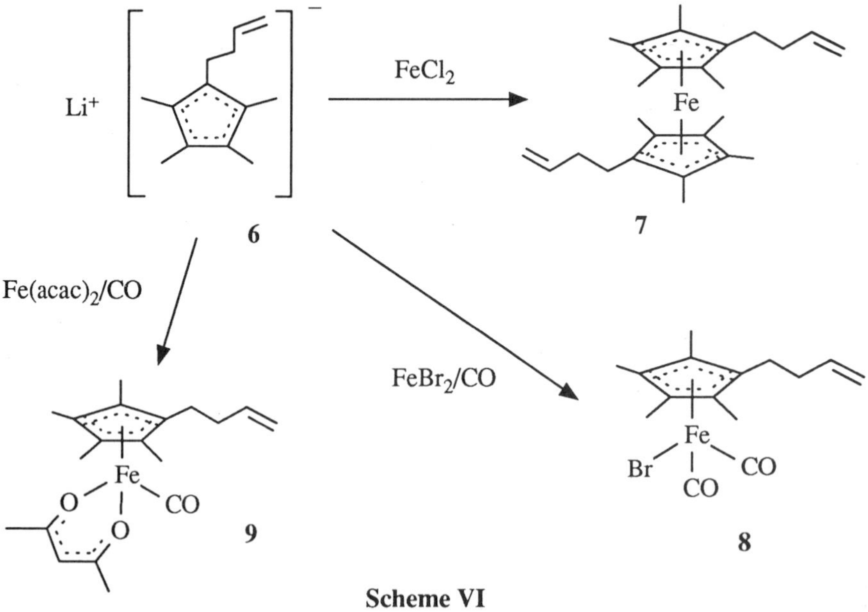

Scheme VI

ω-alkenyl-substituted organometallics **7-9** would seem to be ideal for use in the syntheses of OTM-containing polymers, this issue was not addressed in the original communication.

In several recent papers, Mintz *et al.* reported the reactions of **1** with allylmagnesium bromide,[12] and with vinylmagnesium bromide.[13] Treatment of **1** with allylmagnesium bromide followed by acidic workup was reported to give a mixture of the three expected isomeric forms of $HC_5Me_4(CH_2CH=CH_2)$ along with the tautomeric 3-(2,3,4,5-tetramethylcyclopent-2-en-1-ylidene)-1-propene, although the relative amounts of the different isomers was not reported (Scheme VII). Mintz has a converted this product

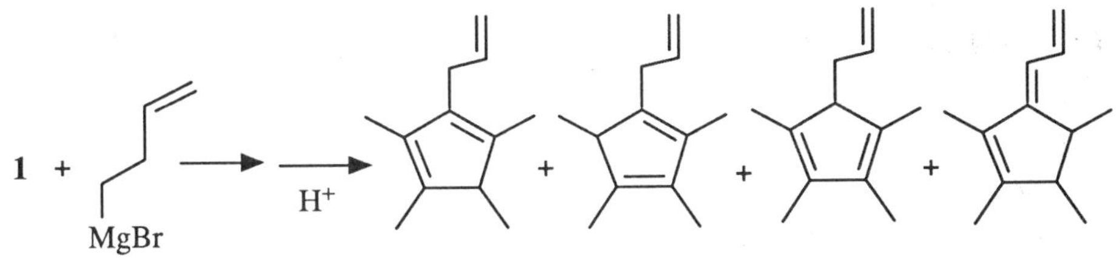

Scheme VII

mixture to the phosphine-substituted $HC_5Me_4(CH_2CH_2CH_2PPh_2)$, which is particularly appropriate for use in the creation of heterobimetallics. The use of this ligand in the syntheses of several functionally-substituted organometallics and heterobimetallics has been investigated.[14]

The reaction of **1** with vinylmagnesium bromide, followed by acidic workup, gave a mixture of the expected vinyltetramethylcyclopentadienes (**11a-b**) and the isomeric pentamethylfulvene (**10**) in essentially quantitative yield (Scheme VIII).[13] Significantly, **10**

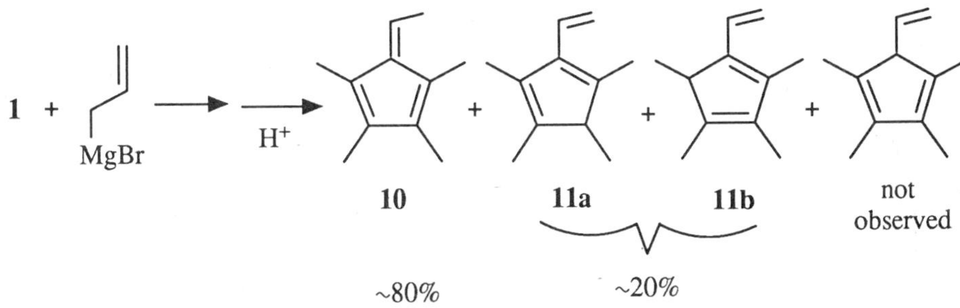

Scheme VIII

was the major product of the reaction. Mintz has capitalized on this result by creating new substituted ligands by reaction of certain nucleophiles with **10** (*e.g.* Scheme IX). One of these addition products, $(Li^+)(C_5Me_4CH(Me)PPh_2^-)$, was used in the synthesis of a phosphine-substituted ferrocene, which was especially designed for use in the creation of heterobimetallics.

While Mintz has shown that nucleophiles, when added to a mixture of **10** and **11a-b**, selectively react with the fulvene **10** to give addition products, we reasoned that reaction with the sterically-encumbered strong base LDA would result in deprotonation of **10** as well as **11a-b**. Indeed, treatment of a mixture of **10** and **11a-b** with LDA resulted in conversion

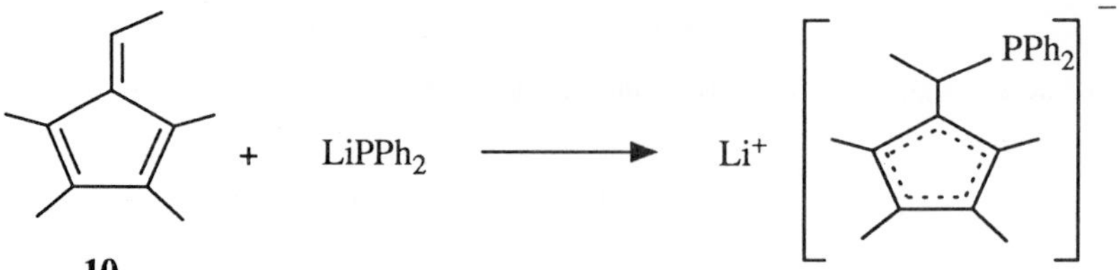

Scheme IX

to the desired lithium 1-vinyl-2,3,4,5-tetramethylcyclopentadienide (**12**), which was isolated 55% yield.[15]

Once synthesized, we used **12** in the creation of several new vinyl-substituted organometallics. The metallocene $(\eta^5\text{-}C_5Me_4CH=CH_2)_2Fe$ (**13**) was synthesized via reaction of **12** with FeI_2 in hot THF (Scheme X; R is -CH=CH$_2$). The product was isolated

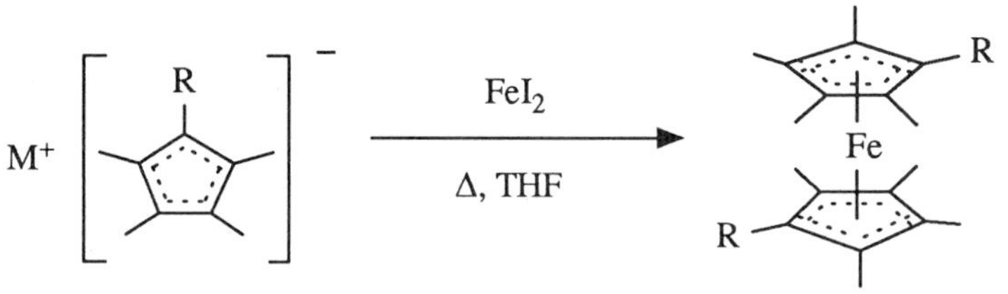

12 (R is -CH=CH$_2$; M is Li)   **13** (R is -CH=CH$_2$)
18 (R is -p-C$_6$H$_4$OCMe$_2$OMe; M is Na)   **19** (R is -p-C$_6$H$_4$OH)

Scheme X

in 50% yield.[16]

The piano-stool compound $(\eta^5\text{-}C_5Me_4CH=CH_2)Mo(CO)_2NO$ (**14**) was also synthesized from **12** (Scheme XI; R is -CH=CH$_2$).[15] In this case, a THF solution of **12** was

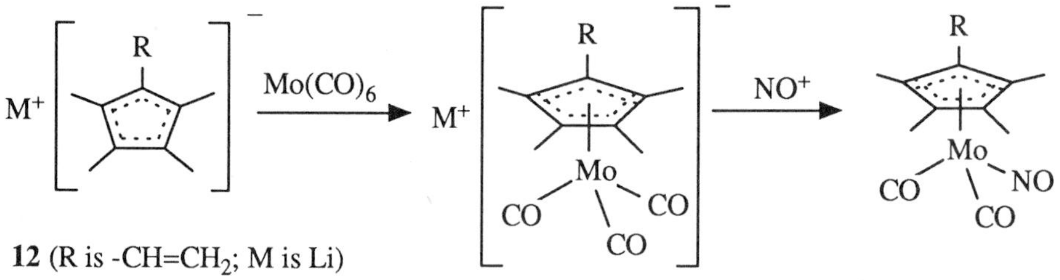

12 (R is -CH=CH$_2$; M is Li)

18 (R is -p-C$_6$H$_4$OCMe$_2$OMe; M is Na)

14 (R is -CH=CH$_2$)

20 (R is -p-C$_6$H$_4$OCMe$_2$OMe)

Scheme XI

treated first with $Mo(CO)_6$, and then with N-methyl-N-nitroso-p-toluenesulfonamide (Diazald, a convenient source of NO$^+$).

Compound **14** was subsequently used in the synthesis of a rare example of a functionally-substituted cluster compound (**15**).[17] This synthesis, which was based on the preparation of the analogous all η^5-C$_5$Me$_5$ containing compound,[18] consisted of photolytically-induced addition of **14** across the Mo-Mo triple bond of (η^5-C$_5$Me$_5$)$_2$Mo$_2$(CO)$_4$ (Scheme XII; R is -CH=CH$_2$). A significant feature of this reaction

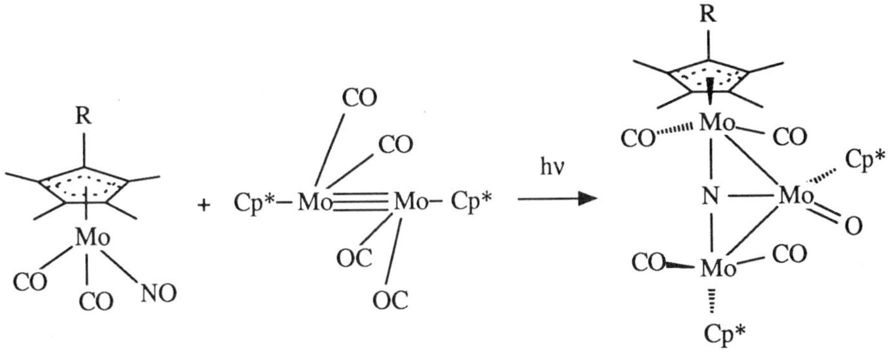

14 (R is -CH=CH$_2$)   Cp* = η^5-C$_5$Me$_5$ **15** (R is -CH=CH$_2$)

20 (R is -p-C$_6$H$_4$OCMe$_2$OMe)   **22** (R is -p-C$_6$H$_4$OH)

Scheme XII

is that the unique η^5-C$_5$Me$_4$R ligand ends up on one of the "basal" molybdenum atoms rather than randomly distributed among the three different sites.

Preliminary investigations have shown that both organomolybdenum compounds **14** and **15** can be used in the syntheses of OTM-containing polymers.[15,17] Free radical initiated polymerization of styrene and **14** (in a 20/1 mole ratio) was accomplished by heating the monomers along with the initiator AIBN in benzene at 80 °C for 48 hours. Methanol was added to the solution in order to precipitate the polymer, which was then purified by repeated precipitations from methylene chloride. Spectroscopic characterization of the resulting polymer confirmed that the organometallic moiety had been incorporated as part of the polymer, with a loading of 1.3 mole per cent.

Attempts to copolymerize **15** with styrene in an analogous manner failed due to decomposition of the cluster.[19] However, copolymerization was accomplished by use of n-butyl lithium as an anionic initiator. In this case, a benzene solution of styrene and **15** (in a 100/1 mole ratio) was treated with a trace of n-butyl lithium. After 20 hours, methanol was added to the mixture in order to quench the reaction and cause precipitation of the resulting polymer, which was then purified by repeated precipitations from methylene chloride. Spectroscopic characterization of the polymer confirmed that the cluster had been chemically incorporated into the polymer, with a loading of 0.67 mole per cent.

In addition to the vinyl-substituted organometallics **13-15**, we have also synthesized p-phenol- and protected p-phenol-substituted analogues (**19-22**).[15,17] The synthesis of the ligand 1-(p-phenol)-2,3,4,5-tetramethylcyclopentadiene (**16**; Scheme XIII) began with the

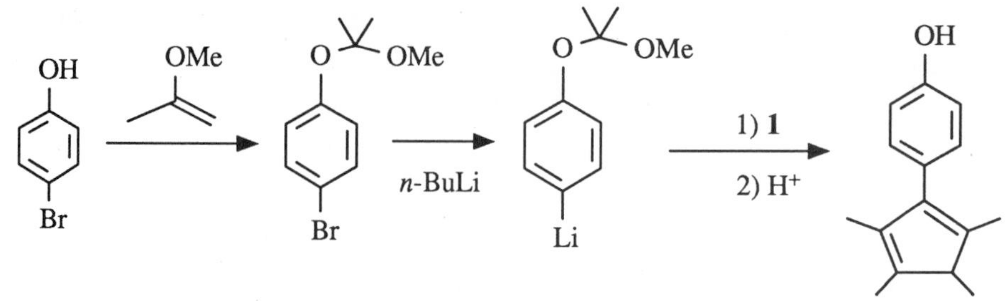

Scheme XIII

16

protection of *p*-bromophenol via reaction with 2-methoxypropene. The protected *p*-bromophenol was then treated with *n*-butyl lithium to give *p*-LiC$_6$H$_4$OCMe$_2$OMe. Subsequent reaction with **1** followed by acidic workup afforded **16** in 80% yield. Interestingly, only a single isomeric form of **16** was normally isolated. Compound **16** was generally reprotected via reaction with 2-methoxypropene to give a product (**17**) which was less susceptible to oxidation, and in a form which was more appropriate for use in subsequent syntheses.

The protected phenol-substituted compound **17** was prepared for use in metallocene syntheses by converting it to the corresponding anion (**18**) via treatment with sodium amide. The reaction of **18** with FeI$_2$ (Scheme X) gave a product mixture which contained crude (η^5-C$_5$Me$_4$-*p*-C$_6$H$_4$OCMe$_2$OMe)$_2$Fe. The product was deprotected and purified by column chromatography on activity III alumina (6% H$_2$O) to give (η^5-C$_5$Me$_4$-*p*-C$_6$H$_4$OH)$_2$Fe (**19**) in 67% yield.[16]

Treatment of **18** with Mo(CO)$_6$ followed by reaction with Diazald (Scheme XI) gave the protected phenol-substituted (η^5-C$_5$Me$_4$-*p*-C$_6$H$_4$OCMe$_2$OMe)Mo(CO)$_2$NO (**20**), which was subsequently treated with aqueous acid to give the deprotected version (η^5-C$_5$Me$_4$-*p*-C$_6$H$_4$OH)Mo(CO)$_2$NO (**21**). The phenol-substituted trimolybdenum cluster **22** (Scheme XII) was synthesized via photolytic reaction of **20** with (η^5-C$_5$Me$_5$)$_2$Mo$_2$(CO)$_4$, followed by deprotection and separation of the products via chromatography on alumina.

OTM-containing polymers were constructed from **21** and **22** by reaction with an excess of a preformed 1/1 styrene/maleic anhydride copolymer.[15,17] Reaction of *ca.* 5 mole percent **21** with the preformed polymer resulted in chemical incorporation of 3.0 mole percent of the substituted organometallic (Scheme XIV; x=3.0). A similar reaction of the preformed polymer with 1.0 mole percent **22** gave a material which contained 0.7 mole percent cluster (Scheme XIV; x=0.7).

A particularly interesting mixed-metal polymer was synthesized from the metallocene **19**. This reaction (Scheme XV) was conducted by first converting the phenol functionalities of **19** to lithium phenoxides via reaction with methyl lithium. Subsequent reaction with (η^5-C$_5$H$_5$)$_2$ZrCl$_2$ gave the desired polymeric product. Detailed studies of the physical and chemical properties of this polymer are currently in progress.

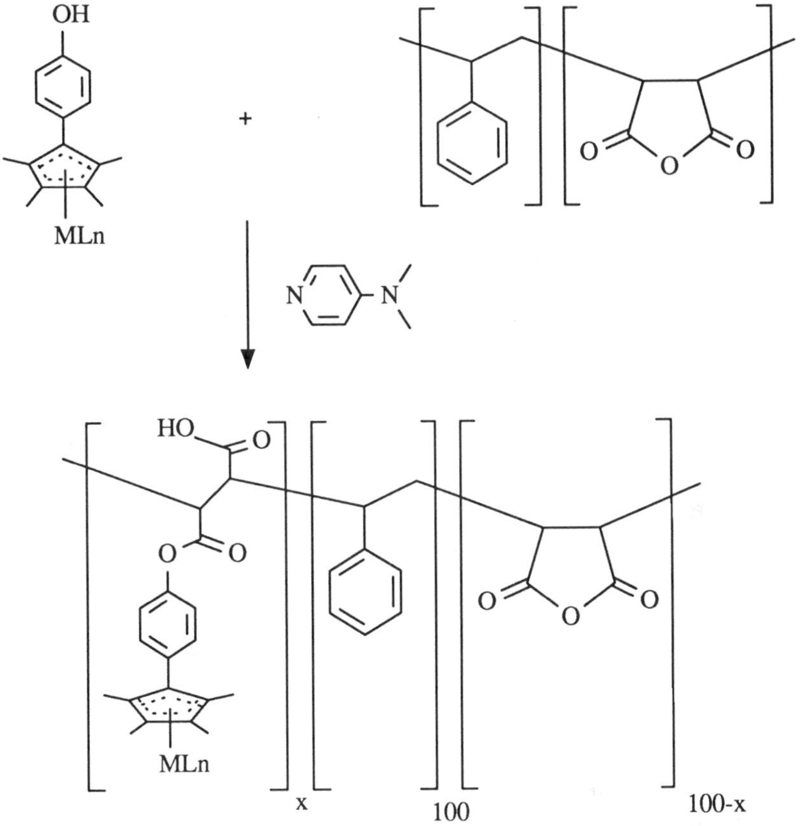

Scheme XIV

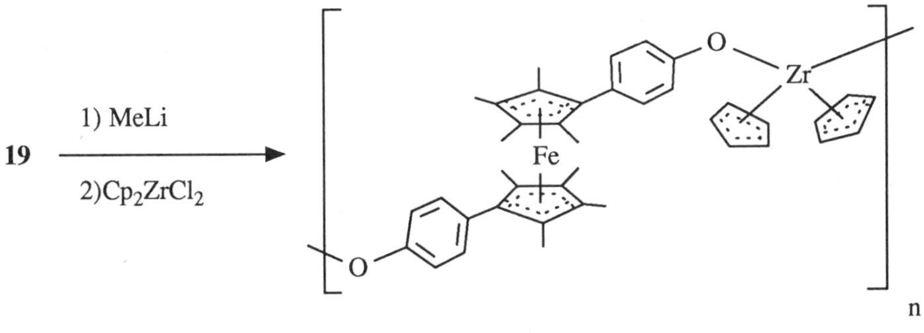

19 $\xrightarrow[\text{2)Cp}_2\text{ZrCl}_2]{\text{1) MeLi}}$

Scheme XV

Conclusion

The syntheses described in this paper illustrate that a variety of functionally-substituted tetramethylcyclopentadienes may be constructed via a general route which consists of the addition of suitable organolithium or Grignard reagents to **1**, followed by acid-catalyzed dehydration. Although the syntheses and subsequent characterization of the products can sometimes be challenging, it is clear that this strategy will lead to the syntheses of a variety of new functionally-substituted organometallics which may be suitable for use in the construction of OTM-containing polymers.

Experimental

General Information. The syntheses of **1-12**, **14**, **16-18**, **20** and **21**, as well as OTM-containing polymers from **14** and **21**, have been reported elsewhere (*vide supra*). The syntheses of **15** and **22**, as well as OTM-containing polymers from **15** and **22** are detailed the M.S. Thesis of J.E. Cortopassi.[17b] Full details will be published elsewhere.[17a]

In the syntheses of the materials listed below, all organometallics were treated as air- and moisture sensitive.[20]

Synthesis of (η^5-C_5Me_4R)$_2$Fe (13, R is -CH=CH$_2$; 19, R is -p-C_6H_4OH). In a typical experiment, a 500 mL flask was charged with 8.0 mmol of FeI$_2$•2THF, 100 mL of THF, and 8.0 mmol of either compound **12** or **18**. The solution was then heated at 60 °C for 12 hours. After cooling the flask and its contents to room temperature, the volatiles were removed *in vacuo* to give a meconium-like crude product. Some alumina (5-10 g) and methylene chloride (*ca.* 30 mL) were added to the crude product, and the resulting slurry was mixed well. After removing the volatiles *in vacuo*, the residue was transferred to the top of a chromatography column which contained activity III alumina (2.5 x 20 cm, packed in hexane).

In the case of the synthesis of **13**, the column was eluted with hexane. The orange band was collected, and the solvents evaporated to give desired product in 50% yield. In the case of the synthesis of **19**, the column was first eluted with hexane, and then with THF. The dark orange band that was eluted with THF was collected. After evaporation of the solvents, the desired product was isolated in 67% yield.

Analytical data for **13**. ^{1}H NMR (CDCl$_3$): δ 1.70 (s, 12H, C-CH$_3$); 1.80 (s, 12H, C-CH$_3$); 5.22 (dd, 2H, $^3J_{cis}$=11.0 Hz, $^2J_{gem}$=2.2 Hz, -CH=CHH); 5.27 (dd, 2H, $^3J_{trans}$=18.3 Hz, $^2J_{gem}$=2.2 Hz, -CH=CHH); 6.34 (dd, 2H, $^3J_{trans}$=18.3 Hz, $^3J_{cis}$=11.0 Hz, -CH=CH$_2$). ^{13}C{H} NMR (CDCl$_3$): δ 9.2 (4C, -CH$_3$); 10.3 (4C, -CH$_3$); 112.5 (2C, -CH=CH$_2$); 133.5 (2C, -CH=CH$_2$).

Analytical data for **19**. ^{1}H NMR (d$_8$-THF): δ 1.72 (s, 12H, C-CH$_3$); 1.75 (s, 12H, C-CH$_3$); 6.57 (d, 4H, phenyl H, J=8.4 Hz); 7.06 (d, 4H, phenyl H, J=8.4 Hz). ^{13}C{H} NMR (d$_8$-THF): δ 10.2 (4C, -CH$_3$); 11.4 (4C, -CH$_3$); 115.0 (4C, aromatic C-H); 132.7 (4C, aromatic C-H); 156.7 (2C, C-OH). Solution IR (THF): 3305 (br, vs), 1701 (w), 1614 (m), 1523 (m), 1379 (w) cm^{-1}.

The use of 19 in the synthesis of a Fe-Zr organometallic polymer. A flask was charged with 1.04 mmol **19** and 40 mL THF. To this solution was added 2.08 mmol CH$_3$Li (1.30 mL of a 1.60 M solution).[21] After allowing the solution to stir for 5 minutes, a solution of 1.05 mmol (η^5-C$_5$H$_5$)$_2$ZrCl$_2$ in 10 mL THF was added. The solution was stirred overnight, and then reduced in volume to *ca.* 10 mL by evaporation of the volatiles *in vacuo*. The polymer was precipitated by adding hexane, and then purified by repeated precipitations from THF. The ^{1}H NMR solution of this material (a dilute solution in d$_8$-THF) exhibited broad resonances at *ca.* δ 1.80 (24 H, *-CH$_3$*), 6.35 (10 H, C$_5$*H$_5$*), 6.65 (4H, phenyl *H*), and 7.05 (4H, phenyl *H*). The product was slightly soluble in THF and acetone, insoluble in hexane.

Acknowledgment

We wish to thank the Mining and Mineral Resources Research Institute (grant number G1184154), the West Virginia University Energy and Water Research Center, and the donors of the Petroleum Research Fund, administered by the American Chemical Society, for financial support of this project.

References

1.   Collman, J.P.; Hegedus, L.S.; Norton, J.R.; Finke, R.G. *Principles and Applications of Organotransition Metal Chemistry*, University Science Books, Mill Valley, CA, 1987.

2.   Macomber, D.W.; Hart, W.P.; Rausch, M.D. *Adv. Organomet. Chem.*, **1982**, *21*, 1.

3.   Pittman, C.U., Jr.; Carraher, C.E.; Reynolds, J. *Encyl. Polym. Sci. Eng, Volume 10*, John Wiley & Sons, New York, 1987.

4.   Carraher, C.E., Jr.; Sheats, J.E.; Pittman, C.U., Jr. *Organometallic Polymers*, Academic Press, New York, 1978.

5.   Carraher, C.E., Jr.; Sheats, J.E.; Pittman, C.U., Jr. *Advances in Organometallic and Inorganic Polymer Science*, Marcel Dekker, New York, 1982.

6.   For the purposes of this chapter, we will consider only those functional groups that participate in the usual types of polymerization reactions. Compounds with alkyl, aryl, and silyl substituents will not be considered.

7.   Burger, U.; Delay, A.; Mazenod, F. *Helv. Chim. Acta.*, **1974**, *57*, 2106.

8. Kohl, F.X.; Jutzi, P. *J. Organomet. Chem.*, **1983**, *243*, 119.

9. Buzinkai, J.F.; Schrock, R.R. *Organometallics*, **1987**, *6*, 1147.

10. Schrock, R.R.; Pederson, S.F.; Churchill, M.R.; Ziller, J.W. *Organometallics*, **1984**, *3*, 1574.

11. Okuda, J.; Zimmerman, K.H. *J. Organomet. Chem.*, **1988**, *344*, C1.

12. Bensley, D.M.; Mintz, E.A.; Sussangkarn, S.J. *J. Org. Chem.*, **1988**, *53*, 4417.

13. Bensley, D.M.; Mintz, E.A. *J. Organomet. Chem.*, **1988**, *353*, 93.

14. D.M. Bensley, Ph.D. Dissertation, West Virginia University (1988).

15. Gibson, C.P.; Bem, D.S.; Falloon, S.B.; Cortopassi, J.E. *Organometallics*, in press.

16. Gibson, C.P.; Bem, D.S.; Falloon, S.B. manuscript in preparation.

17. (a) Gibson, C.P.; Cortopassi, J.E. manuscript in preparation. (b) J.E. Cortopassi, M.S. Thesis, West Virginia University (1990).

18. (a) Gibson, C.P.; Schugart, K.A.; Cortopassi, J.E.; Fenske, R.F.; Dahl, L.F. manuscript in preparation. (b) C.P. Gibson, Ph.D. Dissertation, The University of Wisconsin-Madison (1985).

19. Compound **15** is thermally stable at this temperature.[17] Decomposition of the cluster was evidently caused by reaction with radicals.

20. D.F. Shriver and M.A. Drezdzon *The Manipulation of Air-Sensitive Compounds*, John Wiley & Sons, New York, 1986.

21. The methyl lithium used in this reaction was assayed prior to use. See: (a) Gilman, H.; Cartledge, F.K. *J. Organomet. Chem.*, **1964**, *2*, 447. (b) Turner, R.R.; Altenau, A.G.; Cheng, T.C. *Anal. Chem.*, **1970**, *42*, 1835.

SYNTHESIS OF POLYASPARTAMIDE–BOUND FERROCENE COMPOUNDS[†]

Eberhard W. Neuse and Carol W.N. Mbonyana

Department of Chemistry, University of the Witwatersrand
WITS 2050, Republic of South Africa

ABSTRACT

The di–η^5–cyclopentadienyliron (ferrocene) complex is bound covalently, yet reversibly, to a variety of linear copolyaspartamides *via* short side chains attached to the aspartamide N atoms. The resulting polymer–ferrocene conjugates, purified by dialysis in aqueous solution and isolated by freeze–drying, are characterized analytically and spectroscopically. By virtue of incorporated *tert*–amine functions acting as hydrosolubilizing groups, the polymers are water–soluble despite the lipophilic character of the anchored metallocene. Inherent viscosities typically range from 10 to 15 ml g^{-1}, and the degree of ferrocene incorporation varies from 20 to 75% of theory, depending *inter alia* on the reactant ratios employed. The conjugates are of interest for biomedical applications requiring solubility in aqueous media.

INTRODUCTION

Although the biochemistry of cancerous diseases is complex and, at present, by no means well understood, the leading role played by free radicals, notably in carcinogenesis and tumor promotion, has been commonly accepted (2–4). Dioxygen reduction in the respiring cell leads to the generation of reactive oxygen species, such as superoxide anion radical, hydrogen peroxide, and ultimately, the highly aggressive and invariably destructive hydroxyl radical. Foremost enzymatic components of the host's defence mechanism against the deleterious effects of these species include several forms of superoxide dismutase (SOD), which dismutate superoxide ion into hydrogen peroxide and dioxygen, and of catalase and glutathione peroxidase, which serve to eliminate the hydrogen peroxide formed, thus preventing its transformation by a Fenton–type reaction to the hydroxyl radical (5,6). The delicate balance inherent in this enzymatic cycle can be rudely upset by a superoxide–generating promoter. The problem is compounded by the considerable reduction in SOD activity observed in most transformed cells, concomitantly with an increase in superoxide levels (6), and it is the action of the superoxide and secondary free–radical species, binding to nuclear DNA and causing replication errors during the mitotic stage, which has been implicated as one of the causative factors in malignancy (4). Accordingly, carcinogenic and metastatic processes are inhibited, within limits, by antioxidants and free–radical scavenger agents (2,6,7), and carcinostatic activity is shown also by certain SOD–mimetic metal complexes functioning as superoxide scavengers (4,8). Lastly, with free–radical intermediacy

---

[†] Metallocene Polymers 45. For Part 44, see (1).

implicated in the biological metabolism of numerous anticancer drugs, for some of these a free–radical scavenger function may well be the *modus operandi* (9).

Against this background, it would appear to be a challenging task to investigate the intriguing possibilities likely to be offered by the ferricenium complex system in cancer research. The ferricenium ion, **2**, generated in a one–electron oxidation step from the neutral ferrocene (di–η^5–cyclopentadienyliron) complex, **1**, represents a radical cation species of appreciable stability (10). Reaction of the cation **2** with superoxide anion proceeds with electron transfer and regeneration of dioxygen (Scheme 1a) (11). In a reverse process involving oxidation of the ferrocene to the ferricenium system, ferrocenylcarboxylates may interact with hydroxyl radical, giving rise to the formation of ferricenylcarboxylates in addition to hydroxyl anion (Scheme 1b; ⌒ = alkylene) (12). Lastly, the cation **2** undergoes the typical free–radical recombination by interaction with other free–radical species; after

$$
\begin{array}{ccc}
\underset{\mathbf{2}}{\text{Fe}^+} & \xrightarrow[-\,O_2]{O_2^-} & \underset{\mathbf{1}}{\text{Fe}}
\end{array}
\qquad (1a)
$$

$$
\begin{array}{ccc}
\underset{\text{Fe}}{\overset{\frown}{}}\text{COO}^- & \xrightarrow[-\text{OH}^-]{\bullet\text{OH}} & \underset{\text{Fe}^+}{\overset{\frown}{}}\text{COO}^-
\end{array}
\qquad (1b)
$$

$$
\begin{array}{ccc}
\text{Fe}^+ & \xrightarrow[-\,H^+]{\bullet\,R} & \text{Fe}{-}R
\end{array}
\qquad (1c)
$$

rearrangement from metal–substitution to ring substitution products, followed by deprotonation, this may lead to uncharged, substituted ferrocenes (Scheme 1c) (13). The three reactions represent free radical–scavenging processes with potentially significant biological ramifications in the detoxification of reactive oxygen species and their precursors. In addition, the enzyme–mediated oxidation of ferrocene by hydrogen peroxide has been observed (14), and, conversely, the reduction of ferricenium cation by such biologically important species as NADH (15) and metalloproteins (16) is on record.

This performance pattern of the ferricenium complex provided the incentive for our laboratory to initiate, over the past few years, several collaborative projects with biomedical and pharmacological research groups, involving the testing of ferricenium salts for antineoplastic activity against Ehrlich (17) and Yoshida (18) ascites murine tumor lines, as well as against several human tumor clonogenic cultures (19). While the water–insoluble salts tested all possessed the same inactivity shown by the (very poorly water–soluble) parent complex **1**, several salts distinguished by excellent solubility in water, exemplified by the tetrachloroferrate(III), the picrate, the (solvated) trichloroacetate, and the chloride, were found to be active (17,18), cure rates of up to 100% being obtained under the conditions of the Ehrlich ascites tumor testing protocol. In the human tumor clonogenic assay, even ferrocenylacetic acid, a slightly water–soluble representative of complexes in the reduced (ferrocene) state, showed moderate activity (19). As the limited stability of ferricenium compounds in aqueous solution at pH 7.3 suggests *in vivo* half–lives too short for favorable pharmacokinetic response (18), it appears somewhat questionable whether intravenously or intraperitoneally administered ferricenium salt solutions will effectively survive transport to the target tissue. The observed activity of ferrocenylacetic acid may rather suggest that the redox equilibrium composition of the ferricenium/ferrocene couple in a particular body compartment will largely be dictated by the inherent biological environment in terms of pH and the presence of oxidizing or reducing agents, and not by the compound's initial oxidation state. Hence, it should be of no consequence whether the compound be administered in the reduced (ferrocene) or oxidized (ferricenium) state as long as its water solubility is sufficiently high for rapid dissipation in the central circulation system. This indicated to us that, by tying the inherently hydrophobic ferrocene complex reversibly to a vehicle capable of providing (i) complete water–solubility of the conjugate and, thus, rapid introduction into central circulation, (ii) limited protection against attack by the reticulo–endothelial system while in transit, and (iii) facilitated endocytotic cell entry, one should be able to create optimal pharmacokinetic conditions for aqueous–phase administration without having to resort to prior oxidation of the anchored metallocene complex. Vehicles of this type are conveniently represented by synthetic macromolecules with built–in drug carrier–functions, and several such macromolecular carriers have been prepared in this laboratory as part of an extended program aimed at the development of biocompatible and biodegradable polymer–drug conjugates possessing solubility in aqueous media. In the following, we demonstrate the feasibility of synthesizing ferrocene–containing, water–soluble conjugates possessing linear polymeric backbone structures.

RESULTS AND DISCUSSION

The concept of drug binding to a carrier polymer, based on solid arguments elaborated elsewhere (20–22), is attracting increasing attention in the biochemical and biomedical research community. A number of structural prerequisites must be fulfilled by a polymer designed for drug carrier action. Specifically:

1. The polymer must possess reactive functional groups as suitable binding sites for drug attachment. These sites should be separated from the linear main chain by short (5–15 constituent atoms) spacer segments so as to increase their steric accessibility.
2. One or more biofissionable links, such as amide or ester groups, required for drug release in the biological environment, must be inserted into the spacer. At least one of these must be sufficiently remote from the main chain to permit enzyme approach for cleavage action.
3. The macromolecular main chain should be rendered biodegradable through incorporation of biofissionable intrachain links, such as amide or urethan groups. The main chain and the spacers both should be nontoxic and should possess minimal immunogenicity.
4. The drug–bearing spacers should be sufficiently distanced from each other along the backbone to prevent intramolecular multifunctional drug binding. This requires interposition of main–chain units lacking binding capabilities.
5. The polymer must contain a sufficient number of hydrosolubilizing groups to render the ultimate polymer–drug conjugate water–soluble, and the drug–anchoring chemistry must be designed so as to accommodate this polymer solubility feature, permitting coupling reactions to proceed in aqueous or semi–aqueous solutions.

The carrier polymers selected in accordance with the aforementioned requirements were random copolyamides of the type depicted schematically below, which comprises repeat units bearing unreactive but solubilizing side groups S, as well as repeat units equipped with spacers that are terminated by functional groups F suitable for metallocene binding, and x is chosen to be equal to, or larger than, y.

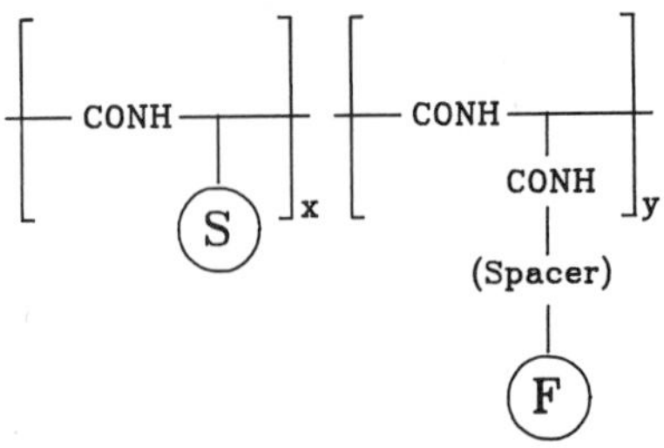

Copolyamides of this type, synthesized from poly–D,L–succinimide in a two–step aminolytic ring opening reaction (Scheme 2; R^i,R'' = amine–functionalized substituents), were made available from a separate program (M. de L. Machado and

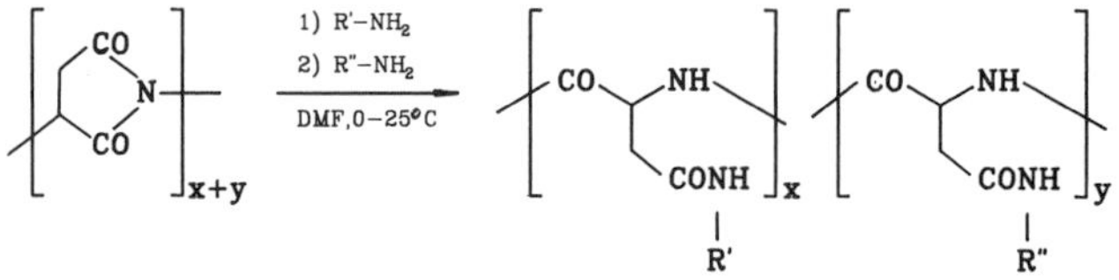

$$(2)$$

E.W. Neuse, unpublished work; see also (23)). Specifically, the polymers listed below, selected from a large number of other copolymer structures, were utilized for the present purpose:

Poly–α,β–D,L–[N–(3–(morpholin–4–yl)propyl)aspartamide–co–N–(2–(piperazinyl)ethyl)aspartamide] (**3**),

Poly–α,β–D,L–[N–(3–(morpholin–4–yl)propyl)aspartamide–co–N–(3,6–diazahexyl)aspartamide] (**4**),

Poly–α,β–D,L–[N–(3–(dimethylamino)propyl)aspartamide–co–N–(2–(piperazinyl)ethyl)aspartamide] (**5**),

Poly–α,β–D,L–[N–(3–(dimethylamino)propyl)aspartamide–co–N–(3,6–diazahexyl)aspartamide] (**6**),

Poly–α,β–D,L–[N–(3–(dimethylamino)propyl)aspartamide–co–N–(2–aminoethyl)aspartamide] (**7**),

Poly–α,β–D,L–[N–(3–(dimethylamino)propyl)aspartamide–co–N–(3–aminopropyl)aspartamide] (**8**), and

Poly–α,β–D,L–[N–(3–(morpholin–4–yl)propyl)aspartamide–co–aspart–hydrazide] (**9**).

For simplicity, all intrachain aspartyl segments in Scheme 2 and subsequent structural representations have been drawn as α–peptide units.

Ferrocene anchoring to the carrier polymers 3–9 was brought about through interaction of the spacer–attached amine functions with ferrocenylcarboxylic acids and ferrocenylcarboxaldehydes. The reactions discussed in the following (Fc = ferrocenyl) are representative.

The N–acylation of **3** with the presynthesized active ferrocenoic acid hydroxysuccinimide ester in a water–acetonitrile–tetrahydrofuran medium gave the polymeric ferrocenoylpiperazide **10** (Scheme 3). The analogous reaction of **6** afforded the ferrocenoylamino–substituted polymer **11** (Scheme 4). Coupling reactions performed with ferrocenylethanoic acid and the copolymers **5** and **8** gave rise to the formation of the conjugates **12** and **13**, respectively (Schemes 5,6), and the conjugate **14** was obtained from the reactant pair 3–ferrocenylpropanoic acid and **7** (Scheme 7). In these reactions, the free carboxylic acids were transformed *in situ* into the reactive hydroxysuccinimide esters, and these were allowed without isolation to interact with the polymeric substrates in aqueous acetonitrile. Under very similar conditions the acylation of **4**, **5**, and **8** with 4–ferrocenylbutanoic acid proceeded

142

smoothly to give the conjugates **15** (Scheme 8), **16** and **17**, respectively.

For the carboxylic acids employed in the foregoing reactions, ferrocenoic acid, 2–ferrocenylethanoic acid, 3–ferrocenylpropanoic acid, and 4–ferrocenylbutanoic acid, the electrochemical half–wave potential $E^1/_2$ (vs. SCE, in MeCN) decreases in

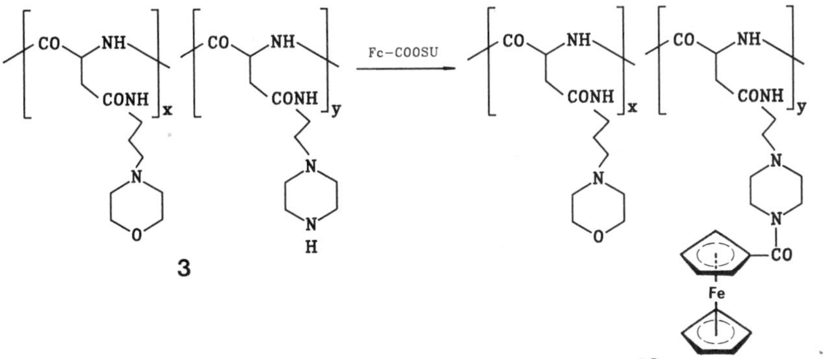

(3)

3 **10**

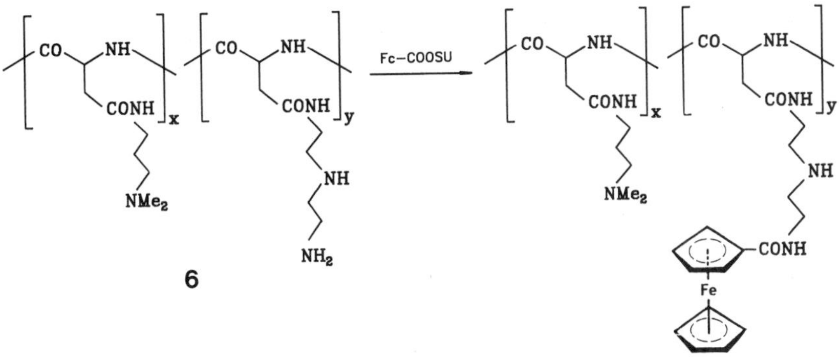

(4)

6 **11**

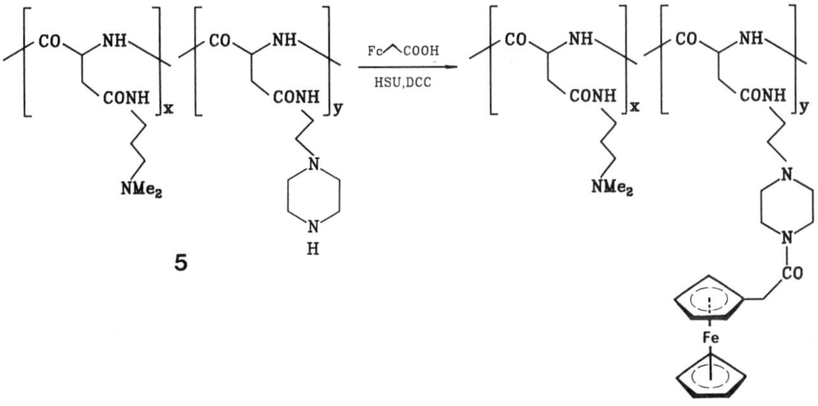

(5)

5 **12**

the given order from a high (0.59V) to a low (0.31V) value (24), the butanoic acid derivative, in fact, being more readily oxidized than the parent complex 1 (0.33V). A similar trend of reduction potentials should become apparent in the respective series of polymer–bound acids (electrochemical work in progress), and this should have some bearing on the relative *in vivo* redox behavior of the conjugates.

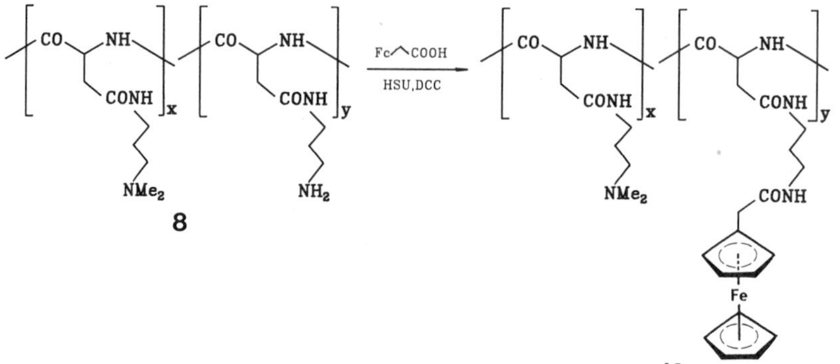

$$\text{8} \quad \xrightarrow[\text{HSU,DCC}]{\text{Fc} \frown \text{COOH}} \quad \text{13} \tag{6}$$

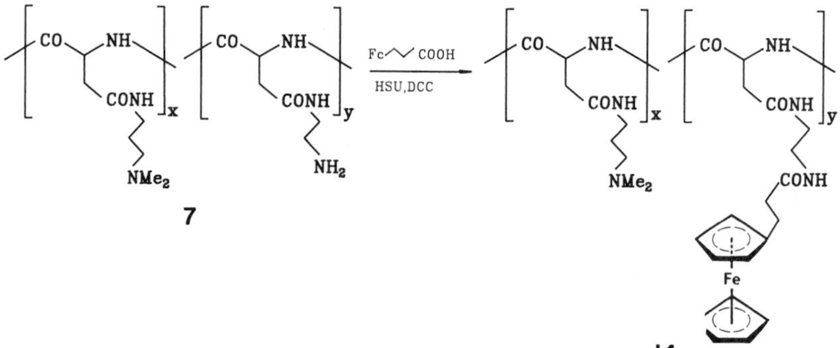

$$\text{7} \quad \xrightarrow[\text{HSU,DCC}]{\text{Fc} \frown\frown \text{COOH}} \quad \text{14} \tag{7}$$

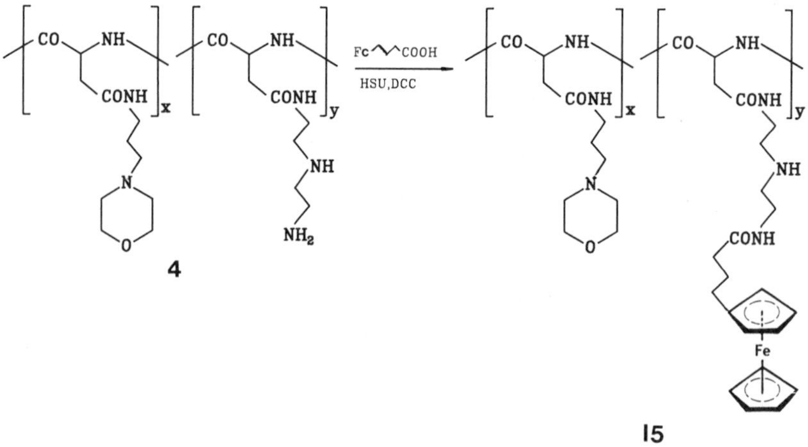

$$\text{4} \quad \xrightarrow[\text{HSU,DCC}]{\text{Fc} \frown\frown\frown \text{COOH}} \quad \text{15} \tag{8}$$

144

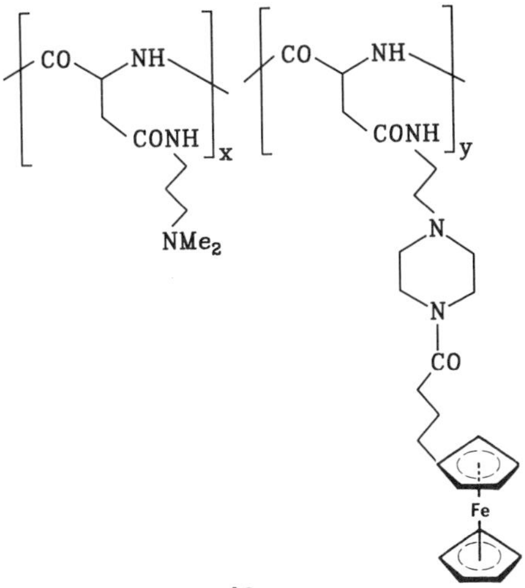

16

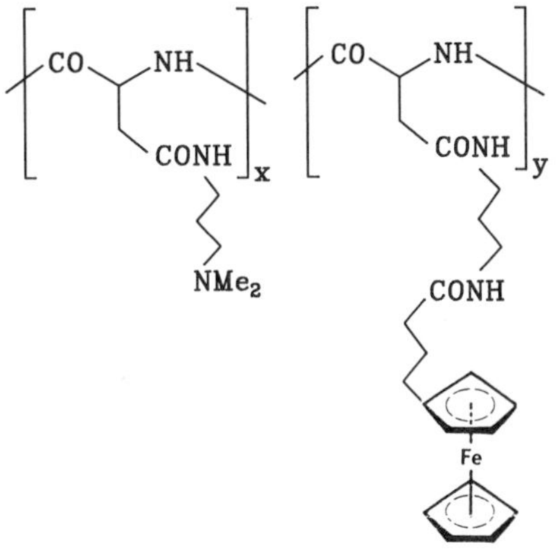

17

The carrier polymer **6** lends itself to ferrocene anchoring by reaction with the formyl function of ferrocenylaldehydes. Although the metallocene attachment is stabilized toward hydrolysis (relative to simple $-N=CH-$ bonding) through imidazoline ring formation as shown in the reaction product **18** (Scheme 9), this

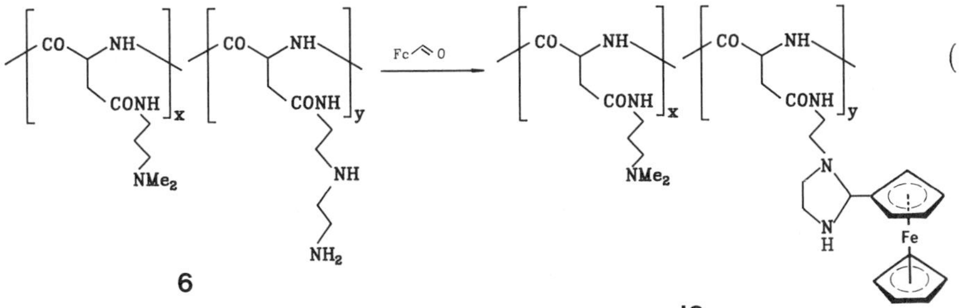

$$(9)$$

6

18

anchoring group was found to be less stable hydrolytically then the amide link in **3–17**. This was indicated by a steady iron depletion on extended aqueous dialysis of the conjugate.

In the last two examples (Scheme 10), the copolymer **9** possessing hydrazide anchoring groups was utilized. Treatment with formylferrocene in water–acetonitrile solution gave the conjugate **19**, in which the metallocene was bound through a hydrazone link. Significantly, soluble products were obtained only whenever $x/y \geq 4$. Lower ratios (i.e. larger proprotions of hydrazide units) in **9** tended to cause crosslinking in contact with the aldehyde, probably *via* ferrocenylmethinyl bridges between hydrazide groups of different chains. A somewhat enhanced tolerance in this respect (probably as a consequence of the greater steric requirements of the ferrocenophane) was observed with **20**, obtained from **9** and 1,1′–(α–oxotri–methylene)ferrocene (OTMF).

The target polymers **10–20**, upon dialysis in aqueous solution and freeze–drying, were obtained as water–soluble solids in yields of 35–90%. Inherent viscosities were typically in the range of 10–15 ml g^{-1}, and the degree of ferrocene incorporation, assessed by elemental analysis and ^{1}H NMR data, ranged from about 20 to 75% of theory, depending critically on reactant ratio and type of anchoring function. The polymers showed the characteristic ferrocene CH out–of–plane deformation and skeletal bands at 810 and 500–490 cm^{-1} in the IR spectra. The

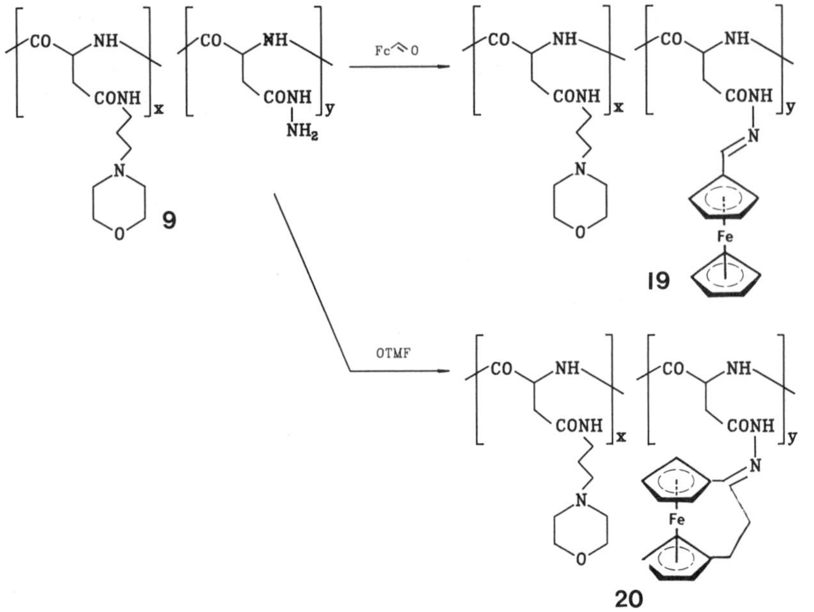

$$(10)$$

cyclopentadienyl ring proton signals appeared at δ 4.2–4.0 ppm (ca. 4.5–4.0 ppm in **10**, **11**, and **18–20**) in the ^{1}H NMR spectra, and the corresponding ring carbon peaks emerged in the ranges of 90–80 ppm (C–1) and 72–69 ppm (remaining C atoms) in the ^{13}C NMR spectra (all data vs. TMS).

In summary, the exemplifying anchoring reactions shown in the foregoing demonstrate the practicability of synthesizing completely water–soluble polymer–ferrocene conjugates for biomedical applications. It remains an objective of future tasks to establish the conditions for hydrolytic and enzyme–mediated release of the metallocene in the biological environment, to explore the feasibility of oxidizing the bound ferrocene to the ferricenium state, and, lastly, to test, in parallel and for comparison, the antineoplastic activities of both the reduced and the oxidized conjugates administered in aqueous dosage forms.

EXPERIMENTAL

Ferrocene Derivatives

Ferrocenoic acid and formylferrocene (ferrocenecarboxaldehyde) (Strem Chemicals, Inc.) were used as received. 2–Ferrocenylethanoic, 3–ferrocenyl–propanoic, and 4–ferrocenylbutanoic acids were prepared as described (25). A literature procedure (26) was also used for the synthesis of 1,1′–(α–oxo–trimethylene)ferrocene. Ferrocenoic acid N–hydroxysuccinimide ester was prepared from the free acid and N–hydroxysuccinimide (HSU) in the presence of N,N′–dicyclohexylcarbodiimide (DCC) in MeCN at −5 to 0°C; it was recrystallized from isopropanol; mp 148–149°C.

Copolyamides 3–9

These were prepared in a separate project (M. de L. Machado and E.W. Neuse, unpublished work) from poly–D,L–succinimide (27) by sequential treatment with R′–NH$_2$ and R″–NH$_2$ (Scheme 2). Briefly, the polyimide, dissolved in N,N–dimethylformamide (DMF), was allowed to react (5–10h, 20–25°C) under anhydrous conditions with the first nucleophile R′–NH$_2$ in the molar ratio $(x + y)/x$. This was followed by treatment of the intermediate with an excess of the second nucleophile R"–NH$_2$ (5–10h, 20–25°C). The ultimate copolymer was purified by dialysis (48h, 12000 molecular mass cut–off) in aqueous phase and was isolated by freeze–drying as a perfectly water–soluble solid. The yield was typically 50–75%, and the inherent viscosity, η_{inh}, ranged from 5 to 25 ml g^{-1}. The synthesis of copolyamide **4** (x/y = 1.0), described below, exemplifies the general procedure. Throughout this experimental part, amounts of polymers are given in base mole units.

4–(3–Aminopropyl)morpholine (R$'$–NH$_2$; 10 mmol) was added dropwise to polysuccinimide (20 mmol) dissolved in DMF (25 ml; redistilled at reduced pressure under N$_2$) and precooled to 0°C. The solution was stirred for 6h at room temperature; it was then added dropwise to a stirred solution of diethylenetriamine (R"–NH$_2$; 20 mmol) in the same solvent (20 ml), again cooled at 0°C. After stirring for another 5h at room temperature, the product polymer was precipitated from the viscous solution by Et$_2$0–petroleum ether (bp 65–70°C) and redissolved in H$_2$O (50 ml). The pH was adjusted to 8 (1M HCl), and the solution was dialyzed and freeze–dried, to give copolyamide **4** in 61% yield; η_{inh}(H$_2$O), 12 ml g^{-1}. The composition (x = y) was ascertained from elemental analytical and ^{1}H NMR data.

Copolyamide–Ferrocene Conjugate 10

The suspension of ferrocenoic acid N–hydroxysuccinimide ester (1.0 mmol) in MeCN–THF (1:1, 6 ml) was added dropwise to a solution of the copolyamide **3** (x/y = 1.0; 2.0 mmol) in H$_2$O (5 ml) at room temperature. The orange mixture was stirred for 24h at the same temperature. EtOH–Et$_2$O (2:3; 25 ml) was added, the resinous precipitate was washed with EtOH and was dissolved in H$_2$O (50 ml). The filtered solution was dialyzed (24h; 6000 molecular mass cut–off) and freeze–dried to give reddish–brown, water–soluble **10**; yield, 64%; η_{inh} (H$_2$O), 10 ml g^{-1}. Anal. found: Fe, 1.9%. Calcd. for **10** (x/y = 1.0; 18% substitution): Fe, 2.0%. Raising the amount of ester to 2.0 mmol under otherwise unchanged conditions gave **10** in 66% yield, and the degree of substitution increased to 31%.

Copolyamide–Ferrocene Conjugate 11

Reaction of ferrocenoic acid hydroxysuccinimide ester (1.0 mmol) with copolyamide **6** (x/y = 1.5; 2.0 mmol) and work–up by dialysis and freeze–drying as in the preceding experiment gave rust–colored, water–soluble **11** in 68% yield; η_{inh}(H$_2$O), 9 ml g^{-1}. Anal. found: Fe, 3.4%. Calcd. for **11** (x/y = 1.5; 35% substitution): Fe, 3.4%.

Copolyamide–Ferrocene Conjugate 12

The active ferrocenylethanoic N–hydroxysuccinimide ester was prepared *in situ* from ferrocenylethanoic acid (1.0 mmol), HSU (1.1 mmol), and DCC (1.2 mmol) in THF (2 ml) at –5°C (1h) and ambient temperature (5h). The ester solution, filtered from the urea by–product, combined with the THF washings of the residue, and reduced in volume to 7 ml, was added dropwise to the precooled (0°C) solution of copolyamide **5** (x/y = 3.0; 0.7 mmol) in H$_2$O (5 ml) with rapid stirring. Following the addition of more THF (1 ml) to dissolve some precipitated active ester, the mixture was stirred for 18h at ambient temperature. Upon the addition of NEt$_3$ (1.0 mmol), stirring was continued for another 6h, the THF component removed under reduced pressure, and the mixture diluted with H$_2$O (8 ml). The filtered solution was dialyzed and freeze–dried as before to give **12** as a water–soluble, brown solid. Yield, 78%; η_{inh}(H$_2$O) 12 ml g^{-1}. Anal. found: Fe, 2.6%. Calcd. for **12** (x/y = 3.0; 45% substitution): Fe, 2.7%.

Copolyamide–Ferrocene Conjugate 13

The conjugate was prepared from copolyamide **8** (x/y = 1.0; 1.0 mmol) and ferrocenylethanoic acid (1.0 mmol) by the method of the preceding experiment. Conjugate **13** was obtained in 90% yield as a water–soluble, brown solid; η_{inh}(H$_2$O),

17 ml g^{-1}. Anal. found: Fe, 3.9$_5$%. Calcd. for **13** (x/y = 1.0; 32% substitution): Fe, 4.0%.

Copolyamide–Ferrocene Conjugate **14**

Treatment of ferrocenylpropanoic acid (1.7 mmol) with HSU (1.8 mmol) and DCC (1.9 mmol) in THF (4 ml) as described above for the ferrocenylethanoic acid ester afforded the active ferrocenylpropanoic acid hydroxysuccinimide ester. Without solute isolation the solution was added to copolyamide **7** (x/y = 1.5; 1.3 mmol). The mixture was stirred for 18h at room temperature, NEt$_3$ (1.7 mmol) was added and stirring continued for another 6h. It was then treated as described for the preparation of **12**, to give conjugate **14** in 67% yield; η_{inh}(H$_2$O), 15 ml g^{-1}. Anal. found: Fe, 6.5%. Calcd. for **14** (x/y = 1.5; 75% substitution): Fe, 6.6%.

Copolyamide–Ferrocene Conjugate **15**

For the *in situ* preparation of the active hydroxysuccinimide ester of ferrocenylbutanoic acid, the latter (1.0 mmol), dissolved in THF (3 ml), was treated with HSU (1.1 mmol), then DCC (1.2 mmol), at 0°C. Work–up as described for the preparation of **12** produced an ester solution, which was added to copolyamide **4** (x/y = 1.0; 1.0 mmol) dissolved in H$_2$O (5 ml). Stirring of the mixture for 18h at room temperature was followed by the addition of NEt$_3$ (1.0 mmol) and further stirring for 8h. After removal of THF, dilution with H$_2$O and filtration as before, the aqueous solution was dialyzed and freeze–dried. Conjugate **15** was collected in 37% yield as a water–soluble mustard–colored solid; η_{inh}(H$_2$O), 7 ml g^{-1}. Anal. found: Fe, 3.9%. Calcd for **15** (x/y = 1.0; 38% substitution): Fe., 3.9$_5$%.

Copolyamide–Ferrocene Conjugate **16**

The conjugate was prepared from copolyamide **5** (x/y = 1.0; 1.0 mmol) and ferrocenylbutanoic acid (1.0 mmol) as in the preceding experiment. This gave conjugate **16** in 58% yield as a mustard–colored, water–soluble solid; η_{inh}(H$_2$O), 11 ml g^{-1}. Anal. found: Fe, 5.1%. Calcd. for **16** (x/y = 1.0; 50% substitution): Fe, 5.0$_5$%.

Copolyamide–Ferrocene Conjugate **17**

The preparation of this conjugate from copolyamide **8** (x/y = 1.0; 1.0 mmol) and ferrocenylbutanoic acid (1.0 mmol) followed the precedure described for the synthesis of **15**. Conjugate **17** was obtained in 82% yield as a mustard–colored, water–soluble solid; η_{inh}(H$_2$O), 14 ml g^{-1}. Anal. found: Fe, 5.3%. Calcd. for **17** (x/y = 1.0; 46% substitution): Fe, 5.3%.

Copolyamide–Ferrocene Conjugate **18**

To the solution of copolyamide **6** (x/y = 1.0; 2.0 mmol) in DMF (15 ml), saturated with N$_2$, was added formylferrocene (4.0 mmol) with stirring and continued introduction of N$_2$. The flask was stoppered, and the solution was stirred for 3 days at room temperature in the dark. After dilution with H$_2$O (100 ml) and filtration, the solution was dialyzed and freeze–dried as before, to give conjugate **18** in 78% yield as a dark red, water–soluble solid; η_{inh}(H$_2$O), 9 ml g^{-1}. Anal. found: Fe, 3.3%. Calcd. for **18** (x/y = 1.0; 27% substitution): Fe, 3.3%. Redialysis (24h) and freeze–drying left polymer (66% recovery), η_{inh}(H$_2$O), 8 ml g^{-1}. Anal. found: Fe, 2.3. Calcd. for **18** (x/y = 1.0; 18% substitution): Fe, 2.3%.

Copolyamide–Ferrocene Conjugate **19**

Formylferrocene (0.5 mmol), dissolved in H$_2$O–MeCN (1:1; 3 ml), was added dropwise to the solution of copolyamide **9** (x/y = 9.0; 0.22 mmol) in H$_2$O (10 ml) saturated with N$_2$ and precooled to 0°C. Introduction of N$_2$ was continued for 15 min, and the clear, orange–brown solution was left to stand for 24h at room temperature in the stoppered flask with protection from light. Upon dilution with H_2O (30 ml) the product solution was dialyzed and freeze–dried. Conjugate **19** was thus obtained in 92% yield; $\eta_{inh}(H_2O)$, 12 ml g^{-1}. Anal. found: Fe, 1.1%. Calcd. for **19** (x/y = 0.9; 45% substitution): Fe, 1.1%. Redialysis (24 h) and freeze–drying gave water soluble polymer (68% recovery) with $\eta_{inh}(H_2O)$, 15 ml g^{-1}. Anal. found: Fe, 0.8%. Calcd. for **19** (x/y = 0.9; 35% substitution): Fe, 0.8%.

Copolyamide–Ferrocene Conjugate 20

1,1′–(α–Oxotrimethylene)ferrocene (0.5 mmol) and copolyamide **9** (x/y = 0.8; 0.25 mmol) were allowed to interact as described in the preceding experiment. Work–up as before gave water–soluble conjugate **20** in 77% yield; $\eta_{inh}(H_2O)$, 0.14 ml g^{-1}. Anal. found: Fe, 1.9%. Calcd. for **20** (x/y = 0.8; 38% substitution): Fe, 1.9%.

The conjugates obtained in the described experiments tended to undergo slow intermolecular solid–state interaction; as a result, extended storage at ambient temperature occasionally led to reduced solubility in H_2O. This aging effect was significantly retarded with samples stored at $-25°C$. In addition, it sometimes proved possible to restore the solubility of aged samples by shaking their acidified (pH 3; 1M HCl) suspensions in H_2O for a few minutes to achieve predominant redissolution. Following readjustment of the pH to 6–7, the filtered solutions were dialyzed and freeze–dried. This gave water–soluble polymers possessing the same analytical, viscometric, and spectroscopic properties as the original conjugates.

ACKNOWLEDGMENT

The authors gratefully acknowledge the financial support provided by SASOL Limited. Mrs. L. Meredith is thanked for her excellent contribution in typing the manuscript.

REFERENCES

1.   E.W. Neuse, and F.B.D. Khan, Appl. Organomet. Chem., *1*, 499 (1987).
2.   W. Troll, G. Witz, B. Goldstein, D. Stone, and T. Sugimura, in "Carcinogenesis", Vol. 7, E. Hecker *et al.*, Eds., Raven Press, New York (1982), p. 593.
3.   T.W. Kensler, D.M. Bush, and W.J. Kozumbo, Science *221*, 75 (1983).
4.   P.J. O'Brien, Environmental Health Perspectives, *64*, 219 (1985).
5.   I. Fridovich, Annu. Rev. Biochem., *44*, 147 (1976).
6.   L.W. Oberley and G.R. Buettner, Cancer Res., *39*, 1141 (1979).
7.   M.D. Sevilla, P. Neta, and L.J. Marnett, Biochem. Biophys. Res. Commun., *115*, 800 (1983).
8.   T.W. Kensler and M.A. Trush, Biochem. Pharmacol., *32*, 3485 (1983).
9.   K.V. Honn and J.R. Dunn, FEBS Lett., *139*, 65 (1982).
10. E.W. Neuse, in "Metal–Containing Polymeric Systems", J.E. Sheats, C.E. Carraher. Jr., C.U. Pittman, Jr. Eds., Plenum Press, New York (1985), p. 99, and refs. therein.
11. M.S. McDowell, J.H. Espenson, and A. Bakač, Inorg. Chem., *23*, 2232 (1984).
12. S.R. Logan and G.A. Salmon, J. Chem. Soc. Perkin Tr. II, 1781 (1983).
13. M. Rosenblum, "Chemistry of the Iron Group Metallocenes", Part I, Wiley, New York (1965), Chap. 6 and refs. therein.
14. R. Epton, M.E. Hobson, and G. Marr, J. Organometal. Chem., *149*, 231 (1978).
15. B.W. Carlson, L.L. Miller, P. Neta, and J. Grodkowski, J. Am. Chem. Soc., *106*, 7233 (1984).
16. J.R. Pladziewicz and M.S. Brenner, Inorg. Chem., *26*, 3629 (1987).
17. P. Köpf–Maier, H. Köpf, and E.W. Neuse, J. Cancer Res. Clin. Oncol., *108*, 336 (1984).
18. M. Wenzel, Y. Wu, E. Liss, and E.W. Neuse, Z. Naturforsch., *43c*, 963 (1988).

19. E.W. Neuse, and F. Kanzawa, Appl. Organomet. Chem., in press.
20. R. Duncan and J. Kopeček, Advan. Polym. Sci., *57*, 51 (1984).
21. M. Poznansky and R.L. Juliano, Pharmacol. Revs., *36*, 277 (1984).
22. R. Duncan, in "Controlled Drug Delivery", 2nd Edn., J.R. Robinson, and V.H. Lee, Eds., Dekker, New York (1987), Chap. 14.
23. E.W. Neuse, and A. Perlwitz, Polym. Prepr., *30*, 361 (1989).
24. N.F. Blom, E.W. Neuse, and H.G. Thomas, Transition Met. Chem., *12*, 301 (1987).
25. E.W. Neuse, F.B.D. Khan, and M.G. Meirim, Appl. Organomet. Chem., *2*, 129 (1988).
26. K.L. Rinehart, Jr., R.J. Curby, Jr., D.H. Gustafson, K.G. Harrison, R.E. Bozak, and D.E. Bublitz, J. Am. Chem. Soc., *84*, 3263 (1962).
27. P. Neri and G. Antoni, Macromol. Synt., *8*, 25 (1982).

POLYCONDENSATION OF CHROMIUMTRICARBONYL-ARENE

SYSTEMS AND ORGANOSTANNANE MONOMERS CATALYZED BY PALLADIUM

Michael E. Wright

Department of Chemistry and Biochemistry
Utah State University
Logan, Utah 84322-0300

INTRODUCTION

Conjugated organic polymers represent an important class of materials
which often have interesting thermal, optical, and electronic properties
[1].   A general limitation to the utility of these polymers is their
insolubility and often high melting points leading to problems in
fabrication.   One solution which has proved quite successful in enhancing
these polymer's processability has been the addition of side-chain groups
(e.g. phenyl groups) to the polymer backbone [2].   Our laboratory is
approaching the problem from an organometallic point of view, that is, we
are using coordination of a transition-metal complex as the side-chain
moiety.   Inorganic and organometallic polymers in the form of coordination
or pendant type are themselves an important class of materials [3].
However, having the transition-metal complexed directly to a conjugated,
organic, polymer backbone produces a unique situation.   The multiple
oxidation states, the coupling of electronic transitions between ligand
and metal, and the accessibility of ligand substitution reactions makes
the organometallic complex an intriguing "side-chain" indeed.

Transition metal–catalyzed cross-coupling reactions have become a
powerful and routine method for the formation of sp^2-sp^2 and sp^2-sp carbon-
carbon bonds [4].   When the technique of cross-coupling is applied to
polymer synthesis the term polycondensation is deemed appropriate because
all transition metal catalyzed reactions produce a small molecule some-
where in the catalytic cycle.   The process will also follow rules which
govern a step-growth polymerization technique.   Step-growth
polymerizations require high reaction efficiency, perfect stoichiometry,
and pure monomers in order to obtain high molecular weight polymers.   In

Inorganic and Metal-Containing Polymeric Materials
Edited by J. Sheats *et al.*, Plenum Press, New York, 1990

eq 1 below we have given the formula describing the relationship of DP and reaction efficiency (P in % reaction) and monomer purity (X_1 = monomer, X_2 = impurity). For example, with a monomer of 100% purity X_2/X_1 = 0 and 98% yield reaction produces a DP of 50. If the monomer is 98% pure (i.e. X_2 = 2%), then the DP drops to 25! Clearly reaction efficiency and monomer purity are critical issues.

$$DP = \frac{1 + X_2/X_1}{1 - P + X_2/X_1} \tag{1}$$

The first utilization of a transition metal-catalyzed cross-coupling reaction for a polymer synthesis was that of Yamamoto and coworkers in the synthesis of poly-p-phenylene (eq 2) [5]. The small molecule produced in the reaction was $MgBr_2$ and molecular weights on the order of 10,000 were obtained.

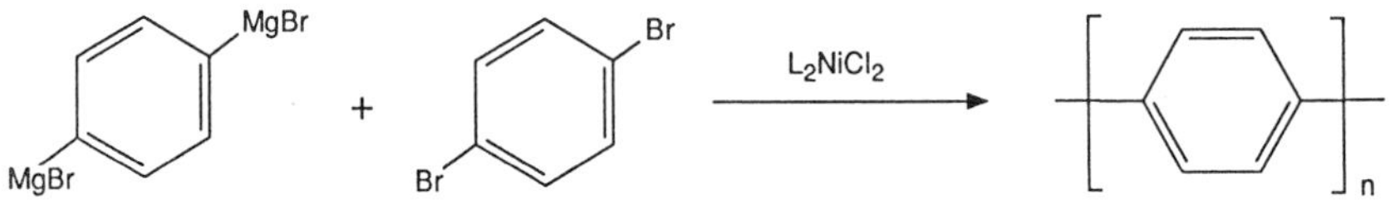

$$\tag{2}$$

The palladium-catalyzed cross-coupling of aryl halides with diamines has also been used in the preparation of polymers [6]. This reaction can be run under an atmosphere of carbon monoxide to produce aromatic polyamides with similar inherent viscosities as polymers prepared via conventional solution polymerization techniques (eq 3) [7].

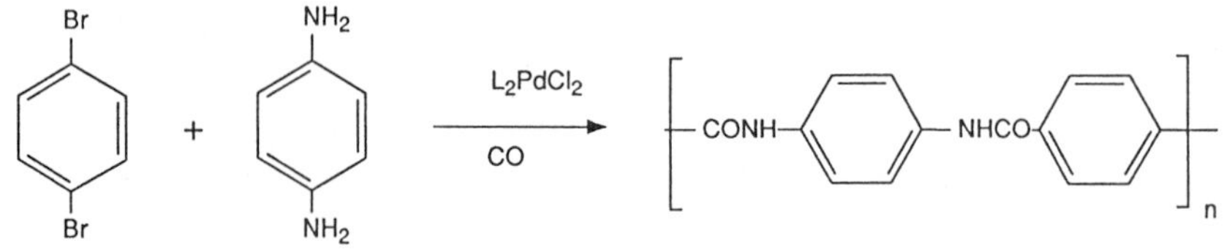

$$\tag{3}$$

The palladium-copper-catalyzed cross-coupling of acetylenic systems with aryl halide has been used successfully to homopolymerize p-bromoethynylbenzene (eq 4) [8]. The poly-p-ethynylbenzene was isolated as a white polymer of relatively low molecular weight. The coupling of acetylenic reagents with aryl halides has been noted to lead to explosions and therefore due care should be exercised.

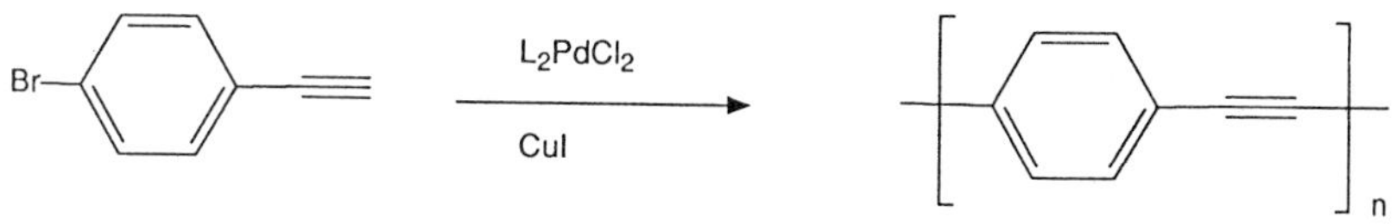

$$\text{(4)}$$

The palladium-catalyzed cross-coupling of thiophenyl zinc reagents with aromatic dihalides was recently described (eq 5) [9]. The method was published in communication form but it did appear this could be a generally useful polymerization technique.

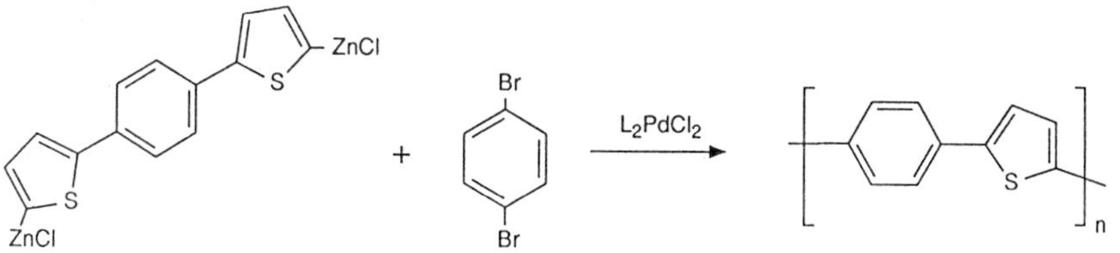

$$\text{(5)}$$

The homopolymerization of boric acid derivatives catalyzed by palladium has very recently appeared by two research groups [10]. This polymerization technique employs cross-coupling chemistry of boric acids developed by Suzuki and coworkers [4]. The yields of the pure carbon polymers were above 89%, (eq 6) [10b].

We recently presented the synthesis of polymer 1 via the palladium-catalyzed polycondensation of $(\eta^6$-arene)$Cr(CO)_3$ and organostannane monomers (eq 7) [11]. This polycondensation technique is based on previous work

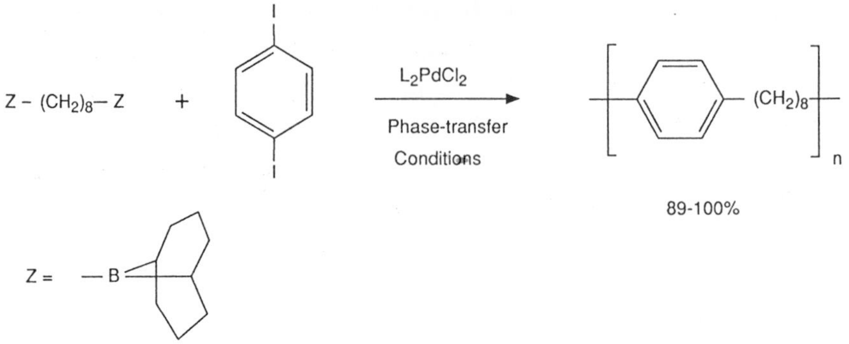

Z = —B

(6)

89-100%

done by the Stille research group [4]. It should be noted that both monomers are isolable and easily obtained in analytically pure form. As mentioned above, monomer purity is critical to the control and success of step-growth polymerizations. An example of a palladium catalyzed polycondensation involving organostannane and organic aryl halides has also been recently reported [12].

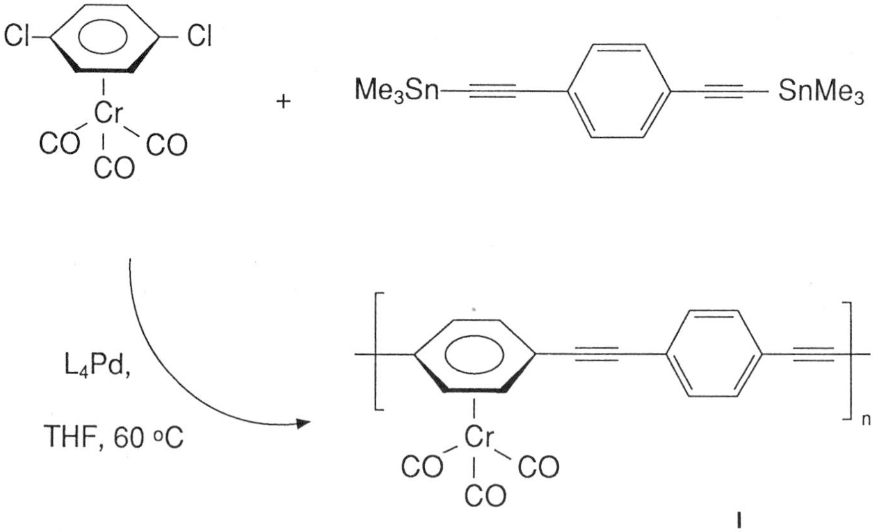

(7)

In this paper I describe the synthesis of a new Group 6 monomer in which a carbonyl group on the metal has been substituted by a tributylphosphine moiety. This new monomer is then utilized in the preparation of new class of conjugated polymers. The palladium-catalyzed cross-coupling reaction is once again a successful method for the polycondensation of organometallic monomers.

RESULTS & DISCUSSION

We recently developed efficient synthetic routes to the required η^6-1,3- and (η^6-1,4-dichlorobenzene)Cr(CO)$_3$ (2) complexes for the preparation of polymers such as 1 [13]. Complex 2 can be further modified by photolysis in the presence of tributylphosphine to produce the new monomer 3 (Scheme I). A relative rate study between 2 and 3 showed the latter complex to be less reactive in the palladium-catalyzed cross-coupling reaction with 1-(trimethylstannyl)-2-phenylethyne. The palladium-catalyzed polycondensation of monomers 3 and 4 resulted in high yields of polymer 5 (>90%). A reaction temperature of 65 °C and the use of THF as solvent proved to be the optimum polymerization conditions. Lower temperatures or a solvent change to toluene afforded little if any polymer after several hours of reaction.

Scheme I

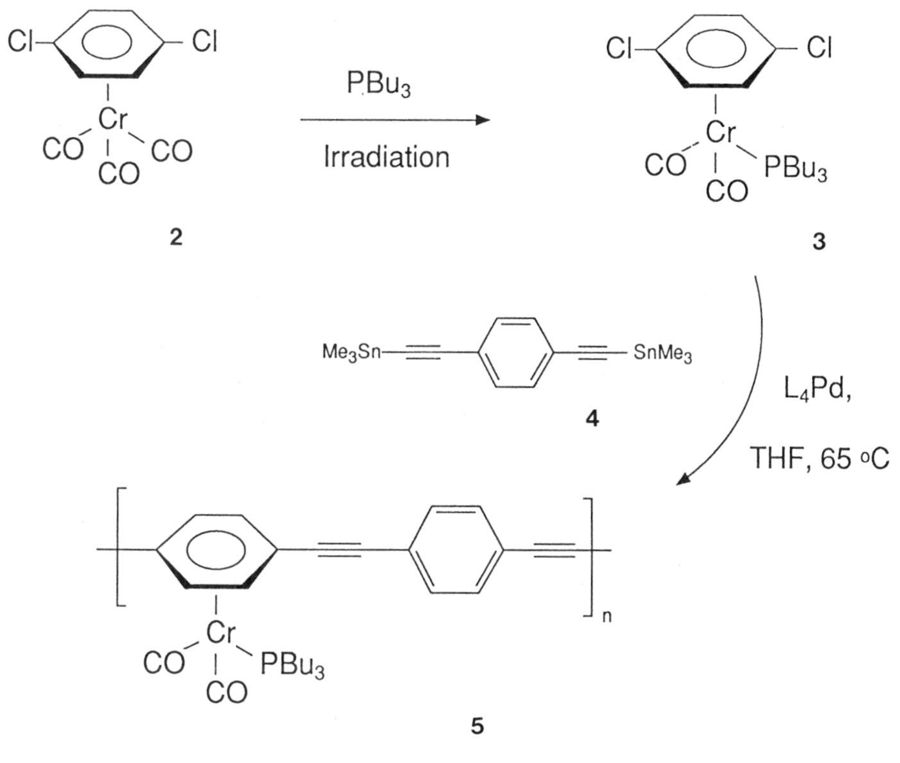

2

3

4

L$_4$Pd,

THF, 65 °C

5

Polymer 5 is isolated as a deep red solid showing limited solubility in chlorinated solvents (e.g. dichloromethane & chloroform). Elemental analysis of a polymer prepared from imbalanced stoichiometry (1:0.95, stannane:chromium monomer, respectively) gave data consistent with an average DP of 8 and the expected 0% chlorine content. Proton NMR data showed a downfield shift in the η^6-arene protons consistent with replacement of the chlorides with acetylenic moieties based on model compounds.

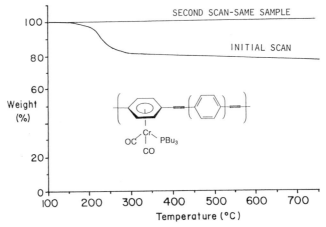

Figure 1.   Thermogravimetric analysis of polymer 5 under an argon atmosphere. A heating rate of 10°C/min was employed.

We have examined the thermal stability of polymer 5 by thermal gravimetric analysis (TGA) and differential scanning calorimeter (DSC). The results are presented in Figures 1 and 2, respectively. As was the case with polymer 1 an initial weight loss occurs at a lower temperature followed by excellent stability thereafter. Notably, the temperature at which 5 begins this initial weight loss is ~50 °C higher than for 1. Using the assumptions that the rate of ligand dissociation and diffusion out of the polymer matrix are similar for both systems at the respective temperatures, this represents approximately a 5 kcal/mol increase in stability towards ligand dissociation for polymer 5.

Infrared data suggests there is loss of tributylphosphine followed by coordination of the $Cr(CO)_2$ moiety to an alkyne unit. The weight loss found in the TGA is consistent with loss of tributylphosphine and the endotherm observed is likely due to evaporation of the free phosphine. A similar process was noted for polymer 1 except this involved loss of carbon monoxide. In essence then, polymers 1 and 5 upon thermal treatment lead to the same highly-crosslinked and thermally stable polymer.

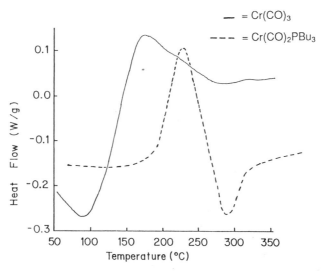

Figure 2.   Differential scanning calorimeter of polymer 5 (dashed line) and polymer 1 (solid line) under an argon atmosphere. Heating rate was 10 °C/min.

CONCLUDING REMARKS

The incorporation of a tributylphosphine group on the chromium metal center produced polymers which thermally set at a higher temperature and with a greater weight loss; although, the new thermally set polymers demonstrated excellent thermal stability thereafter up to 750 °C. The initial loss of mass corresponds to the %-weight of the tributylphosphine in the polymer and leads us to believe that the thermal setting process involves ejection of the phosphine from the chromium center. This degradation mechanism could be confirmed by elemental analysis of the process after heating.

EXPERIMENTAL

General. All manipulations of compounds and solvents were carried out by using standard Schlenk techniques. Solvents were degassed and purified by distillation under nitrogen from standard drying agents. Spectroscopic measurements utilized the following instrumentation: ^{1}H NMR, Varian XL 300; ^{13}C NMR, Varian XL 300 (at 75.4 MHz). The 1,4-bis(trimethylstannylethynyl)benzene, (η^6-1,4-dichlorobenzene)Cr(CO)$_3$, and 4,4'-bis(trimethylstannylethynyl)biphenyl were prepared by literature methods [11]. Polymer analyses were performed using a duPont 9900 thermal analysis data station. Elemental analyses were performed at Atlantic Microlab Inc.

Preparation of Monomer 3. A benzene (20 mL) solution containing 2 (0.50 g, 1.8 mmol) and tributylphosphine (0.50 g, 2.5 mmol) was irradiated in water jacketed quartz reaction vessel with a 175 watt medium pressure mercury lamp for 6 h. The solvent was removed under reduced pressure and crude product subjected to column chromatography on alumina. Gradient elution with pet ether/benzene (final 2/1, v/v) produced a yellow-orange band which was collected and the solvents removed to afford pure 3 (0.65 g, 81%). ^{1}H NMR (CDCl$_3$) δ 4.90 (d, J = 3 Hz, 4 H), 1.66-1.60 (m, 6 H), 1.43-1.37 (m, 12 H), 0.94 (t, J = 7 Hz, 9 H); ^{13}C NMR (CDCl$_3$) δ 238.1 (d, J = 21 Hz), 104.8 (ipso η^6-arene carbon), 85.3 (η^6-arene CH), 29.4 (d, J = 21 Hz), 25.5, 24.5 (d, J = 12 Hz), 13.8. IR (CH$_2$Cl$_2$) ν_{CO} 1906 and 1851 cm^{-1}. Anal. Calcd for C$_{20}$H$_{31}$CrCl$_2$O$_2$P: C, 52.5; H, 6.8. Found: C, 52.0; H, 6.7.

Preparation of Polymer 5. A Schlenk flask was charged with THF (5 mL), 3 (100 mg, 0.23 mmol), 4 (133 mg, 0.25 mmol), and (PPh$_3$)$_2$PdCl$_2$ (3.2 mg, 2 mol-%), in that order, and allowed to react at 65 °C for 36 h. The mixture was diluted with ether and the polymer collected on glass frit. The polymer was washed with ether, dichloromethane, and finally benzene. The polymer was dried under reduced pressure at 60 °C for 48 h to afford 140 mg (93%) of a deep red polymer. IR (CH$_2$Cl$_2$) ν_{CO} 1899 and 1848 cm^{-1}. Anal. Calcd for {C$_{36}$H$_{39}$CrO$_2$P)$_n$} (SnMe$_3$ end-capped): DP = 8; C, 71.30; H, 6.52. Found: C, 71.27; H, 6.35.

ACKNOWLEDGEMENT

I wish to express my sincere gratitude to donors of the Petroleum Research Fund, administered by the American Chemical Society, for financial support of this research.

REFERENCES

1. "Nonlinear Optical and Electroactive Polymers", Eds. P. N. Prasad, D. R. Ulrich, Plenum Press, New York (1988). G. Wegner, Macromol. Chem., Macromol. Symp. 1:151 (1986). P. E. Cassidy, "Thermally Stable Polymers: Syntheses and Properties", Marcel Dekker, New York (1980). C. Arnold Jr, J. Polym. Sci., Macromol. Rev. 14:265 (1979). W. W. Wright, The Development of Heat Resistant Organic Polymers, in: "Degradation and Stabilization of Polymers," Wiley, New York (1975). J. G. Speight; P. Kovacic; F. W. Koch, J. Macromol. Sci. Rev. Macromol. Chem., 5:295 (1971). G. K. Noren; J. K. Stille, J. Polym. Sci. D.: Macromol. Rev., 5:385 (1971). H. F. Mark, J. Appl. Polym. Sci., Appl. Polym. Symp., 35:13 (1979).

2. J. K. Stille; F. W. Harris; R. O. Rakutis; H. Mukamol, J. Polym. Sci. B. 791 (1966). J. K. Stille; R. O. Rabutis; H. Mukamol; F. W. Harris, Macromolecules 1:431 (1968). J. K. Stille; G. K. Noren, Polym. Lett. 7:525 (1969).

3. M. Zeldin, K. J. Wynne, Allcock, H. R., Eds., "Inorganic and Organometallic Polymers: Macromolecules Containing Silicon, Phosphorus, and other Inorganic Elements," ACS Symp. Ser., Washington D. C. (1987). C. U. Pittman, jr., M. D. Rausch, Pure Appl. Chem. 58:617 (1986). J. E. Sheats, C. E. Carraher, C. U. Pittman, jr., Eds., "Metal-Containing Polymer Systems," Plenum, New York (1985).

4. Grignard cross-coupling: T. Hayashi, M. Kumada, Acc. Chem. Res. 15:395 (1982). K. Tamao, K. Sumitani, Y. Kiso, M. Zembayashi, A. Fijioka, S.-I. Kodama, I. Nakajima, A. Minato, M. Kumada, Bull. Chem. Soc. Jpn. 49:1958 (1976). Boronic acid cross-coupling: A. Suzuki, Acc. Chem. Res. 15:178 (1982). A. Suzuki, Pure Appl. Chem. 57:1749 (1985). Organostannane cross-coupling: J. K. Stille, Pure Appl. Chem. 57:1771 (1985). J. K. Stille, Angew. Chem., Int. Ed. Engl. 25:508 (1986).

5. T. Yamamoto, Y. Hayashi, A. Yamamoto, Bull. Chem. Soc. Jpn. 51:2091 (1978).

6. This includes other Heck type cross-coupling reactions: A. Greiner, W. Hietz, Makromol. Chem., Rapid Commun. 9:581 (1988). M. Suzuki, K. Sho, J. C. Lin, T. Saegusa, Polym. Bull. 21:415 (1989). W. Heitz, W. Brugging, L. Freund, M. Gailberger, A. Greiner, H. Jung, U. Kampschulte, N. Niessner, F. Osan, H.-W. Schmidt, W. Wicker, Makromol. Chem. 189:119 (1988).

7. M. Yoneyama, M. Kakimoto, Imai, Y. Macromolecules 21:1908 (1988).

8. D. L. Trumbo, C. S. Marvel, <u>J. Polym. Sci. Chem. Ed.</u> 24:2311 (1986). S. J. Havens, P. M. Hergenrother, <u>J. Polym. Sci., Polym. Lett. Ed.</u> 23:587 (1985).

9. A. Pelter, M. Rowlands, I. H. Jenkins, <u>Tetrahedron Lett</u>. 28:5213 (1987).

10. M. Rehan, A.-D. Schluter, G. Wegner, W. J. Feast, <u>Polymer</u> 30:1060 (1989). (b) E. Cramer, V. Percec, <u>Polym. Preprints</u> 31(1):516 (1990).

11. M. E. Wright, <u>Polymer Preprints</u> 29(1):294 (1988). M. E. Wright, <u>Macromolecules</u> 22:3256 (1989).

12. M. Bochmann, K. Kelly, <u>Chem. Commun.</u> 532 (1989).

13. M. E. Wright, <u>Organometallics</u> 8:407 (1989).

METAL COORDINATION POLYMERS AS POTENTIAL HIGH-ENERGY

LITHOGRAPHIC RESISTS

Ronald D. Archer, Valentino J. Tramontano, Ven O. Ochaya,
Paul V. West, and William G. Cumming

Department of Chemistry, University of Massachusetts
Amherst, MA 01003

ABSTRACT

Linear metal coordination polymers can provide materials which are very sensitive to high-energy radiation. Such polymers should make good lithographic resists. In addition to our previously reported uranyl and vanadyl polymers, we have extended this work to cobalt and chromium coordination polymers which have thio, dithio, sulfoxo, and sulfone bridges between two β-diketonato ligands on adjacent metal centers. The cobalt (III) polymers with two 2,4-pentanedionato and one leucinato ligand per metal have been thoroughly investigated and are the focus of this chapter. The thio-, dithio-, and sulfone-bridged polymers are extremely efficient in their scission reactions under cesium-137 radiation with G_s values of greater than 10 in each case. All bond breaking occurs at C-S and/or S-S bonds, not at the cobalt centers. Progress with similar chromium(III) polymers which have a third β-diketonato ligand is briefly noted.

INTRODUCTION

During the past few years, a number of new applications for metal coordination polymers have appeared as detailed by Pittman and others earlier in this volume (1). The use of metal coordination polymers for lithographic purposes dates back to the nineteenth century when blueprint and chromium lithographic procedures were developed (2). In both of these cases, low molecular weight coordination species are polymerized by light, which provides less soluble photo products. The alternative situation in which metal coordination polymers are both soluble enough to allow good film formation and can then be degraded by light is not that common inasmuch as soluble metal coordination polymers are not that overly abundant.

On the other hand, high energy radiation is well known to provide more soluble organic polymers by causing degradation of long chain polymers which have lower solubility. One particularly interesting application is the use of polymers as photo resists or as electron-beam resists in lithography (3). Higher density integrated circuit chips have allowed metal oxide semiconductor random access memory device capacity to increase by several orders of magnitude in the recent past with minimum feature size of less than one micron at present. Further reductions are in progress, but diffraction effects will eventually limit the resolution of visible and ultraviolet-sensitive resists. However, higher energy electron beams, x-ray, gamma ray, and ion-beam systems can provide the necessary resolution required for submicron resolution. But organic polymers have limited sensitivity to high energy radiation with G values (chemical events/100 eV absorbed) typically not much greater than one, though sulfur containing organic polymers with G values of about 10 are known. G_s = G for scission.

Heavy metal atoms provide an increased absorption of energy because of the photoelectric effect absorption, which has a greater than Z^4 power dependence, where Z = atomic number (4,5). See Figure 1. The uranyl dicarboxylate polymers which we have studied (6,7) have G_s values of 50 and more at 662 keV, where the photoelectric effect is quite low, even for uranium (Figure 2). In the usual 10 – 20 keV range of electron beam studies, 3d-transition metal polymers should have a tremendous advantage over simple organic polymers. Poly(vinylferrocene) shows a 10 to 20 fold increase in sensitivity to x-ray radiation relative to simple vinyl polymers, but even poly(vinylferrocene) is not sufficiently sensitive to be useful (8). Miramura (9,10) used pendant ferrocene groups to enhance the x-ray sensitivity of lithographic resists. However, mixtures of ferric oxide with lithographic resists did not enhance the sensitivity of lithographic resists (11). Enhanced sensitivity of PMMA [poly(methyl methacrylate)] has been obtained through the incorporation of cesium (Z = 55) and thallium (Z = 81) ions into copolymer of methacrylic acid and methyl methacrylate (12). One thallium (28 % by wt) copolymer provided an enhancement factor of about 40 relative to PMMA.

Therefore, in order to get the advantage of the heavier atoms together with the sensitivity of the organic polymers containing sulfide and sulfone bridges, we have synthesized several cobalt(III) coordination polymers, one of which was briefly described earlier (13,14). The general synthesis for three of the polymers derived from leucinatobis(2,4-pentanedionato)cobalt(III), Co(leu)(acac)$_2$, is shown in equations (1) - (5):

$$[Co(NO_2)_6]^{3-} + 2\ acac^- \longrightarrow trans\text{-}[Co(acac)_2(NO_2)_2]^- + 4\ NO_2^- \qquad (1)$$

where acac$^-$ is the anion of 2,4-pentanedione,

$$[Co(acac)_2(NO_2)_2]^- + leu^- \underset{Norit\ \mathbf{A}}{\xrightarrow{\hspace{1cm}}} Co(leu)(acac)_2 + 2\ NO_2^- \qquad (2)$$

where leu$^-$ is the anion of leucine. Derivatives of simpler amino acids such as glycine are more difficult to purify. Note that a rearrangement of ligands takes place as the product of (2) has three bidentate rings--not the trans arrangement observed in the product of (1).

$$Co(leu)(acac)_2 + SCl_2 \longrightarrow [-acacCo(leu)acac\text{-}S\text{-}]_n + 2\ HCl \qquad (3)$$

$$Co(leu)(acac)_2 + S_2Cl_2 \longrightarrow [-acacCo(leu)acac\text{-}S\text{-}S\text{-}]_n + 2\ HCl \qquad (4)$$

$$Co(leu)(acac)_2 + SOCl_2 \longrightarrow [-acacCo(leu)acac\text{-}\underset{O}{S}\text{-}]_n + 2\ HCl \qquad (5)$$

These three polymerization reactions can be summarized as shown in **Scheme 1**, where Q = S, S$_2$, or SO. Q bridges the β-diketones from adjacent units in the polymer chains. The corresponding sulfone polymer can be synthesized by the oxidation of the sulfoxide polymer with hydrogen peroxide.

$$[-acacCo(leu)acac\text{-}\underset{O}{S}\text{-}]_n \xrightarrow[H_2O_2]{} [-acacCo(leu)acac\text{-}\overset{O}{\underset{O}{S}}\text{-}]_n \qquad (6)$$

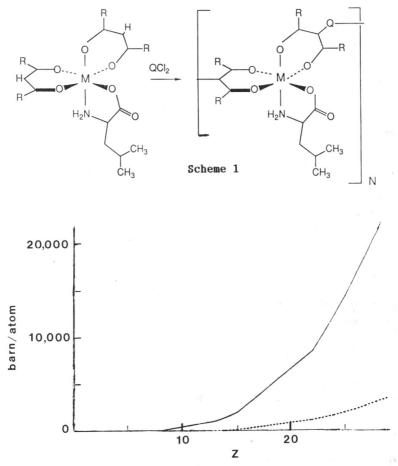

Scheme 1

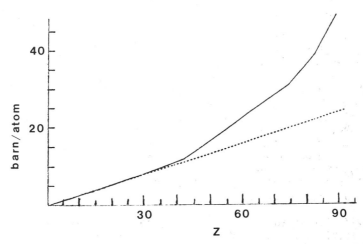

Fig. 1. Mass absorption coefficient of gamma radiation at 10 (———) and 20 (----) keV *vs.* atomic number Z (15). Photoelectric effect absorption dominates at these energies.

Fig. 2. Total (———) and Compton (----) mass absorption at *ca.* 600 keV *vs.* atomic number Z (15). Photoelectric effect absorption is the difference.

As noted below, the cobalt(III) coordination polymers have excellent G_s values. On the other hand, they appear to undergo cleavage only at the C-S and S-S bonds when irradiated at high energy, rather than at the cobalt centers [which would be expected for radiation in the ultraviolet region, where cobalt(III) to cobalt(II) charge transfer chemistry occurs for cobalt(III) β-diketone complexes (16)]. In fact, other than the sulfoxo species, they all appear to have G_s values at least as high as any organic polymers, even at 662 keV, where cobalt has no particular advantage over ligher elements other than the Compton effect linear increases with increases in atomic number. Coupled with the higher quantity of energy absorbed for identical films and their good adhesion to silica-coated silicon, these species have excellent resist potential. In response to the concern of resist specialists about residual metal ions (17), it is gratifying to note that no detectable free cobalt ions can be observed after several megarads of irradiation at high energies, either for the monomer or the polymer. This observation led us to consider the more volatile fluorinated chromium(III) β-diketonates. Progress with the chromium species is also noted herein, though this work has not proceeded as far as the cobalt(III) studies.

EXPERIMENTAL

Solvents and ligands were carefully dried prior to use. Sulfur dichloride and sulfur monochloride were purified just prior to use by appropriate literature methods (18,19).

(S-Leucinato-*N,O*)bis(2,4-pentanedionato-*O,O'*)cobalt(III).-This neutral complex was prepared by the method of Laurie (20). Yields of 20 - 29% of pure Co(leu)(acac)$_2$ were obtained from the reaction between Na[*trans*-Co-(acac)$_2$(NO$_2$)] and Na[$\bar{S}$-(leu)] {S in R,S-nomenclature is equivalent to *l* in *d,l*} after column chromatography, rotary evaporation and recrystallization.

Anal. Calcd for C$_{16}$H$_{26}$NO$_6$Co: C, 49.6; H, 6.8; N, 3.6. Found: C, 49.5; H, 6.7; N, 3.6.

(S-Leucinato-*N,O*)bis[2,2,6,6-tetramethyl-3,5-heptanedionato-*O,O'*)-cobalt(III).--An analogous reaction between the corresponding 2,2,6,6-tetramethylheptanedionate (or dipivaloylmethanate, dpm), but at much lower concentrations due to limited solubility in water, provided a 15% yield of the intermediate product and a 25% yield of the final Co(leu)(dpm)$_2$ product. FT-NMR carbon-13 results are in agreement with the formulation given.

Anal. Calcd for C$_{28}$H$_{50}$NO$_6$Co: C, 60.5; H, 9.1; N, 2.5. Found: C, 60.3; H, 9.0; N, 2.5.

Poly{(S-leucinato-*N,O*)cobalt(III)-u-[3,3'-dithiobis(2,4-pentanedion-ato-*O,O'*)]}.--Initially, 1.7460 g disulfur dichloride (12.93 mmol) in 6 mL dry dichloroethane was added dropwise to 5.0088 g Co(leu)(acac)$_2$ (12.93 mmol) in 20 mL dimethylacetamide (DMAC) with 0.7095 g sodium carbonate (6.69 mmol) as a slurry in the dimethylacetamide solution under argon and with vigorous stirring. The mixture was stirred for 24 - 36 h and then precipitated with diethyl ether. Alternatively, the solvent was removed *in vacuo* at 45°C. The product was collected on a fritted funnel, washed with water, and dried *in vacuo* at 100°C; yield, 5.5 g; 95%. The polymer was fractionated (three times) by a DMAC/acetone or dimethyl sulfoxide/acetone solvent/nonsolvent precipitation to remove low molecular-weight material.

Anal. Calcd for [C$_{16}$H$_{24}$NO$_6$S$_2$Co]$_n$: C, 42.8; H, 5.4; N, 3.1; S, 14.3. Found: C, 42.5; H, 5.1; N, 2.9; S, 14.2.

Poly{(S-leucinato-*N,O*)cobalt(III)-u-[3,3'-thiobis(2,4-pentanedionato-*O,O'*)]}.--Initially, 0.7592 g sulfur dichloride (7.37 mmol) in 6 mL dry dichloroethane was added dropwise to 2.8514 g Co(acac)$_2$(leu) (7.36 mmol) in 60 mL dichloroethane or dichloromethane under argon with vigorous stirring.

164

After 24 h of stirring HCl and the solvent were removed *in vacuo*, washed with water, and dried *in vacuo* at 100°C; yield, 2.76 g; 90%. The polymer was fractionated by an acetone/water fractional precipitation.

Anal. Calcd. for $[C_{16}H_{24}NO_6SCo]_n$: C, 46.0; H, 5.8; N, 3.4; S, 7.7. Found (from $C_2H_4Cl_2$): C, 45.4; H, 5.7; N, 3.0; S, 7.3. Found (from CH_2Cl_2): C, 45.9; H, 5.6; N, 2.8; S, 7.5.

(The use of more polar solvents such as acetonitrile with an acid acceptor proved less satisfactory with SCl_2 than with S_2Cl_2 above.)

Poly{(S-leucinato-*N,O*)cobalt(III)-u-[3,3'-sulfoxobis(2,4-pentanedio-nato-*O,O'*)]}.--Initially, 0.9213 g thionyl chloride (7.74 mmol) in 4 mL dry dichloroethane was added dropwise to 2.9925 g Co(acac)$_2$(leu) (7.74 mmol) in 10 mL dimethylformamide and 15 mL dichloroethane under argon with vigorous stirring at 0°C. The mixture was then allowed to stir for at least 24 h. The solvent and any residual HCl was removed at 45°C *in vacuo*, and the product was dried at 100°C; yield, 3.13 g; 93%. The polymer was fractionated with an acetone/water fractional precipitation.

Anal. Calcd for $[C_{16}H_{24}NO_7SCo]_n$: C, 44.3; H, 5.6; N, 3.2; S, 7.4. Found: C, 44.1; H, 5.3; N, 3.8; S, 6.9.

Poly{(S-leucinato-*N,O*)cobalt(III)-u-[3,3'-sulfonebis(2,4-pentanedio-nato-*O,O'*)]}.--Initially, 0.10 mL hydrogen peroxide was added dropwise to 0.0603 g fractionated sulfoxide polymer in 10 mL acetonitrile with vigorous stirring. The mixture was heated to 40°C and stirred for 24 h. The solvent was removed *in vacuo* at 40°C, and the sample was then dried *in vacuo* at 100°C.

Anal. Calcd for $[C_{16}H_{24}NO_8SCo]_n$: C, 42.8; H, 5.4; N, 3.1; S, 7.1. Found: C, 41.0; H, 5.6; N, 3.4; S, 6.9.

Poly{(1,1,1-trifluoro-2,4-pentanedionato-*O,O'*)chromium(III)-u-[3,3'-thiobis(2,4-pentanedionato-*O,O'*)}, [Cr(tfa)(acac$_2$S)]$_n$.--The synthesis of this polymer has been based on the reaction of Cr(tfa)(acac)$_2$, which was prepared by the method of Palmer, Fay and Piper (21) with SCl_2 (exact 1:1 mole ratio) in dichloromethane at -10°C. The infrared spectra and the intrinsic viscosity of 0.17 dL/g for the product are consistent with polymerization, though the GPC results were less conclusive.

Physical Characterizations.--Gel permeation chromatography was conducted in N-methylpyrrolidone (NMP) using a 10^3 A ultrastyragel column (Waters) calibrated with polystyrene standards. The viscosity measurements were made in NMP at 30.0°C; scribe-stripping was performed according to a standard scribing method (22); nuclear magnetic resonance measurements were recorded (as δ ppm values relative to tetramethylsilane) with Varian XL-200 and 300 MHz Fourier-transform spectrometers with 128 to 5000 transients on 5 - 10 mg/mL in DMSO-d$_6$; infrared measurements (KBr pellets) were made with a Mattsen Cygnus 100 Fourier-transform spectrometer, and ultraviolet/visible measurements [dimethylacetamide (DMAC) solutions] were recorded using a Perkin-Elmer lambda array spectrophotometer coupled to an IBM-PC.

Electron spin resonance spectroscopy measurements were made with an IBM ESP 300 instrument with quartz tubes sealed *in vacuo* prior to irradiation. To remove color centers from the quartz, one end of the tube was heated with a torch while the other end was immersed in liquid nitrogen. The sample was then reversed and the process repeated.

Polymer Irradiation.--Samples of the polymers sealed *in vacuo* were exposed to cesium-137 gamma irradiation (662 KeV) at doses ranging from 0.02 to 0.07 Mrad/h in a Radiation Machinery Corporation Gammator model B.

NMR Molecular Weight Determinations.--For exact 1:1 sulfur halide: cobalt chelate ratios, *on average* one end group per chain will be a cobalt chelate (which still has a vinylic proton) and the other end will be a sulfur halide. From the ratio of one-sixth the total intensity of the two leucinato methyl NMR doublets to the total intensity of the vinylic proton NMR peak(s), the number of repeating units is obtained, which in turn can be multiplied by the emperical weight of the repeating unit to obtain the molecular weight.

RESULTS AND DISCUSSION

Linear coordination polymers of cobalt(III) have been prepared from the reactions of Co(leu)(acac)$_2$ with SCl$_2$, S$_2$Cl$_2$, and SOCl$_2$. An earlier attempt to prepare polymers by an analogous procedure, but with labile divalent metal ions, had been attempted elsewhere (23). The result of that study was the preparation of metal sulfides! Our approach has been to use inert metal chelates. No evidence of sulfide formation was obtained in our syntheses. The solvents had to be varied with the different sulfur halides and were chosen from a combination of knowledge of the reactivity of the sulfur halides, the need for solubility throughout the polymerization reaction, and trial and error. Attempts to remove all of the low molecular weight byproducts through recrystallization were not always successful.

The cobalt sulfone polymer was prepared by hydrogen peroxide oxidation of the corresponding sulfoxide polymer.

Several properties of the polymers are provided in Table 1. The number-average molecular weights may be higher than the NMR values because the vinylic proton signals (one proton per chain is expected) are still at the noise level after several thousand transients. See Figure 3. However, the viscosity of the dithio polymers (including several other lower molecular samples) are closer to those expected for the NMR molecular weight values than for the GPC molecular weight values though low molecular weight material in many of the polymer preparations causes the viscosities to be lower than ideal. Whereas, the average chain should have one cobalt chelate end group, statistically one-fourth should have two such end groups and one-fourth should have none.

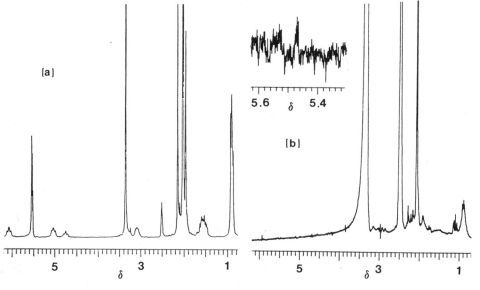

Fig. 3. Proton NMR of (a) Co(leu)(acac)$_2$ and (b) [-acacCo(leu)acac-SO-]$_n$.

Table 1. Linear cobalt(III) polymers derived from Co(leu)(acac)$_2$

Bridge[a]	M_n(GPC)[b]	M_n(NMR)[c]	Visc[d]	UV/Vis[e]	TGA[f]	Infrared[g]
-S-S-	34,000	$\geq$13,000	0.11	280/553	180°	h,i
-S-	13,500	$\geq$15,000	0.10	279/546	160°	h,i
-SO-	16,000	-	-	276/553	-	1025[h,i]
-SO$_2$-	17,000	$\geq$10,000	0.084	279/548	190°	1160[h,i]

[a]replaces vinylic-hydrogens of acetylacetonate ligands.
[b]NMP solvent, 1000 A ultrastyragel column, & polystyrene standards.
[c]based on FT-NMR of weak vinylic-proton signal in DMSO-d$_6$.
[d]intrinsic viscosity, dL/g in NMP at 30°C.
[e]nm, molar extinction coefficients about 10^4 and 10^2 $M^{-1}cm^{-1}$/
 repeating unit; DMAC solvent.
[f]temperature of first major weight loss.
[g]cm^{-1}.
[h]780 cm^{-1} monomer vinylic C-H mode not observed.
[i]1550 cm^{-1} peak replaces two peaks observed at 1570 and 1520 cm^{-1}
 in monomer.

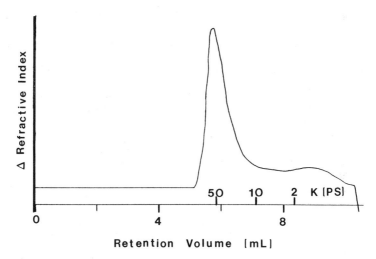

Fig. 4. GPC of [-acacCo(leu)acac-S-S-]$_n$ in NMP with polystyrene (PS)
M_n calibrations shown in thousands (K).

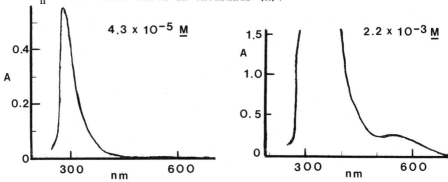

Fig. 5. The ultraviolet/visible spectrum of [-acacCo(leu)acac-S-S-]$_n$ in
DMAC.

167

The ultraviolet/visible spectra of the polymers provide no surprises. The stronger ultraviolet peak is quite solvent dependent, which is consistent with its assignment as a charge-transfer peak.

The thermogravimetric results show that these polymers are not extremely thermally stable, unlike our zirconium Schiff-base polymers, which show thermal stability to about 500°C (24). All of the cobalt(III) β-diketonate polymers show decomposition before 200°C is reached.

The infrared spectra of the polymers are characterized by a coalescence of the 1570 and 1520 cm^{-1} C=O and C=C peaks, the disappearance of the vinylic C-H mode at 780 cm^{-1} and the appearance of S=O and SO$_2$ peaks that do not exist in the Co(leu)(acac)$_2$ monomer.

The irradiation of these cobalt polymers with 662 keV gamma rays provides approximate G$_s$ values of 18, 10, 0.4, and 60 for the dithio, thio, sulfoxo, and sulfone polymers, respectively, when Charlesby's equation (25) is used together with the GPC molecular weight results. Whereas these values might seem quite high, polybutylene sulfone (PBS) has been reported to have G$_s$ values of 12.2 and 23.7 in $vacuo$ and in air, respectively (26). However, because the mass absorption coefficient for the more massive cobalt unit is 3.6 times that of PBS, the value we have obtained is reasonable. The electron spin resonance spectra of the irradiated polymers suggest radicals centered primarily on the sulfur atoms, with about 5% delocalization on the cobalt centers based on the hyperfine structure of the dithio and thio polymers. No hyperfine splitting is observed for the sulfone polymer, and the monomer shows no ESR signal at all. This verifies the observation that no spectral change is observed for extensive cobalt-60 irradiation of the monomer in methanol solutions. Evidence for the S-H stretching vibration at about 2500 cm^{-1} has been obtained by Fourier-transform infrared spectroscopy after irradiation and exposure to moist air. Also, a loss in the intensity of the symmetrical SO$_2$ stretch near 1160 cm^{-1} after irradiation has been observed for the sulfone polymer. These results coupled with our ability to make films of the dithio polymer on silicon wafers which have oxidized surfaces look encouraging. Unfortunately, the films peel from the surface when treated with the solvent combinations we have used in an attempt to determine the sensitivity to electron beams.

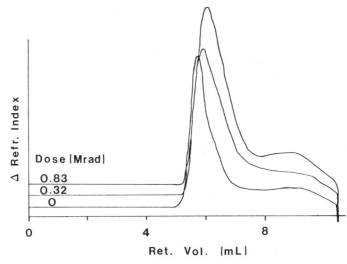

Fig. 6. The GPC results for irradiated [-acacCo(leu)acac-S-S-]$_n$.

Metals have also been used to enhance the sensitivities of negative resist materials for which radiation causes cross-linking (27,28,29). As noted initially, the low solubility of most inorganic polymers makes negative resists with metal containing species quite feasible. However, to date, all attempts to incorporate metal ions or organometallics in resist polymers have encountered one or more serious drawbacks. A recent variation involves the use of a cobalt(III) chelate, which has also been shown (30) to provide photoinitiated crosslinking and image formation with ultraviolet radiation (254 nm). The process involves photoreductive labilization of the $[Co(NH_3)_5Br]^{2+}$ ion, which yields three species capable of crosslinking: the cobalt(II) cation, ammonia, and the bromine radical, which should be capable of Lewis acid, base, and radical catalysis, respectively. The process includes radiation of the cobalt chelate perchlorate in films of a copolymer of glycidyl methacrylate and ethyl acrylate (COP), which has crosslinkable epoxide groups capable of acid or base catalysis.

We have also expanded our studies to include chromium(III) systems. However, for the chromium centers we have chosen the more volatile trifluoro (tfa$^-$) and hexafluoro (hfa$^-$) derivatives of 2,4-pentanedione. Since the fluorinated derivatives are deactivated relative to the electrophilic attack used in these polymerizations, the $Cr(acac)_2(tfa)$ mixed ligand complex has been used initially to obtain polymeric products with sulfur halides. Whereas the intrinsic viscosity of one of the polymers so obtained was 0.17 dL/g, the GPC results suggest only oligomeric products.

We have also nitrated the $Cr(acac)_2(tfa)$ complex and attempted to reduce the nitro groups to amine groups in order to provide links with organic oligomers. Unfortunately, the first nitro together with the trifluoro group deactivates the system such that no dinitro product can be obtained. Collman (31) had noted similar problems earlier in his attempts to nitrate all three rings on tris(β-diketonates). This deactivation may also help to explain the problems we appear to be having in getting good polymers with the chromium(III) systems to date.

SUMMARY

Four cobalt polymers have been synthesized, three of which are very sensitive to gamma irradiation with resulting G_s values which exceed the analogous organic polymers. Furthermore, the cobalt is **not** released during the scission reactions. Films which adhere well enough for commercial use have not yet been obtained for the cobalt systems. Extension to chromium is in progress.

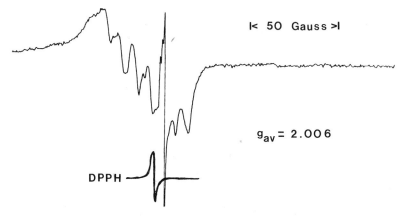

|< 50 Gauss >|

$g_{av} = 2.006$

DPPH

Fig. 7. ESR spectrum of irradiated $[-acacCo(leu)acac-S-S-]_n$.

Acknowledgement.--The authors acknowledge the support of the Office of Naval Research for this reseach.

REFERENCES

1. C. U. Pittman, Jr., J. E. Sheats, C. E. Carraher, Jr., M. Zeldin, and B. Currell, **this volume**, Chapter 1.

2. J. Kosar, "Light Sensitive Systems," Wiley, New York, 1965, Chapters 1 and 2.

3. M. J. Bowden and S. R. Turner, "Electronic and Photonic Applications of Polymers," Amer. Chem. Soc., Washington, D.C., 1988, Chapter 1, 55 ref.

4. G. Friedlander, J. W. Kennedy, E. S. Macias, and J. M. Miller, "Nuclear and Radiochemistry," Wiley, New York, 1981, p. 225.

5. A. H. Compton and S. K. Allison, "X-Rays in Theory and Experiment," Van Nostrand, Princeton, 1935.

6. R. D. Archer, C. J. Hardiman, R. Grybos, and J. C. W. Chien, U. S. Patent No. 4,693,957 (Sept. 1987).

7. Archer, R. D., Hardiman, C. J., and Lee, A. Y. in "Photochemistry and Photophysics of Coordination Compounds," H. Yersin and A. Vogler, eds., Springer Verlag, Berlin, 1987, pp 285-90.

8. L. F. Thompson, E. D. Feit, J. J. Bowden, P. V. Lenzo, and E. G. Spencer, *J. Electrochem. Soc.*, **121**, 1500 (1974).

9. M. Miramura, A. Muira, and O. Tada, *Japan Kokai*, **77**, 146223-224 (Dec. 1977).

10. M. Miramura, *Japan Kokai,* **77**, 146220 (Dec. 1977).

11. T. Kitakoji, Y. Toneda, and M. Fujimore, *Japan Kokai,* **76**, 148417-418 (Dec. 1976).

12. I. Haller, R. Feder, M. Hatzakis, and E. Spiller, *J. Electrochem. Soc.,* **126**, 154 (1979).

13. R. D. Archer, V. J. Tramontano, B. Wang, and P. V. West, *Proc. 11th Conf. Coord. Chem., Smolenice, CSSR,* 1987, pp. 1 - 6.

14. R. D. Archer, B. Wang, V. J. Tramontano, A. Y. Lee, and V. O. Ochaya, in "Inorganic and Organometallic Polymers," M. Zeldin, K. J. Wynne, and H. R. Allcock, eds., Amer. Chem. Soc., Washington, DC, 1987, pp 463-68.

15. R. A. Mann, "Gamma Ray Cross Section Data," General Electric, 1960, Report DC 60-9-75.

16. R. L. Lintvedt, in "Concepts of Inorganic Photochemistry," A. W. Adamson and P.D. Fleischauer, eds., Wiley, NY, 1975, p. 299.

17. S. Fujimura and H. Yano, *J. Electrochem. Soc.,* **135**, 1195 (1988).

18. R. J. Rosser and F. R. Whitt, *J. Appl. Chem.,* 1960, **10**, 229.

19. D. D. Perrin, W. L. Armarego, and D. R. Perrin, "Purification of Laboratory Chemicals," Pergamon Press, Oxford, 1980.

20. S. H. Laurie, *Austr. J. Chem.*, **21**, 679 (1968).

21. R. A. Palmer, R. C. Fay, and T. S. Piper, *Inorg. Chem.*, **3**, 875 (1964).

22. W. R. Tooke, Jr. and J. R. Montalvo, *J. Paint Technol.* **38**, 18 (1966).

23. R. D. G. Jones and L. F. Power, *Austral. J. Chem.*, **24**, 735 (1971).

24. R. D. Archer and B. Wang, *Inorg. Chem.*, **29**, 39 (1990).

25. M. Dole, "The Radiation Chemistry of Macromolecules," Vol. II, Academic Press, New York, 1973, p. 101.

26. J. R. Brown and J. H. O'Donnell, *Macromolecules*, **3**, 265 (1970); *ibid.*, **5**, 109 (1972).

27. G. N. Taylor, *Solid State Technol.*, **May 1980**, 73-80; **June 1984**, 124-31 --reviews x-ray resists and target effects on x-ray absorption.

28. H. Curtis, E. Irvin, and B. F. G. Johnson, *Chem. Brit.*, **22**, 327 (1986).

29. D. M. Allen, *J. Photograph. Sci.*, **24**, 61 (1976).

30. C. Kutal, and C. G. Willson, *J. Electrochem. Soc.*, **134**, 2280 (1987).

31. J. P. Collman, R. L. Marshall, W. L. Young III, and S. D. Goldby, *Inorg. Chem.*, **1**, 704 (1962).

Ti-B-N-C CONTAINING POLYMERS: POTENTIAL

PRECURSORS FOR CERAMICS

Kenneth E. Gonsalves

Stevens Institute of Technology
Department of Chemistry and Chemical Engineering
Hoboken, New Jersey 07030, USA

ABSTRACT

Titanium containing fibers were synthesized via the modification of
organic materials. Thermal processing of these chemically modified
organic fibers in reactive/inert atmospheres produced materials containing Ti(C,O). Materials analysis was performed using SEM-EDAX, and Auger
spectroscopy. Novel monomeric titanium precursors were synthesized for
the modification of organic materials such as polyacrylonitrile. Polyureidoborazines were also synthesized by the reaction of tri-isocyanatotrimethyl borazine with linear aliphatic diamines. The insoluble
oligomers were characterized by infra-red spectroscopy and the effect
of the number of methylene units on thermal transitions determined by
DSC. The oligomers were pyrolyzed in argon and ammonia atmospheres and
the residues analyzed for boron, nitrogen and carbon.

INTRODUCTION

There is considerable interest in the development of high temperature, high strength, ceramic fibers and composites, which exhibit high
resistance to oxidation, corrosion and thermal shock. A major component
of this interest lies in the development of innovative technologies that
will be required. Traditional methods such as sintering cannot be
readily applied to high temperature structural materials like SiC to
produce complicated shapes, especially in fiber form. Various methods
for producing SiC articles are chemical vapor deposition on a carbon
filament core, hot pressing of SiC powder with metallic additives and
conversion of an organometallic polymer precursor.

The shaping of polymers into forms, which can subsequently be con-
served during pyrolysis is a technology which has been extensively
employed in the manufacturing of graphite fibers. Graphite technology
has suggested heating of organometallic polymers to obtain skeleton
structures composed of SiC, Si_3N_4 and other carbides, borides, nitrides
especially in shapes not readily amenable by other manufacturing
methods [1].

Inorganic and Metal-Containing Polymeric Materials
Edited by J. Sheats *et al.*, Plenum Press, New York, 1990

A number of organometallic polymers including poly(silazane), poly-
(carbosilane), poly(phenylborazole), poly(silastyrene) and poly(titano
carbosilane) have been synthesized [2,3]. Some of these organosilicon
polymers i.e., poly(carbosilane) and poly(titano carbosilane), could be
shaped into fibers by melt spinning at temperatures of 250 to 350 $^{\circ}$C in
an inert atmosphere, and subsequent drawing at wind-up velocities of
several hundred meters per minute. To increase the melting point of
the "green" fibers thus produced and to set the structure, the "green"
fibers need to be cured usually in air at 25 to 150 $^{\circ}$C. Then the cured
fibers can be pyrolyzed at 1200 to 1400 $^{\circ}$C to generate the ceramic fiber.
The art of producing complex shaped ceramics through the organometallic
polymeric precursor route outlined above therefore involves the synthesis
of precursors, which can be shaped by employing production scale opera-
tions followed by pyrolysis to produce the desired properties within the
shape.

An important area of application of ceramic fibers is in ceramic-
fiber composites. Substantial increase in fracture toughness of
composites, in which a significant volume fraction of fine unidirectional
fibers are introduced, have been reported. Thus, refractory ceramic
fibers have the potential to toughen refractory ceramic matrices, which
would allow higher temperature applications of such composites than
silicate-based glass matrices, but also provide a broad range of other
desirable physical and chemical properties. SiC fibers from polymer
pyrolysis have been encouragingly used in reinforcing glass and ceramic
matrices. As a consequence, developing fibers of different compositions
have been cited as a particularly important need and opportunity to
broaden the scope of ceramic composites. The enhanced interest in
polymeric precursors to ceramics is therefore based on the potential
capability of obtaining fibers from these materials, which can then be
pyrolyzed to ceramic fibers.

The potential for further extensive development in ceramic-fiber
composites is thus dependent upon the production of a wide array of
fiber compositions as a significant existing problem is of property mis-
match between the fiber and matrix. It has, therefore, been emphasized
that the challenge is to either obtain new fibers with improved
thermoxidative stability in comparison to the limited number currently
available, or to suitably modify existing fiber materials.

We have focused on the Ti-B-C-N Systems in our attempts to obtain
fibers of such materials. Our approach to produce such fibers is
analogous to the processing of organic polymers into carbon fibers [4].
Synthesizing the above mentioned ceramics by the pyrolysis of organometallic polymers, or other precursors [2,3], has particularly
facilitated the fabrication of such ceramic fibers.

I. Ti-C-N SYSTEMS

In our attempt towards such fibers we felt that if we utilized
organic polymers (as a source of carbon) and suitably modified it with
an organotitanium complex at the molecular level, we would have a
system containing the right ingredients i.e. Ti and C. Such a polymeric
system could conceivably be processed into a fiber [5].

Two probable systems $TiCl_4$ and alkyl cyanides form complexes of
the type $(R-C{\equiv}N)_2TiCl_4$ [6] and titanium amides $(R_2N)_4Ti$ also complex
with the $-C{\equiv}N$ group [5] as below:

$$Ti-NMe_2 \ + \ N{\equiv}C-R \ \longrightarrow \ -Ti-N{=}C\underset{R}{\overset{NMe_2}{<}}$$

Polyacrylonitrile (PAN) has pendant nitrile groups and therefore was selected as a prime candidate for modification by organotitanium complexes. PAN also can be easily fabricated into fibers and processed at elevated temperatures to produce "carbon" which can simultaneously undergo a "solid state" reaction with titanium. Under appropriate processing conditions it was envisaged that Ti-C would be the main constituent of the end product.

Thus far we have synthesized organometallic monomers and modified organic polymers; fabricated such modified organic polymers into fibers followed by their thermal processing into ceramics. These features are summarized in Scheme I.

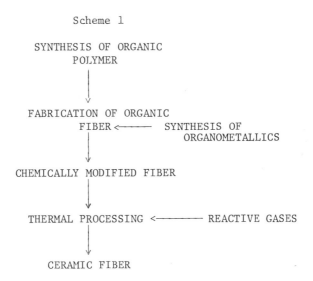

Scheme 1

SYNTHESIS OF ORGANIC
POLYMER

↓

FABRICATION OF ORGANIC
FIBER ←——— SYNTHESIS OF
ORGANOMETALLICS

↓

CHEMICALLY MODIFIED FIBER

↓

THERMAL PROCESSING ←——— REACTIVE GASES

↓

CERAMIC FIBER

Synthesis: Model Systems

Titanium containing monomers, tetrakis(dialkylamino)titanium $(R_2N)_4Ti$ ($R=C_2H_5$, CH_3) were synthesized by a slight modification of the procedure reported in the literature [7,8]. In general, first the lithium dialkylamides (LDA) were obtained by reacting methyl lithium in diethyl ether with the appropriate dialkylamine at low temperatures ($-20^\circ C$). To the resulting LDA was then added $TiCl_4$ in benzene, the temperature being maintained below $10^\circ C$. These reactions were conducted in an argon environment to exclude moisture and air. In both instances the titanium monomers were high boiling viscous liquids. The above sequence of reactions are presented below in Scheme 2

Scheme 2

$$R_2NH + CH_3Li \xrightarrow[\text{Temp.} < -20^\circ C]{\text{diethyl ether}} R_2NLi + CH_4$$

$$R_2NLi + TiCl_4 \xrightarrow[\text{Temp.} < 10^\circ C]{\text{benzene}} (R_2N)_4Ti + LiCl$$

$$(R = CH_3, C_2H_5-)$$

Polyacrylonitrile (PAN) was prepared by known procedures [5]. Purified acrylonitrile was polymerized with a redox initiator system in water by precipitation polymerization. The intrinsic viscosity, $\eta = 0.98$ dl/g, determined at $25°C$ in DMF was suitable for fiber spinning.

Preliminary investigations regarding the feasibility of these monomers as potential sources of "Active Ti" in solid state reactions were conducted prior to fiber fabrication and processing studies. The titanium monomer $[(CH_3)_2N]_4Ti$, on pyrolysis from ambient to $1200°C$, in an argon atmosphere left a black residue. On chemical analysis it was observed that the residue contained 61 % titanium, whereas the un-pyrolysed sample contains only 27 %. The residue was also found to contain carbon and nitrogen. In a parallel experiment, polyacrylonitrile powder (PAN) and the monomer $Ti[N(CH_3)_2]_4$ in a 10:1 ratio by weight, mixed under argon for 24h, was similarly pyrolysed. A black residue was obtained which contained 5.1 % Ti, 87.5 % C and 5.2 % N. It can be assumed that a mixture of Ti(C,N) and carbon are present. The total weight loss in the first experiments was ca 40 %. The above pyrolyses are summarized in Scheme 3.

Scheme 3

$$Ti(NR_2)_4 \xrightarrow[\text{ambient} - > 1200°C]{\text{argon}} Ti(C,N)$$

$$Ti(NR_2)_4 + -[CH_2-CH]_n^{CN} \xrightarrow[\text{ambient} - > 1200°C]{\text{argon}} Ti(C,N) + C$$

From these initial experiments it was concluded that these monomers could provide the titanium, a crucial source for reaction with carbon. The production of titanium with excess carbon in the latter experiment was critical as it was anticipated that an analogous solid state reaction would occur in the titanium modified PAN fibers.

Fiber Processing

PAN powder was dissolved in dimethylformamide (DMF) and the "dope" extruded into a chilled water bath to coagulate the polymer. The as-spun fibers were drawn in a hot water bath before drying. The drawn fibers were dried overnight in an oven at $60°C$, under tension. It may be mentioned that the PAN fibers were porous to enable the titanium monomers to diffuse sufficiently into the inner core to provide a homogeneous fiber. The titanium monomer was expected to intercolate through these pores and channels and form molecular level complexes with the abundant reactive pendant cyano groups of PAN.

Modification of the PAN fiber was done after drawing and drying. The fibers were immersed in solution of $Ti(NEt_2)_4$/benzene for 48h at $10°C$. The concentration of $Ti(NEt_2)_4$ in benzene was varied to determine the effect on fiber structure. The modified fibers were pyrolyzed under different conditions, which are summarized in Table I. Fig. 1 is an SEM micrograph of the surface of such a fiber, thermally processed in argon. Table II summarizes the modification of PAN fibers by the monomer $Ti(NMe_2)_4$. Here no solvent was used as the monomer was less viscous and easier to handle. Fig. 2 is an SEM micrograph of a cross section of a fiber pyrolysed in ammonia. The fibers were grayish with a metallic lustre.

Table 1. Processing Conditions & Thermal Analysis of PAN fibers modified by Ti(NEt$_2$)$_4$

Sample No.	Ti(NEt$_2$)$_4$ Benzene	Thermal Processing Conditions	IDT[b]	
1	Conc.	Directly heated to 1000° in argon	519	18% at 850°C
2		Sample 1 reheated to 560°C for 1 hr and then 1150°C in argon	654	16% at 850°C
3	Conc.	Heated to 1050°C for 2h in argon	-	No Weight Loss
4	Dilute	Directly heated to 560°h for 1 h and then to 1150° in argon	552	64% at 850°C
5	Dilute	Directly heated to 1000°C in argon	510	18% at 600°C
7	PAN powder + dilute Ti(NEt$_2$)$_4$	Heated to 560°C for 1h and 1050°C for 1h	620	7%

a. conc.: neat Ti(NEt$_2$)$_4$; dilute: Ti(NEt$_2$)$_4$ + benzene

b. IDT: Initial decomposition temperature; c. Det. by thermogravimetric analysis

Table II. Processing Conditions & Thermal Analysis of PAN modified by $Ti(NMe_2)_4$

Sample No.	Sample description	Thermal Processing Conditions	% Weight loss
8	$Ti(NMe_2)_4$ (neat)	Heated in argon to $1000^{o}C$ for 1 h and then to $1200^{o}C$	No weight loss
9	PAN powder + $Ti(NMe_2)_4$ (neat)	Heated in argon to $500^{o}C$ for 1h, then to $1000^{o}C$ for 1h and finally to $1200^{o}C$	9.2%
10	PAN powder $Ti(NMe_2)_4$ (neat)	Heated in ammonia for 1h, then to $900^{o}C$ also in ammonia. Finally from $900^{o}C$ to $1200^{o}C$ in nitrogen	
11	PAN fiber + $Ti(NMe_2)_4$ (neat)	Heated in ammonia to $560^{o}C$ for 1h; then to $900^{o}C$ also in ammonia. Finally from $900^{o}C$ to $1200^{o}C$ in nitrogen	No weight loss

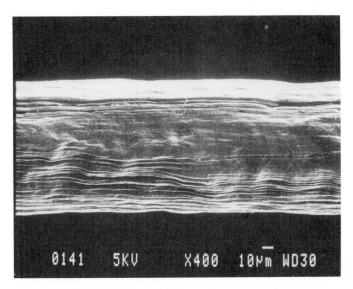

Fig. 1. SEM micrograph (400X): Surface of PAN fiber
chemically modified by $Ti(NEt_2)_4$

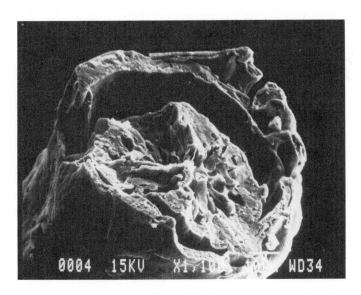

Fig. 2. SEM micrograph(1100X): Cross-section of fiber
modified by $Ti(NMe_2)_4$ at room temperature

Thermal Analysis

The results of thermogravimetric analyses (TGA) of the fibers and powders are also summarized in Tables I and II. A perusal of the thermal stability of these fibers indicates that processing conditions are vital not only to ultimate mechanical fiber properties but also to their thermal stability. Sample 3 showed no weight loss till 800°C in a nitrogen atmosphere. Similarly samples 8 and 11 showed no weight losses till 850°C. Sample 8 showed a weight gain at 430-550°C, possibly nitridation. The TGA traces are shown in Fig. 3.

Materials Characterization

Scanning Auger electron spectroscopy (AES) was performed on a finely powdered single fiber mounted on a copper backed adhesive tape. The AES survey of the powder is given in Figure 4. The Auger transitions of the major elemental constitutents of the fiber, i.e., C and Ti, are evident. Contributions from the carbon and oxygen of the adhesive are possible as copper signals were observed in the ESCA survey. From these initial AES and ESCA surveys, it appears that the fibers are composed of Ti(C,O).

Semiquantitative elemental analysis was performed on the fiber samples by SEM-EDAX. The Ti content on the surface ranged around 90-95% and in the interior from 65 to 90%, depending on the conditions of PAN fiber modification. The inner core contained a higher percentage of titanium when the modifications were performed at 20°C, compared to -10°C. Thus temperature possibly affects the rate of titanium monomer diffusion as well as reaction rate of the -C≡N group and the titanium monomer.

We are currently attempting to densify the fiber by conducting the pyrolysis in reducing hydrocarbon atmospheres and increasing the maximum processing temperature to 1700-2000°C. Tensile testing of these fibers will then be carried out.

II. BORON CONTAINING POLYMERS

We recently reported the synthesis of transition metal containing polyamides and polyureas [9], utilizing low temperature interfacial and solution techniques. We have now attempted to extend the solution method to the synthesis of boron containing polymers, viz. [10]. These polymers are of increasing interest as potential precursors for B-N-C type ceramic materials [10], particularly in fiber form.

Monomers and Polyureidoborazines

The monomer tri-isocyanto-trimethyl-borazine (TITMB) was synthesized from trichloro-trimethyl-borazine (TCTMB) according to the method reported in the literature [12]. The intermediate TCTMB was synthesized via a modification of the procedure of Brown and Laubengayer [13]. Methyl-amine hydrochloride and excess of BCl_3 were refluxed in chlorobenzene for 6h and then maintained for 16h at 80°C. The reaction mixture was filtered warm, vacuum dried and subsequently sublimed under reduced pressure. A slight increase in yield (12%) of pure TCTMB was achieved. The pure TCTMB was then reacted with silver cyanate to obtain TITMB. The above synthesis is given below in Scheme 4. The IR spectra of TITMB was similar to that reported previously.

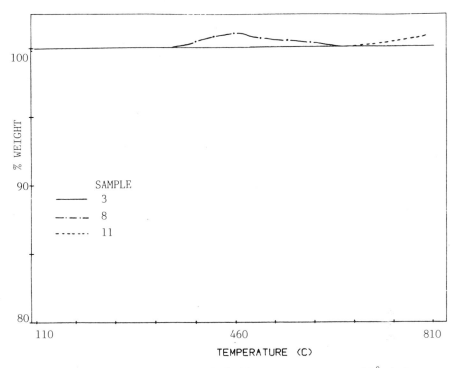

Fig. 3. TGA of samples 3,8,11. Heating rate 20°C/min in nitrogen.

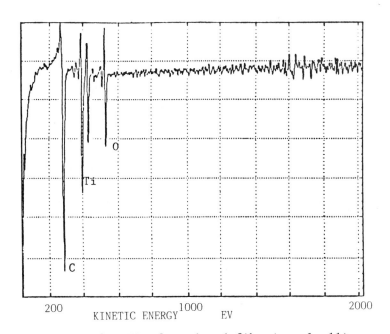

Fig. 4. AES of powdered fiber(sample 11).

SCHEME 4

$$BCl_3 + CH_3NH_2 \cdot HCl \xrightarrow[130\,^{\circ}C]{\text{chlorobenzene}}$$

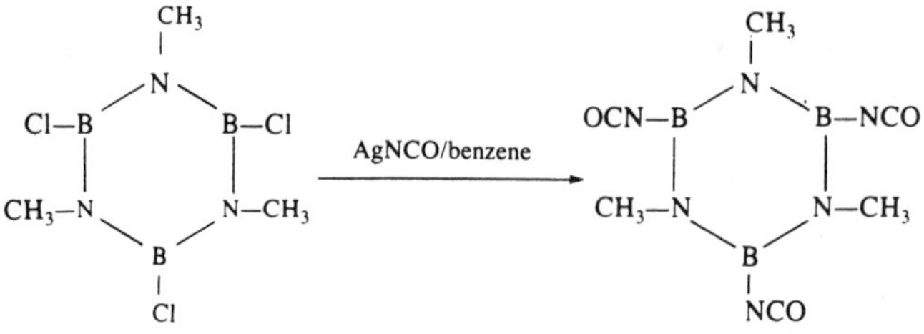

TCTMB   TITMB

Polyureidoborazines (PUB) were synthesized by reacting various aliphatic diamines with TITMB. In general, the monomer TITMB was dissolved in benzene and reacted with the diamine at ambient temperature. Reaction appeared to be immediate as evidenced by the solution turning cloudy followed by the formation of a white precipitate. The concentrations of the reactants were also varied to determine the effects on polymer formation. The synthesis of PUB is shown in Scheme 5. Details are given in Table III.

The polymers were insoluble in most organic solvents making their structural and molecular weight determinations difficult. However, on the basis of infrared spectroscopy studies and thermal analysis we have deduced that the above reactions do result in higher molecular weight materials. Elemental analysis in conjunction with IR spectroscopy data indicate that the repeat unit in the PUBs have the possible structure I shown in Scheme 5.

SCHEME 5

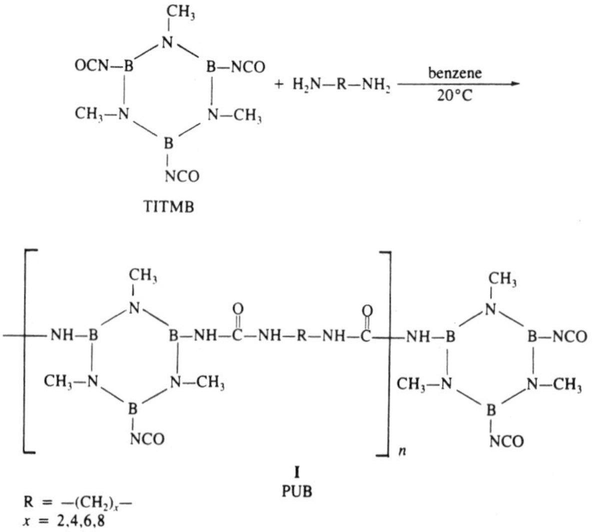

$R = -(CH_2)_x-$
$x = 2,4,6,8$

Infrared Spectra

IR spectra for the polyureidoborazines (PUB) were recorded in the solid state. In all the PUBs, B-N ring vibrations were present at 1400-1450 cm^{-1} and the strong NCO absorption at 2300 and also possibly at 2170 cm^{-1}. PUBs synthesized by reacting equimolar amounts of TITMB and diamine showed these two possible NCO absorptions compared to a single absorption at 2310 cm^{-1} for the NCO group in the monomer TITMB.

As mentioned above, due to the insolubility of the polymers in organic solvents, we have attempted to determine the "linear" structure of these polymer and their degree of polymerization by using IR spectroscopy techniques. To provide a basis for structure I we deduced that there is approximately one NCO group in the repeat unit thereby limiting extensive chain branching. For this purpose the extinction coefficient ε of the monomer TITMB was determined. The net absorbance for the NCO peak was measured for peak maxima at 2298 cm^{-1} at varying concentrations. The samples were obtained by finely mixing and grinding TITMB in KBr using a Wig-L-Bug and obtaining a pressed pellet. From a plot of absorbance versus concentration, ε was determined to be 506. Assuming that the NCO absorption in the polymer approximates those in TITMB, the extinction coefficient ε' for the polymer is one third ε, i.e., 169. From ε' and absorptions for polymer NCO peaks at 2298 cm^{-1} it was assessed that the number of NCO groups per repeat unit is 0.6. This value can be considered to approximate unity if scattering losses in the polymer as well as losses in reflection are accounted for.

On the basis of the above, the area under the NCO peaks in the DP polymers were compared with the corresponding area in TITMB. The DP calculated ranged between 3-5, indicating oligomers.

DSC

Endothermic melting transitions were observed at 155°C and 74°C for TCTMB and TITMB respectively. The observed transitions for the PUB's were significantly higher than those of the monomer TITMB which is indicative of a reaction between the monomer and the diamines resulting in a higher molecular weight material. A distinctive trend in the thermal transitions of the polymers with respect to the number of methylene units in the backbone is also evident. In the case of the ethylene diamine polymer a transition was observed at 244°C with possible ensuing degradation (endothermic peak at 331°C). For butylene, hexamethylene and octamethylene diamines two endothermic transitions were observed between 230-160°C and 245-226°C.

Pyrolysis

The PUB's were pyrolyzed from ambient to 900°C in an NH_3 stream and then from 900-1200°C in a nitrogen stream. The char yield was approximately 80%. The residue from polymer 1 was white while the remaining were brownish beige. All the residues gave broad IR bands centered at 1400 cm^{-1}, indicative of B-N. No carbonyl or NCO absorptions were observed. These results are reasonable since B-N is reportedly synthesized by heating the condensation product of orthoboric acid and urea in a stream of ammonia till 1650°C [14]. However, when the pyrolyses were conducted in argon, a black residue was left, indicative of significant amount of carbon in the char. The weight loss was approximately 70 %. Initial characterization by XRD of the former residues indicate the presence of a graphite phase, a small amount of BN and other complex B-N-C phases.

Table III. Polymerization Conditions of Triisocyanato Borazines and Diamines

Polymer Number	M_1	M_2	0.0025 mol M_1 in benzene (ml)	0.0025 mol M_2 in benzene (ml)	Yield %	Thermal Transitions* °C
1	Triisocyanato Timethylborazine	Ethylene-diamine	40	60	90	244,331
2	Triisocyanato Trimethylborazine	Butylene-diamine	40	60	95	230,245
3	Triisocyanato Trimethylborazine	Hexamethylene-diamine	40	60	94	203,227
4	Triisocyanato Trimethylborazine	Hexamethylene-diamine	250	5	92	186,228
5	Triisocyanato Trimethylborazine	Octamethylene-diamine	40	60	92	160,226
6	Triisocyanato Trimethylborazine	Ethylene-diamine	40	0.0038 mol in 60 ml	Quantitative	—

*Determined by DSC- Heating Rate 20°C/min

CONCLUSION

We have demonstrated that PAN can be suitably modified by organo-metallics and thermally processed to metal carbides under appropriate conditions. Similarly Polyureidoborazines have the potential of being processed into ceramic fiber materials.

ACKNOWLEDGMENTS

Partial support of the National Science Foundation (Grant number MSM-8612801) is gratefully acknowledged. Thanks to Dr. K.T. Kembaiyan for his assistance in the X-ray diffraction studies. Acknowledgment is made to the Longman Group, UK for permission to include the article "Polyureidoborazines" by K. Gonsalves and R. Agarwal, J. Appl. Orgmet. Chem. (1988) 2, 245-249 and also to John Wiley & Sons. USA for permission to include the article "Modification of PAN by Titanium Dialkylamides: Precursors for Titanium Carbide" by K. Gonsalves and R. Agarwal, J. Appl. Polym. Sci (1988) 36, 1659-1665.

REFERENCES

[1]   K.J. Wynne and R.W. Rice, Ann. Rev. Mater. Sci. 14, 297 (1989).
[2]   "Inorganic and Organometallic Polymers" ACS Symposium Series 360 M. Zeldin, K.J. Wynne and H.R. Allcock Eds., Washington, D.C. 1988.
[3]   G. Pouskouleli Ceramics International 15, 213 (1989.
[4]   M.K. Jain and A.S. Abhiraman J. Mater. Sci. 22, 278 (1987).
[5]   K. Gonsalves and R. Agarwal, J. Appl. Polym. Sci. 36, 1659 (1988).
[6]   H.J. Coerver and C. Curran J. Am. Chem. Soc. 80, 3522 (1958); Chem. Commun. 116 (1967).
[7]   D.C. Bradley and I.M. Thomas, J. Chem. Soc. 3587 (1960).
[8]   A.D. Jenkins, M.F. Lappert and R.C. Srivastava, Polym. Journal 7, 289 (1971).
[9]   K. Gonsalves, R.W. Lenz and M.D. Rausch, Appl. Orgmet. Chem. 1 81 (1987); K. Gonsalves, Z. Lin and M.D. Rausch, J. Am. Chem. Soc. 106, 3862 (1984).
[10] K. Gonsalves and R. Agarwal, Appl. Orgmet. Chem. 2, 245 (1988).
[11] B.A. Bender, R.W. Rice and J.R. Spann Ceram. Proceedings: Engineering and Science page 1171, July-Aug. 1985.
[12] M.F. Lappert and H. Pyszora J. Chem. Soc. 1733 (1963).
[13] C.A. Brown and A.W. Laubengayer, J. Am. Chem. Soc. 77, 3699 (1955).
[14] T.E. O'Connor J. Am. Chem. Soc. 1753 (1962).

NOVEL INORGANIC MATERIALS AND HETEROGENEOUS CATALYSIS: PREPARATION AND

PROPERTIES OF HIGH SURFACE AREA SILICON CARBIDE AND SILICON OXYNITRIDES

Peter W. Lednor[*] and René de Ruiter

Koninklijke/Shell-Laboratorium, Amsterdam
(Shell Research B.V.),
Badhuisweg 3, 1031CM Amsterdam, The Netherlands

ABSTRACT

The synthesis and stability of some high surface area forms of silicon carbide and silicon oxynitride are described. Such materials may be of interest in the context of heterogeneous catalysis. Porous silicon carbide was prepared, following a recent report, by pyrolysis of the organosilicon polymer $(PhSiO_{1.5})_n$. The porous structure of this material was found to show superior thermal stability to that of silica. The same polymer undergoes reactions with flowing, gaseous water vapour, air or ammonia at 500°C to give high surface area inorganic materials. The silicon oxynitride product obtained from reaction of the polymer with ammonia is unstable to the atmosphere at ambient temperature. However, silicon oxynitride powders of high surface area can also be obtained by reaction of an amorphous silica with ammonia at 1100°C, and these materials are of much higher stability, showing only a moderate decrease in surface nitrogen content after exposure to air, hydrogen or water vapour at 750°C.

INTRODUCTION

The general aim of the present work is to extend the range of inorganic materials which find application in heterogeneous catalysis. The inorganic materials most often used in catalysis are silica, alumina or mixed alumino-silicates such as amorphous silica–alumina or zeolites; these compounds can be used as catalysts or as supports for catalytically active species such as metal particles. The possibility of achieving novel catalytic transformations of hydrocarbons or petrochemicals would be increased by the availability of materials with properties not displayed by the above classes of materials. The important properties of inorganic materials with respect to catalysis are (i) the surface chemistry, i.e. the number, nature, and reactivity of groups on the surface of the material, (ii) the texture, by which is meant the surface area and the type of porosity present (note that high surface area, for example > 50 m^2 g^{-1}, can be obtained from the external surface of small particles, but in practice catalysts derive high surface areas from a network of interconnected pores), and (iii) the stability, which refers to oxidative, thermal or hydrothermal stability of surface species or of the texture.

Inorganic and Metal-Containing Polymeric Materials
Edited by J. Sheats *et al.*, Plenum Press, New York, 1990

Our approach to finding new types of heterogeneous catalysts is based on making use of recent developments in the chemistry of inorganic materials (ceramics, glasses and polymers). Relevant developments include the wide-ranging interest in (i) compositions other than oxides, for instance, nitrides, carbides or borides, and (ii) new preparative routes such as organo-sol-gel chemistry, polymer pyrolysis and deposition from the gas phase (1).

The properties relevant to catalysis (surface chemistry, texture, stability) which might be displayed by materials such as nitrides, carbides or borides cannot be estimated with any confidence, and are also unlikely to be trivial extensions of what is known for oxides. A limiting factor in developing work in this area has undoubtedly been the absence of readily available syntheses of high surface area forms of such compositions. For example, porous oxides are easily obtained through aqueous or non-aqueous sol-gel chemistry, but comparable solution chemistry for non-oxides does not exist, although it is certainly a challenging and worthwhile target. However, the new preparative routes referred to above may offer opportunities for synthesising high surface area materials.

Interest in new compositions and new synthetic routes in the context of catalysis is growing, and recent examples of the synthesis and use of non-oxidic, ceramic compositions in catalysis include SiC as a support for Ni or Pt in CO hydrogenation (2), SiC as a support for Co and Mo for thiophene hydrodesulphurisation (3), transition metal (Ti, Ta, Mo or W) carbides for methanol decomposition (4), early transition metal carbides, nitrides or borides for hydrodenitrogenation of quinoline (5), and the synthesis of high surface area molybdenum carbide (6).

The present paper discusses the preparation and properties of high surface area silicon carbide and oxynitride with respect to possible application in catalysis. The synthetic work includes new routes to high surface area forms of these materials. Regarding properties, an important aspect is stability. This refers both to the stability of a pore system in a non-oxide material, on which there is very little information available, and to the stability of the surface composition; in the case of the latter, oxidation to the oxide will be thermodynamically preferred in most cases. We report data on the textural stability of porous silicon carbide and on the surface stability of high surface area silicon oxynitrides. Some of the work reported in the present paper has been described at recent conferences (7,8) and in a communication (9).

POROUS SILICON CARBIDE PREPARED BY PYROLYSIS OF THE ORGANOSILICON POLYMER $(C_6H_5SiO_{1.5})_n$

Silicon carbide has been manufactured commercially since 1891 and the current world market is about 500 000 tons. This material is dense and crystalline. It is only recently, however, that a porous form has been reported. These two forms can be regarded as the analogues of quartz (dense, crystalline silicon oxide) and silica gel (porous, amorphous silicon oxide). We were interested in the properties of the porous silicon carbide, and in particular its stability. It is not improbable that this be higher than that of silica in view of the four-fold coordination of carbon compared to the two-fold coordination of oxygen. Although data on the stabilities of dense forms are available, the information is not necessarily relevant to the properties of porous forms.

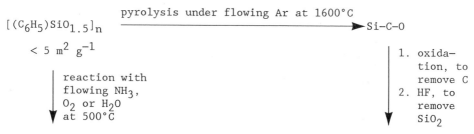

$[(C_6H_5)SiO_{1.5}]_n$

$< 5 \ m^2 \ g^{-1}$

pyrolysis under flowing Ar at 1600°C $\longrightarrow$ Si-C-O

reaction with
flowing NH_3,
O_2 or H_2O
at 500°C

1. oxida-
   tion, to
   remove C
2. HF, to
   remove
   SiO_2

High surface area
silicon oxynitride
or oxide
(see Table 3)

High surface area
silicon carbide
(see Tables 1 and 2)

Fig. 1. Conversion of an organometallic polymer
to high surface area inorganic materials

Table 1. Characterisation of High Surface
Area SiC[a]

Composition

Surface: XPS[b] (atom %) : Si, 46; C, 42; O, 7; F, 5

Bulk : IR : Strong band at 840 cm^{-1}
               No band at 1125 cm^{-1}[c]

Solid State NMR[d]

^{29}Si: −15.3, −16.0,
           −17.9, −19.9, −20.8
           −24.1, −25.2, −26.8

^{13}C: 15.6, 20.6, 23.7

Texture[d]

Surface Area	: 172	$m^2 \ g^{-1}$
Average Pore Diameter	: 13	nm
Pore Volume	: 0.34 ml	g^{-1}
Micropore Volume	: None	

a. Sample prepared by pyrolysis of 140 g of polymer (see text)
   in a tube furnace under a dried argon flow of 10–15 Nl h^{-1}.
   The product was heated in an air flow at 750°C until no
   further CO was produced, and then stirred in a concentrated
   HF solution for several hours followed by evaporation to
   dryness. Yield : 24.6 g, 33 mol%; appearance: beige powder.
b. X-ray photoelectron spectroscopy; the figures represent an
   average composition for the first few surface layers.
c. The Si-O vibration in amorphous silica was found to occur
   at 1125 cm^{-1}.
d. All shifts relative to tetramethylsilane. Literature values
   (11) for the 6H polytype of SiC are:
   ^{29}Si : −13.9, −20.2, −24.5
   ^{13}C : 15.2, 20.2, 23.2
e. By nitrogen adsorption (BET method).

Synthesis

The preparation of a porous silicon carbide has been described by Fox and co-workers (10). The synthesis is based on heating the organosilicon polymer $(C_6H_5SiO_{1.5})_n$ at 1600°C under argon (see Fig. 1). The pyrolysis reaction results in an intramolecular carbothermic reduction, i.e. the carbon bonded to silicon is used to remove oxygen and to form the carbide (the commercial manufacture of silicon carbide uses an external source of carbon: for example, by mixing quartz sand and petroleum cokes). The product is purified by oxidation to remove excess carbon, followed by treatment with HF to remove silica.

Characterisation of the material we obtained is reported in Table 1. The chief points to note regarding composition are (i) the atomic ratio of Si:C in the surface, which is very close to one, and (ii) the fact that both infra-red spectroscopy and solid state ^{29}Si nmr show the absence of silica in the bulk. The solid state ^{29}Si nmr spectrum is similar to that recently reported (11) for one of the many modifications of SiC, but consists of three groups of peaks rather than simply three peaks; it was found (11) that ^{29}Si nmr is quite sensitive to the local environment in silicon carbides. With respect to texture, the data show the material to be mesoporous; such a texture is eminently suitable for catalysis.

Stability

The stability of the texture was determined by exposing the porous silicon carbide, and a reference silica, to (i) a thermal treatment (flowing argon or nitrogen) or (ii) a steam treatment (nitrogen containing 2 vol% water vapour), both treatments being carried out for 4 hours. The data (Table 2) show that the pore structure of porous silicon carbide is superior in terms of thermal stability, but not as regards hydrothermal stability, to that of silica. As far as we are aware, improved thermal stability has not previously been demonstrated for a porous non-oxide compared to the analogous oxides.

Table 2. Textural Stability of High
Surface Area SiC and SiO_2[a]

	Surface Area, $m^2 g^{-1}$	
	SiO_2	SiC
Initial Surface Area	300	172
Thermal Stability		
1200°C	<5	103
1000°C	245	145
Hydrothermal Stability		
1000°C	40	28
750°C	284	111

a. Treatments carried out for 4 h under a N_2 or Ar flow (thermal stability) or N_2 + 2 vol% H_2O (hydrothermal stability); gas flow rates ca. 10 Nl h^{-1}, 2 g of solid used. The reference silica was a commercially available Shell sample, code S980, mesoporous, 300 $m^2 g^{-1}$, 1.5 mm spheres.

OTHER HIGH SURFACE AREA MATERIALS FROM REACTIONS OF THE ORGANOSILICON
POLYMER $(C_6H_5SiO_{1.5})_n$ AT 500 °C

The title polymer has a ladder structure in which each silicon atom is
bonded terminally to a phenyl group and by three oxygen bridges to other
silicon atoms. It occurred to us that it should be possible to substitute
inorganic groups for the phenyl group under relatively mild conditions (see
Fig. 1). This should allow the preparation of inorganic materials at lower
temperatures than the 1600°C used in the pyrolysis described in the previous
section. Furthermore, elimination of small molecules in a solid can be used
to create porosity (in the preparation of porous magnesium oxide, for
example, by elimination of carbon dioxide from magnesium carbonate). These
possibilities were examined by treatment of the solid polymer with flowing
ammonia, air and water vapour.

Synthesis

Reactions were carried out by flowing gases over the polymer, contained
in a quartz glass tube and heated by a cylindrical oven. Preliminary experiments with ammonia at 350, 400 or 500°C for periods of 24 h showed that a
temperature of 500°C was necessary for complete removal of the phenyl group,
as monitored by the absence of characteristic bands in the infra-red spectrum. This time period and temperature were then used for other reactions.

The data obtained on these reactions are summarised in Table 3. The
product from the reaction with ammonia showed a surface composition, in atom
%, of Si, 32; O, 49; N, 10; C, 8, some of the carbon signal observed being
from the sample holder. The Si:N:O atomic ratio is 1:0.32:1.5, which corresponds very closely to the ratio of 1:0.33:1.5 which would be found in
$[N(SiO_{1.5})_3]_n$, resulting from replacement of each phenyl group with an NH_2
group, with subsequent condensation reactions of the latter to $N(Si\equiv)_3$
units.

Table 3. Reactions of $[C_6H_5SiO_{1.5}]_n$ with Flowing Gases at 500°C,
24 h[a]

Reactant	Complete Reaction[b]	Product		
		Surface Atomic Ratio[c]		Surface Area[d], $m^2 g^{-1}$
NH_3	Yes	Si:N:O	1:0.3:1.5	300
H_2O	Yes	Si:O	1:2	30
O_2	Yes	Si:O	1:2	300
Ar	No			

a. Reactions carried out in a quartz glass tube
b. Indicated by the absence of characteristic phenyl vibrations
   in the IR spectrum
c. By XPS
d. By nitrogen adsorption (BET method)

Similar reactions were carried out using air or water vapour, and in both cases (i) FT-IR spectroscopy showed complete elimination of the phenyl groups, and (ii) surface Si:O atomic ratios were 1:2 (carbon was between 6 and 7 atom %, but again, this is at least in part due to the sample holder).

Finally, it should be noted that all three reactions do indeed lead to creation of surface area; the starting polymer, as indicated in Fig. 1, was found to have a surface area of less than 5 m^2 g^{-1}.

Stability

The product of the reaction between the polymer and ammonia was not stable. The surface composition, described above, was measured by XPS on a sample immediately after preparation. After exposure to air at ambient temperature for one hour the surface nitrogen concentration had declined from 10 to 8 atom %. This suggests a sensitivity towards hydrolysis, although this does not go to completion: two months after preparation the bulk nitrogen content of a sample was 3.9 wt%. The texture of the product was also unstable: the initial surface area was 320 m^2 g^{-1}, with subsequent values of 140 m^2 g^{-1} and < 5 m^2 g^{-1} being measured over a period of weeks.

This work establishes that it is indeed possible to synthesise high surface area inorganic materials through reaction of organometallic polymers at moderate (compared to pyrolysis) temperatures. One advantage of this method is the experimental simplicity of gas-solid reactions. An additional advantage might be the use of polymer technology to shape the material prior to reaction. The products of the above reactions are of no particular interest for our purposes as high surface area silica is widely available and the oxynitride is unstable. However, this instability is not an intrinsic property of high surface area silicon oxynitrides: in the next section we show that a stable form can be prepared by a different route.

SILICON OXYNITRIDES

Silicon nitride is receiving considerable attention as a structural ceramic, which requires a fully dense material, but it is also of interest to consider whether it could be prepared in high surface area forms. Our work in this direction has led to a synthesis of a high surface area silicon oxynitride. We summarise here the synthetic work and then report data on the stability of the surface.

Synthesis

Low temperature routes. The reaction of a molecular silicon compound with ammonia to give silicon nitride is formally similar to the reaction of a silicon compound with water to give silica. This latter reaction readily gives silica gels: for example, from silicon alkoxides. The gels can be dried to give porous solids. There do not appear, however, to be any examples of the formation of a porous silicon nitride using solution chemistry. (Note that with respect to catalysis we are concerned with open porosity, based on interconnected channels, and not a closed porosity consisting of isolated voids in the material; the latter situation is usually the case when ceramic compositions of less than theoretical density are prepared.)

A report has appeared (12) on the formation of a high surface area amorphous silicon nitride from a solution phase reaction of tetrachloro-silane with liquid ammonia. In our hands (7) this reaction gave a material with a surface area of 90 m^2 g^{-1}; both the form of the BET isotherm and transmission electron microscopy indicated that the surface area was derived from the external surface of small particles, i.e. a non-porous material. An interesting question which remains is whether modification of the solution chemistry could give a porous material. There is increasing understanding of the factors which control structural evolution in oxides formed in solution (13,14), but no equivalent information, as far as we are aware, on non-oxides.

High temperature routes. In an alternative approach (9) we have investigated the reaction of high surface area silica phases with gaseous ammonia at about 1100°C. The literature on reactions of silica with ammonia shows large differences in the degree of nitrogen incorporation (substitution for oxygen) in the silica (9). In our work we used silica samples of differing crystallinity and porosity. The materials used and the degree of nitrogen incorporation are shown in Tables 4 and 5. The most interesting result from this work is that one particular silica phase, amorphous and with a surface area (external) of 300 m^2 g^{-1}, could be reacted with ammonia at 1100°C to give a silicon oxynitride containing up to 28 wt% N in the surface and 27 wt% in the bulk (calculated for the known compound Si_2N_2O: 28 wt%) and with a surface area of 150 m^2 g^{-1}. Additional information on prior literature work and on characterisation has been reported (7,9). We have since learnt of similar work, in which a dried silica sol was used for reaction with ammonia to give high surface area silicon oxynitrides (15).

Table 4. Types of Silica Used for Reaction
with Ammonia

	Porous	Non-Porous
Crystalline	Silicalite[a]	Quartz Wool, Quartz Powder
Amorphous	Silica Gel[b]	Aerosil-300[b]

a. An Al-free structural analogue of
   zeolite ZSM-5
b. Surface area ca. 300 m^2 g^{-1}

Table 5. Reaction of Various Types of SiO_2 with Flowing NH_3 (30 $l.h^{-1}$)
at 1100°C[a]

SiO_2	Bulk N, wt%	Surface N, wt%	Surface N, atom %	Surface Area, m^2 g^{-1}
Quartz Wool	<1	14	20	
Quartz Powder	2	10	14	
Silicalite/15 h	2	7	10	
Silicalite	4	7	11	
Silica Gel (10 g)	10	13	19	34
Aerosil	22	24	34	
Aerosil/12 h	25	28	40	132
Aerosil/24 h	27	28	40	150

a. All reactions carried out for 4 h using 1-5 g silica, unless otherwise stated. The products obtained from Aerosil and silica gel are
   amorphous, as determined by X-ray powder diffraction.

Table 6. Stability of Silicon Oxynitride Surfaces to Air, Steam
and Hydrogen[a]

	Surface N, wt%	Bulk N, wt%	Surface Area, $m^2 g^{-1}$
Silicon Oxynitride (A)	28	25	132
After Steam at 680°C	19	21	140
After Hydrogen at 750°C	20		
Silicon Oxynitride (B)	22	18	152
After Steam at 750°C	15	17	
After Hydrogen at 750°C	15		
After Air at 750°C	17		

a. All treatments for 4 h at 1 bar using a gas flow of 5 $1.h^{-1}$
   (steam = nitrogen + 2% H_2O) and 0.1 - 1.0 g SiON.

Stability

We wished to establish the stability of the silicon oxynitride surface
under conditions relevant to catalysis and have used the high surface area
oxynitride prepared by the reaction of silica with gaseous ammonia described
above. Information was obtained by treating samples with flowing gases, and
subsequently re-measuring the surface nitrogen concentration by XPS.

The results are shown in Table 6, and establish that the high surface
area silicon oxynitrides have good stability, particularly as regards the
retention of significant amounts of surface nitrogen under both reducing and
oxidising conditions. This makes further investigation of application of
these materials in the field of catalysis worthwhile.

With respect to the extent of oxidation found, it should be noted that
bulk silicon nitride is protected against oxidation by the formation of a
thin film of silica, which is a barrier to diffusion, and it is possible
that a very thin film is providing stability in our case (note that XPS
measurements give an average over the first few outermost atomic layers).
This might also explain the difference between the hydrolytic stability of
the silicon oxynitride obtained from silica and the silicon oxynitride
obtained from the polymer (see above), if differences in texture determine
the effectivenes of such a film. However, it should also be noted that the
hydrolytic stability of Si-N bonds can vary widely in molecular compounds.
Thus hexamethyldisilazane is resistant to boiling water or aqueous alkali
but is hydrolysed immediately by aqueous methanol (in which it dissolves),
whereas hexaphenyldisilazane can be recrystallised from ethanol, and is
destroyed only slowly when dissolved in boiling aqueous-alcoholic alkali
(16). Thus the reactivity of the Si-N bond is largely dependent on the local
environment, and this might also be the basis for the observed differences.

CONCLUSIONS

The present work has contributed to the growing number of routes to
high surface area non-oxide materials, through the reaction of an organometallic polymer with flowing gases under relatively mild conditions and
through the reaction of silica phases with ammonia at high temperature.

Furthermore, results obtained with respect to the thermal stability of the pore structure in porous silicon carbide and the stability towards air, hydrogen or steam of the surface of a silicon oxynitride powder indicate that the stability of high surface area non-oxidic materials can be promising with respect to potential application in catalysis.

REFERENCES

1. C.J. Brinker, D.E. Clark and D.R. Ulrich (Eds.), "Better Ceramics Through Chemistry. III", Materials Research Society Symposium Proceedings, Vol. 121, 1988.
2. M.A. Vannice, Y.-L. Chao and R.M. Friedman, Appl. Catal., 20, 91 (1986).
3. M.L. Ledoux, S. Hantzer, C.P. Huu, J. Guille and M.P. Desaneaux, J. Catal., 114, 176 (1988).
4. M. Orita, I. Kojima and E. Miyazaki, Bull. Chem. Soc. Jpn., 59, 689 (1986).
5. (a) S.T. Oyama, J.C. Schlatter, J.E. Metcalfe and J.M. Lambert, Ind. Eng. Chem. Res., 27, 1639 (1988); (b) J.C. Schlatter, S.T. Oyama, J.E. Metcalfe and J.M. Lambert, Ind. Eng. Chem. Res., 27, 1648 (1988).
6. J.S. Lee, L. Volpe, F.H. Ribeiro and M. Boudart, J. Catal., 112, 44, (1988).
7. P.W. Lednor and R. de Ruiter, in: Ref. (1), p. 497.
8. P.W. Lednor and R. de Ruiter, in: "Euro-ceramics", G. de With, R.A. Terpstra and R. Metselaar (Eds.), Elsevier Applied Science, 1.53 (1989).
9. P.W. Lednor and R. de Ruiter, J. Chem. Soc., Chem. Commun., 320 (1989).
10. D.A. White, S.M. Oleff and J.R. Fox, Adv. Ceram. Mater., 2, 53 (1987).
11. G.R. Finlay, J.S. Hartman, M.F. Richardson and B.L. Williams, J. Chem. Soc., Chem. Commun., 159 (1985).
12. T. Yamada, T. Kawahito, amd T. Iwai, J. Mater. Sci. Lett., 2, 275 (1983); see also Chemical Abstracts 92:131547f.
13. C.J. Brinker, J. Non-Cryst. Solids, 100, 31 (1988).
14. D. Schaefer, Science, 243, 1023 (1989).
15. (a) J. Sjöberg, K. Rundgren, J. Osten-Sacken, R. Pompe and B. Larsson, Sci. Ceram., 14, 205 (1988); (b) J. Sjöberg, K. Rundgren, R. Pompe and B. Larsson, in "High Tech Ceramics", (P. Vincenzini, Ed.), Elsevier Science Publishers, 535 (1987).
16. C. Eaborn, "Organosilicon Compounds", Butterworths, 1960.

THE POLYSILANES: THEORETICAL PREDICTIONS OF THEIR CONFORMATIONAL CHARACTERISTICS AND CONFIGURATION-DEPENDENT PROPERTIES

William J. Welsh

Department of Chemistry, University of Missouri-St. Louis, St. Louis, MO 63121, USA

ABSTRACT

The poly(organosilanes) [-SiRR'-] represent a novel class of semi-inorganic polymers whose unusual physical and chemical properties find applications in numerous technological applications. Among their most unusual properties, the poly(organosilanes) and more recently the analogous poly(organogermanes) exhibit large shifts in λ_{max} of their electronic spectra as a function of the size and nature of their substituents R and R'. Some examples exhibit thermochromism, again as a function of the size and nature of their substituents. Explanations for this behavior commonly invoke the conformational characteristics of the chains' backbone. The author reports here his applications of theoretical computational chemistry techniques to investigaté the conformational and configuration-dependent properties of the polysilanes and polygermanes. Recent preliminary calculations are also reported which address the unexpected observation of thermochromism and an anomalous blue-shifted λ_{max} in the poly(bis(p-alkoxyphenyl)silanes).

INTRODUCTION

Substituted poly(organosilanes) [-SiRR'-] have recently become the focus of intense scientific and commercial interest, and synthetic efforts have been successful in obtaining high molecular weight organosilane polymers and copolymers containing both alkyl and aryl substituents.[1-5]

These materials are soluble in common organic solvents and can be processed into a variety of shapes, films, and fibers. They also exhibit strong UV absorption whose λ_{max} varies markedly with molecular weight and with the nature and size of the substituents R and R'. These

materials are also highly radiation sensitive, being easily degraded by
UV initiators or by heating to high temperatures. As a consequence of
these unusual properties, organosilane polymers have found commercial
applications as UV photoresists in photolithography, as radical
photoinitiators, as impregnating agents for strengthening ceramics, and
as precursors for silicon carbide fibers. Examples have also shown
promise as dopable electrical conductors and semiconductors.[1-5] More
recently, Miller and Sooriyakumaran[6] have synthesized a series of
poly(organogermanes) [-GeRR'-] which exhibit many physical and
electronic properties similar to the analogous polysilanes.

Much of the scientific interest in the polysilanes stems from the
observation that the position of λ_{max} in their electronic spectra
depends strongly on the nature and size of the substituents R and R' and
on the conformation of the chain backbone.[1-10] Specifically, UV
spectroscopic data reveal a marked bathochromic shift in λ_{max}
associated with an increase in the bulk of the substituents. In
addition, certain polysilanes and polygermanes, such as poly(di-n-
hexylsilane) (PDHS) and higher alkyl homologs, are highly and reversibly
thermochromic: below approximately 40°C PDHS absorbs red-shifted at
λ_{max} = 370-380 nm whereas above 40°C bands near λ_{max} = 325-320 nm
dominate.[7,8]

The unusual conformational dependence of these electronic effects
has stimulated both basic and applied research activity in several
industrial and academic laboratories.[1-5] The growing body of
experimental[1-10] and theoretical[11-22] evidence suggests that these
spectroscopic phenomena are attributed to long-range conformational
transitions of the silicon backbone, specifically from sequences of
helical or random coil configurations to sequences corresponding more
nearly to the planar zigzag, all-_trans_, conformation. In turn, the
conformational transitions appear to be induced by a concerted process
involving side-chain crystallization and the occurrence of thermally
induced _trans-gauche_ conformational transitions along the silicon
backbone.[3] Viewed at the molecular level, melting of the side chains
ensuing above a certain temperature allows for conformational
transitions away from the _trans_ state (which in general is the one
preferred from the standpoint of both side-chain packing efficiency and
conformational energy considerations) to the conformationally higher
energy _gauche_ states. As temperature increases the _gauche_ states become
more populated relative to the alternative _trans_ state, hence the chain
experiences a gradual and characteristic blue shift in λ_{max} in
accordance with these rod-to-coil transformations.

Several theoretical studies,[11-22] using both molecular mechanics (MM)[11-13,15,16,18,20] and molecular orbital (MO)[14,17,19,21,22] approaches, have already revealed important information regarding the structural, electronic, and conformational properties of the polysilanes. Regarding their conformational properties, the earlier MM modeling studies were confined to the simple polysilanes such as $[-SiH_2-]$ and $[-Si(CH_3)_2-]$.[11,12,15] They have since been extended to include bulkier R and R' substituents such as the higher alkyl homologs (e.g., n-pentyl) and even aryl groups.[13,16,18,20] Likewise, most MO approaches[14,17,19,22] have been restricted to model compounds for only the simplest member of the polysilanes, i.e., $[-SiH_2-]$. Furthermore, the polygermanes have remained virtually untouched from the standpoint of theoretical studies. Only recently have MO calculations been extended to model compounds beyond $[-SiH_2-]$ to the simplest poly(organosilane) $[-Si(CH_3)_2-]$.[21] This latter study[21] also included results applicable to the analogous $[-Ge(CH_3)_2-]$ chain.

In the present paper, the author reviews his work concerned with calculating the structural, conformational, and configuration-dependent properties of the polysilanes.[11,15,20,21] The structures and conformational energies are derived in large part from MM and semiempirical MO calculations, sometimes carried out in tandem to afford comparisons between the various computational approaches. The structural and energetic data thus provided have served in some cases as input for rotational isomeric state (RIS) calculations[23] of these chains' unperturbed dimensions (mean-square end-to-end distance), a configuration-dependent property of great interest since it reflects the spatial configuration of the polymer in solution under theta conditions. Moreover, quantitative comparisons of these theoretically-derived values of the chain dimensions with those obtained via experimental determination provide a valuable test of the theoretical assumptions adopted in the computations.

METHODOLOGY

For details of the individual methodologies employed in each study reviewed herein the reader is directed to the particular reference cited. One feature common to all of these studies is the use of model compounds to represent a particular polymer chain. In general these model compounds are devised to incorporate all or at least most of the essential features of the parent polymer chain. At the same time, these model compounds should be constructed conservatively from the standpoint of length of segment and number of atoms (e.g., size of substituents) so as to economize on computer resources and to ensure reasonable CPU turnaround times. Typical model compounds employed in the studies discussed herein range from five to seven backbone silicon atoms depending on the size of the substituents. Examples are depicted in Figure 1.

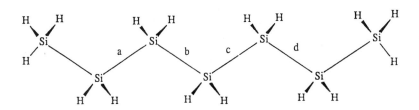

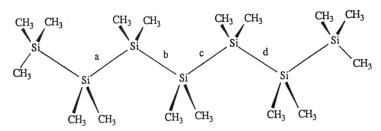

Figure 1. Examples of Polysilane Model Compounds.

RESULTS AND DISCUSSION

The Simple Polysilanes: The Parent Polysilane [-SiH$_2$-] and Poly(dimethylsilane) [-Si(CH$_3$)$_2$-]

Molecular Mechanics Studies. Damewood and West, using full-relaxation (FR) MM techniques, investigated the structure and conformational energies of model compounds for both polysilane [-SiH$_2$-] and poly(dimethylsilylene) [-Si(CH$_3$)$_2$-].[12] Later Welsh et al. [15] applied MM methods incorporating approximations referred to as NR for

"no relaxation" and PR for "partial relaxation" in a similar study of
the same polymers. The NR, PR, and FR force fields differ basically in
the degree to which each method allows relaxation of the molecule's
internal degrees of freedom in order to achieve energy minimization.
The NR force field included both steric (nonbonded) and torsional terms,
and conformational energies were computed without allowance for any
geometry optimization (molecular relaxation or deformation) following
torsional rotations about the backbone bonds. The PR force field as
applied in the study of $[-Si(CH_3)_2-]$ was nearly identical to the NR
force field except that the former optimized the geometry with respect
to torsional rotations about the pendant $Si-CH_3$ bonds for each backbone
conformation considered.

Contour maps of the conformational energy E vs. torsion angles ϕ_b
and ϕ_c (see Figure 1) were generated using energies calculated by
scanning the entire 0-360° conformational energy space in increments of
5° for the NR calculations and 20° for the PR calculations. These
calculated E vs. (ϕ_b, ϕ_c) data sets were also used to evaluate the
conformational partition funcion z, the average energy <E>, and the
average torsional angles $<\phi_b>$ and $<\phi_c>$ for each conformational state s
of interest (e.g., TT, GT, GG), using[24]

$$z_s = \sum_{\phi_b} \sum_{\phi_c} \exp(-E_k/RT) \tag{1}$$

$$<E>_s = z_s^{-1} \sum_{\phi_b} \sum_{\phi_c} E_k \exp(-E_k/RT) \tag{2}$$

$$<\phi_j>_s = z_s^{-1} \sum_{\phi_b} \sum_{\phi_c} \phi_j \exp(-E_k/RT) \tag{3}$$

where the subscript k refers to each conformation (ϕ_b, ϕ_c) and j is b or
c.

Some readers may be perplexed at the prospect of applying
computationally less sophisticated MM methods, such as NR and PR, *after*
analysis by the more rigorous FR method. One might ponder: "What is to
be gained?" The response is twofold: (1) the computational speed and
affordability of these calculations permit use of fine grids in scanning
conformational space, and (2) these methods traverse the entire
conformational terrain (rather than seeking energy minima only) and in
doing so quantify domain sizes and rotational barriers. Such

conformational information is invaluable for a detailed analysis of dynamic flexibility and configuration-dependent properties. These methods[11,15] can thus be regarded as complementary to the computationally more rigorous full-relaxation methods such as that employed by Damewood and West,[12] which are intended to locate energy minima (global or local) within conformational-energy space.

The results of both the NR and PR conformational energy calculations are conveniently depicted graphically in terms of maps which plot the conformational energy as contour lines with the torsion angles ϕ_b and ϕ_c along the axes. A general feature of all polysilanes when compared with their analogous hydrocarbon chains is the former's considerably greater conformational flexibility, specifically in terms of more energetically accessible conformational domains and lower barriers between energy minima. This is due primarily to the greater length of the Si-Si bond (2.34Å) compared with the C-C bond (1.53Å)[25] which results in less severe repulsions for most conformations.

Considering the MM calculations on model compounds of polysilane [-SiH$_2$-], the NR and PR force fields yield virtually identical conformational energy maps (Figure 2). Associated values of <E>, z, and <ϕ_b, ϕ_c>, gleaned from the energy map for states TT, TG (GT), G$^\pm$G$^\pm$, and G$^\pm$G$^\mp$, are listed in Table 1. It is seen that the TG state is preferred over the TT state by ca. 0.2 kcal mol^{-1} and that G$^\pm$G$^\pm$ states are preferred over the alternative TT state by ca. 0.5 kcal mol^{-1}. Even the G$^\pm$G$^+$ states, typically found prohibitively repulsive for many chains (including the alkanes[23]), are preferred over the TT state by ca. 0.4 kcal mol^{-1}. However, the domain size of each conformational state, as determined by the relative magnitudes of z, follows the more typical order TG > TT » G$^\pm$G$^\pm$ » G$^\pm$G$^+$. By comparison, in the analogous n-alkanes TT states are preferred over G$^\pm$G$^\pm$ states by ca. 1.0 kcal mol^{-1}, and G$^\pm$G$^+$ states are almost prohibitively high in energy.[23,26] Given the preference for conformations corresponding to sequences of _gauche_ states of the same sign (and based on the assumption that _inter_molecular energies generally have only a small effect on conformation[23]), the crystalline-state configuration of [-SiH$_2$-] is predicted to be helical rather than similar to the all-_trans_ planar zigzag conformation of polyethylene [-CH$_2$CH$_2$-]. For polysilane, nearly all regions of conformational-energy space are found within 2.0 kcal mol^{-1} of the energy minima; this is in sharp contrast to the relatively high

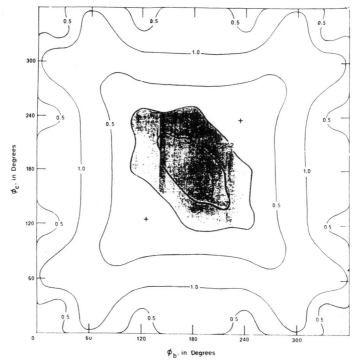

Figure 2. Conformational Energy Map for a [-SiH$_2$-] Model
Compound, as derived from the NR Force-Field Calculations.[15]
The <u>trans</u> conformation corresponds to $\phi = 0°$.

Table 1. Pertinent Conformational Parameters Determined by the NR Force
Field Calculations on the [-SiH$_2$-] Model Compound

	TT	TG (GT)	$G^{\pm}G^{\pm}$	$G^{\pm}G^{\mp}$
$z_s{}^a$	1.00	1.22	0.63	0.31
$\langle E_s \rangle^b$	-0.204	-0.408	-0.711	-0.612
$\langle \phi_b, \phi_c \rangle_{min}{}^c$	(0°,0°)	(0°,125°)	(±125°,±125°)	(±120°,+120°)

^{a}Expressed relative to $z_s = 1.00$ for the TT state. ^{b}In kcal mol^{-1}. ^{c}In
which (0°,0°) corresponds to the all-<u>trans</u>, planar zigzag, conformation.

torsional barriers (>6 kcal mol^{-1}) and large regions of prohibitively
high energy found in the case of the n-alkanes.[23]

Values of the relative energies and associated conformations ϕ for
polysilane, as calculated by the FR[12] and NR[15] force fields, are
compared in Table 2. Agreement is satisfactory, as one would expect for
such a structurally simple and conformationally flexible molecule as
[-SiH$_2$-].

Table 2. Relative Energies[a] and Torsional Angles[b] of Various Conformers of the [-SiH$_2$-] Model Compound as Calculated by the NR and FR[c] Force Fields

	NR[c]		FR	
	$<E>^a$	$<\phi>^b$	E^a	ϕ^b
TT	0.51	0.0	0.7	0.0
TG	0.21	125	0.4	121.4
GG	0.00	125	0.0	125.3
G^+G^-	0.41	-120	0.4	-111.2

[a]In kcal mol^{-1}, given relative to E = 0.0 kcal mol^{-1} for the minimum-energy conformation. [b]In degrees, given relative to ϕ = (0°, 0°) for the all-_trans_, planar zigzag conformation. [c]Values of E and ϕ represent Boltzmann-weighted averages taken from the generated energy maps.

The corresponding conformational energy map for poly(dimethylsilylene) [-Si(CH$_3$)$_2$-] based on the PR force field is given in Figure 3.[15] Associated values of $<E>$, z, and $<\phi_b, \phi_c>$ are listed in Table 3 along with those derived from the NR force field. A summary of the relative energies and their associated conformations for poly(dimethylsilylene) as calculated by the NR, PR, and FR force fields is given in Table 4. In this case, substantial qualitative differences are noted among the results given by the three force fields. The most striking difference is that, whereas the GG state is found by NR to be high in energy, it is the preferred state according to the FR calculations. Since values of the bond lengths and backbone bond angles used in both methods are nearly identical, the discrepancy is attributed largely to the torsional relaxation of the pendant Si-CH$_3$ groups afforded by the FR method. That torsional relaxation of the pendant groups constitutes a dominant influence is corroborated by the results of the PR calculations. Specifically, the PR results show a marked reduction in the relative energy for the GG conformation, although not sufficient to render it the preferred conformation. Of course, other forms of structural relaxation (e.g., deformation of the Si-C-H and Si-Si-C bond angles, which the PR force field did not include) would be expected to contribute but less so.

Notwithstanding this discrepancy, the three force fields do yield some similarities in terms of modeling the conformational characteristics of poly(dimethylsilylene).[12,15] Specifically, all three find the $G^±G^+$ states prohibitively high in energy, and the relative conformational energy preferences of TT and TG states are in reasonable agreement. Still, although corresponding values of φ given by the NR and FR force fields are close to each other and to conventional values (i.e., ±120°), the PR calculations located <u>gauche</u> states at torsion angles quite smaller in magnitude (thus corresponding to a less compact chain).

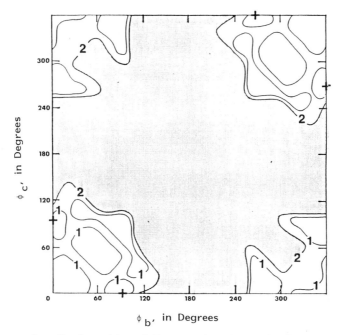

Figure 3. Conformational Energy Map for a [-Si(CH₃)₂-] Model Compound, as derived from the PR Force Field Calculations.[15] The <u>trans</u> conformation corresponds to $\phi = 0°$.

Table 3. Pertinent Conformational Parameters for the [-Si(CH₃)₂-] Model Compound As Determined by the NR and PR Energy Calculations

	NR			PR		
	z_s[a]	$<E_s>$[b]	$<\phi>$[c]	z_s[a]	$<E_s>$[b]	$<\phi>$[c]
T(T)	1.00	−5.99	0	1.00	−16.7	0
T(G)	0.26	−6.04	125	0.82	−17.1	96
G(G)	~0.0	(~−2)	125	0.38	−17.0	89
G⁺(G⁻)	~0.0	>6	−115	~0.0	−	−

[a]Expressed relative to z = 1.00 for the TT state. [b]In kcal mol⁻¹, given as the Boltzmann-weighted averages derived from the conformational energy maps. [c]In degrees, in which 0° corresponds to the all-<u>trans</u>, planar zigzag, conformation.

Values of z obtained from both the NR and PR calculations for
$[-Si(CH_3)_2-]$[15] indicate that the size of the domain associated with TT
is larger than that for TG or GG. Figure 3 is noted for its large
regions of prohibitively high energy, in contrast to that (Figure 2) for
polysilane itself for which nearly all regions are within 2 kcal mol^{-1}
of the energy minima. The difference is attributable to the greater
steric bulk of the methyl groups in poly(dimethylsilylene) relative to
that of the H atoms in polysilane.

Table 4. Relative Energies[a] and Associated Torsion Angles[b] of Various
Conformers of the $[-Si(CH_3)_2-]$ Model Compound As Calculated by the NR,
PR, and FR[c] Force Field Methods

	NR[c]		PR[c]		FR	
	$<E>$[a]	$<\phi>$[b]	$<E>$[a]	$<\phi>$[b]	E[a]	ϕ[b]
T(T)	0.1	0.0	0.42	0	0.9	0.0
T(G)	0.0	125	0.00	98	0.8	120.3
G(G)	~4	125	0.08	89	0.0	125.3
$G^+(G^-)$	>12	-115	d	-	38.4	-107.7

[a] In kcal mol^{-1}, given relative to the energy of the minimum-energy
conformation. [b] In degrees, given relative to $\phi = 0°$ corresonding to the
trans conformation. [c] Values $<E>$ and $<\phi>$ correspond to Boltzmann-
weighted averages derived from the respective energy maps. [d] Calculated
energy was prohibitively high.

Calculation of the Chains' Unperturbed Dimensions. Inspection of
the conformational energy maps (Figures 2 and 3) reveals that adoption
of the familiar three-state (i.e., T, G^+, G^-) scheme is sufficient for
application of the RIS theory to these chains.[23] The corresponding
statistical weight matrix U, inclusive of both first-order and second-
order interactions (those interactions depending on, respectively, one
and two skeletal-bond rotations), is represented by[23]

$$U = \begin{bmatrix} 1 & \sigma & \sigma \\ 1 & \sigma\psi & \sigma\omega \\ 1 & \sigma\omega & \sigma\psi \end{bmatrix} \tag{4}$$

In the case of the NR and PR calculations, the statistical weights σ,
ψ, and ω were determined from the respective values of z derived from
the conformational energy maps; for example

$$\sigma = z_{TG}/z_{TT} = \sigma_0 \exp(-E_\sigma/RT) \tag{5}$$

Values of the statistical weight parameters determined in this manner
account explicitly for the relative size of the domains for each state,

as denoted by the so-called "entropy factor" σ_0 in eq. 5. In the case of the FR calculations,[12] for which the absence of conformational energy maps precludes computation of z values, the statistical weight parameters were taken as simple Boltzmann factors in the conformational energy, for example, $\sigma = \exp(-E_\sigma/RT)$. Such a procedure embodies the tacit assumption of equal domain sizes (i.e., $\sigma_0 = \psi_0 = \omega_0 = 1.0$). Characteristic ratios $C_{n\to\infty} = \lim_{n\to\infty} [r^2]_0/nl^2$, where $[r^2]_0$ represents the mean-square end-to-end distance unperturbed by excluded volume effects, n the number of backbone bonds, and l the backbone bond length, were then calculated according to established methods described in the literature.[23]

The calculated torsion angles listed in Tables 2 and 3 for polysilane and poly(dimethylsilylene), respectively, were used in conjunction with eqs. 4 and 5 to calculate values of $C_{n\to\infty}$ for the two polymers. The statistical weight parameters σ, ψ, and ω computed at 25°C based on the NR, PR, and FR calculations are listed in Table 5.

Table 5. Values of the Statistical Weight Parameters Computed at 25°C Derived from Results of the NR, PR, and FR Force Field Calculations

	[$-SiH_2-$]			[$-Si(CH_3)_2-$]		
	NR	PR	FR	NR	PR	FR
σ	1.6	–	1.6	0.27	0.82	1.2
ψ	1.5	–	2.0	0.00	0.56	3.8
ω	0.52	–	1.0	0.00	0.00	0.0

For polysilane, the corresponding $C_{n\to\infty}$ values at 25°C are 4.1 and 3.9 based on the NR and FR calculations, respectively. The nearly identical values reflect the similarity in conformational characteristics obtained by the two different force fields for this chain. The relatively small value of $C_{n\to\infty}$ for polysilane is indicative of its rather low chain extensibility. This feature is consistent in turn with the chain's high overall conformational flexibility (i.e., no overwhelming preferences for any one particular conformational state), with its identifiable preferences for the more compact TG and $G^\pm G^\pm$ states over the alternative and more chain-extending TT state and moreover, and with its allowance for $G^\pm G^\mp$ states whose occurrence typically leads to reversals in chain direction. For comparison, accepted values of $C_{n\to\infty}$ at 25°C for two other flexible polymers, polyethylene and poly(dimethylsiloxane), are 6.7 and 6.4, respectively.[23,27,28]

Values of $C_{n \to \infty}$ at 25°C for poly(dimethylsilylene) are 15.0, 13.2, and 12.5 based on the NR, PR, and FR calculations, respectively.[15] Given the differences in conformational preferences predicted by these three force-field methods, the small range of values is surprising. Specifically, the value $C_{n \to \infty} = 12.5$ obtained based on the FR calculations is only slightly smaller than the corresponding value of 15.0 based on the NR results. Yet the FR results indicate strong preferences for GG states over the alternative TT and TG states while, in contrast, the NR results yield precisely the opposite preferences (i.e., TT and TG states over GG states). These quantitative and qualitative differences are obviously not strongly reflected in the calculated $C_{n \to \infty}$ values.

An explanation for the close agreement in calculated $C_{n \to \infty}$ values, irrespective of the MM force field applied, can be found from a detailed analysis of the allowable sequences of conformational states for [-Si(CH$_3$)$_2$-] elucidated by the separate MM calculations. For example, the helix associated with ...G$^\pm$G$^\pm$G$^\pm$G$^\pm$G$^\pm$... sequences is predicted by the FR calculations as having a high probability of occurrence. This sequence is spatially more compact than that associated with either the ...TTTTT... or ..TG$^\pm$TG$^\pm$T... sequences predicted by the NR calculations. However, this compactness is compensated for somewhat by the fact that the NR calculations also indicate high probability for sequences such as ...TG$^\pm$TG$^+$T... and ...TG$^\pm$TTG$^+$T... Inclusion of these sequences (i.e., those characterized by the presence of nearby gauche states of opposite sign) will tend to divert the chain direction and thus foreshorten the dimensions of the chain. The FR calculations predict relatively low probability of occurrence of such sequences and, in general, of any sequences corresponding to reversal or redirection of chain propagation. Thus in different but nearly compensatory ways the NR and FR calculations allow for more compact sequences of states than the fully extended all-trans configuration. If in fact the poly(dimethylsilylene) chain adopted exclusively an all-trans conformation, then as $n \to \infty$ the calculated value of C_n would approach ∞ rather than the values (12.5-15.0) obtained.[15]

As would be expected from the intermediate nature of its calculated conformational energies, the PR force field yielded $C_{n \to \infty} = 13.2$[15]. This is nearly midway between the $C_{n \to \infty}$ values obtained from using the NR and FR calculations. The value of $C_{n\to\infty}$ obtained based on the PR results would have been somewhat closer to that (12.5) obtained based on the FR results except that the locations of the _gauche_ states obtained in the PR calculations (i.e., 90-100°) are more extended than those found in the NR or FR calculations (120-125°).

The calculated $C_{n\to\infty}$ values for [-SiH$_2$-] increase slightly as the temperature increases, reflecting the greater accessibility of the higher energy and chain-extending _trans_ states. The NR results yielded $C_{n\to\infty}$ values of 3.85, 4.02, 4.15, and 4.25 at 0, 25, 50, and 75°C, respectively, for polysilane. In contrast, the calculated $C_{n\to\infty}$ values for poly(dimethylsilylene) are affected negligibly in this temperature range. This is expected given that all states except TT and TG are virtually excluded according to the NR calculations.

<u>Molecular Orbital Studies</u>. Conformational energy calculations on [-SiH$_2$-] and [-Si(CH$_3$)$_2$-] model compounds have also been carried out using MNDO[29,30] semiempirical MO techniques. These studies[21] also included some preliminary results for a model compound of the analogous germanium-containing polymer [-Ge(CH$_3$)$_2$-].

The calculated conformational energies ΔE applicable to [-SiH$_2$-] are plotted vs. ϕ in Figure 4. The results indicate a preference for the _trans_ state (TT) over the alternative _gauche_ states (TG$^\pm$) by 0.54 kcal mol^{-1}. The _trans_ state appears to be associated with the global energy minimum with respect to this rotation. The chain is found to exhibit high rotational flexibility with a maximum energy barrier of only 0.80 kcal mol^{-1} located at $\phi = 0°$ (_cis_).

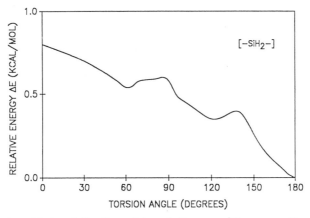

Figure 4. Plot of Conformational Energy ΔE versus Torsion Angle ϕ for a [-SiH$_2$-] Model Compound, as derived from MNDO Calculations.[21] The _trans_ conformation corresponds to $\phi = 180°$.

Extended calculations were carried out for conformations associated with the $G^\pm G^\pm$ and $G^\pm G^\mp$ states for purposes of application of RIS theory to compute the configuration dependent properties of these chains.[15,23] The calculated energies for the $G^\pm G^\pm$ and $G^\pm G^\mp$ states are only 0.86 kcal mol^{-1} and 1.56 kcal mol^{-1}, respectively, above that for the TT state. The MNDO results thus indicate a highly flexible chain.

These MNDO results are consistent with the earlier MNDO calculations of Bigelow and McGrane[22] although the trans-gauche energy differences appear slightly smaller. They are also qualitatively consistent with the results of recent ab initio calculations[19] which indicate preferences for trans over gauche by ~0.17 kcal mol^{-1} at both the 3-21G* and 6-31G* levels but which indicate a preference for gauche over trans by a slight 0.04 kcal mol^{-1} when correlation is included in the calculations (Table 6: MP2/6-31G*).

The above MNDO results disagree however with the earlier MM results[12,15] which consistently indicate a preference for gauche over trans for the [-SiH$_2$-] model compounds. Values of conformational energies obtained from various theoretical studies are compared in Table 6, which also includes some ab initio results calculated by the author at the 3-21G and 3-21G* levels.[21]

Table 6. Results of Conformational Energies[a] Taken from various Computational Studies on Model Systems of [-SiH$_2$-].

| Conformation[b] | Present Work | | | | Other Work | | | |
	MNDO	3-21G	3-21G*	MM[c]	MNDO[d]	3-21G*[e]	6-31G*[e]	MP2/6-31G*[e]
TT	0.00	0.000	0.000	0.0	0.00-0.421	0.00	0.00	0.00
TG	0.54	0.093	0.150	-0.3	0.59[f]	0.17	0.19	-0.04
TE	0.38	-	-	-	-	0.75	0.62	0.58
TC	0.80	-	-	-	-	1.26	1.65	1.25
$G^\pm G^\pm$	0.86	-	-	-0.7	-	0.17	-	-
$G^\pm G^\mp$	1.56	-	-	-0.3	-	-	-	-

[a]Energies are given in units of kcal/mol, all relative to the TT state. [b]T=trans, G=gauche, E=eclipsed, C=cis. [c]Molecular mechanics results taken from References 12 and 15. [d]Taken from Reference 22, the variation for the TT state being due to whether TT is defined as "rigid" (0.00 kcal mol) or "skewed" (0.421 kcal/mol). [e]Taken from Reference 19. [f]Value represents 1.764 kcal/mol (Table IV of Reference 22) divided by 3 assumed as the number of TG states contained in the Si$_{10}$H$_{22}$ model compound.

The MNDO results corresponding to the [-Si(CH$_3$)$_2$-] chain[21] are plotted in Figure 5. The results indicate a preference for _trans_ over _gauche_ by 1.4 kcal mol^{-1}. Another, local energy minimum was located near ϕ = 90° about 1.1 kcal mol^{-1} above the _trans_ energy minimum. A skewed _gauche_ state was found similarly in earlier MM calculations on these model compounds.[15] The energy minimum associated with the _trans_ state is also skewed, in this case toward values near ϕ = 170°. The maximum energy barrier was located at ϕ = 0° (_cis_), about 2.75 kcal mol^{-1} higher in energy than _trans_. Hence this chain is predicted to be rotationally much less flexible than the corresponding [-SiH$_2$-] chain, a conclusion in qualitative agreement with the MM results.[12,15]

Based on these MNDO results,[21] the [-Si(CH$_3$)$_2$-] model compound is depicted as a chain possessing distinct _trans_ and _gauche_ energy minima but whose locations differ considerably from the typical values of ϕ = 180° and ϕ = 60°, respectively. In particular, at ϕ = 90° the "gauche" minimum is found to be skewed some 30° (toward the more extended _trans_ conformation) relative to the location of the typical _gauche_ state (i.e., 60°). These MNDO results[21] agree with the conclusions of earlier MM calculations[15] indicating a preference for _trans_ over _gauche_, although MNDO gives an energy difference nearly twice as high (1.4 vs. ~0.9 kcal mol^{-1}).

Preliminary MNDO calculations on the [-Ge(CH$_3$)$_2$-] model compound[21] yielded a conformational energy profile that is remarkably similar in shape to that obtained for the analogous [-Si(CH$_3$)$_2$-] model compound

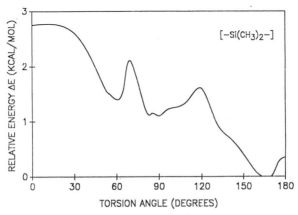

Figure 5. Plot of Conformational Energy ΔE versus Torsion Angle ϕ for a [-Si(CH$_3$)$_2$-] Model Compound, as derived from MNDO Calculations.[21] The _trans_ conformation corresponds to ϕ = 0°.

except that, for the former, energy barriers to rotation are uniformly lower. This is reasonable since the Ge-Ge bonds along the chain backbone are longer than the corresponding Si-Si bonds by about 0.3Å (MNDO gives average bond lengths for Si-Si and Ge-Ge of 2.35Å and 2.65Å, respectively).[25] This additional length leads to a considerable decrease in steric congestion and a concomitant increase in rotational flexibility.

Polysilanes with Bulky Aromatic Substitutents:

Poly(phenylmethylsilylene) [-SiPhMe-] and Poly(silastyrene) [-SiPhH-SiH$_2$-]

MM calculations with full geometry optimization have been carried out[20] to calculate the molecular structure and conformational energies of model compounds for poly(phenylmethylsilylene) [-SiPhMe-] and poly(silastyrene) [-SiPhH-SiH$_2$-]. In each case the calculations considered both the isotactic and syndiotactic stereochemical isomeric forms of the model compounds. The structures considered are illustrated in Figures 6 and 7.

The structural and conformational energy data thus provided were used to calculate, again by application of RIS theory,[23] the unperturbed dimensions $<r^2>_0$ and its corresponding temperature coefficient $d(\ln <r_2>_0)/dT_K$ for each chain. The results, given as usual in terms of the characteristic ratio $C_n = [r^2]_0/nl^2$, can be interpreted in terms of the extensibility and conformational flexibility of the chains dissolved in a θ solvent. In addition, the results have allowed comparison with experimental values of C_n obtained for a number of organopolysilanes from solution light-scattering measurements.[31]

MM Conformational Energy Calculations. Values of the MM calculated conformational energies and associated torsion angles are summarized in Tables 7 and 8 for the iso and syn forms of the [-SiPhMe-] and [-SiPhH-SiH$_2$-] model compounds, respectively.[20]

On the basis of the results given in Table 7, both iso and syn [-SiPhMe-] show an overwhelmingly strong preference for the _trans_ (planar zigzag) conformation. In fact, given their high relative energy, _gauche_ states of any type or in any sequence would be virtually excluded at room temperature. This result reflects the influence of the steric bulk of the substituents. Specifically, steric congestion renders all but the all-_trans_ conformation highly unfavorable. As expected, $G^{\pm}G^{\mp}$ states give rise to the familiar "pentane effect"[23] and, as such, represent domains of extraordinarily high conformational

MODEL COMPOUND FOR [-SiPhMe-]

ISOTACTIC   SYNDIOTACTIC

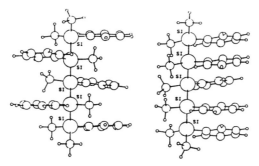

Figure 6.   [-SiPhMe-] Model Compounds used for Molecular
Mechanics Study.[20]

MODEL COMPOUND FOR [SiPhH-SiH₂-]

ISOTACTIC   SYNDIOTACTIC

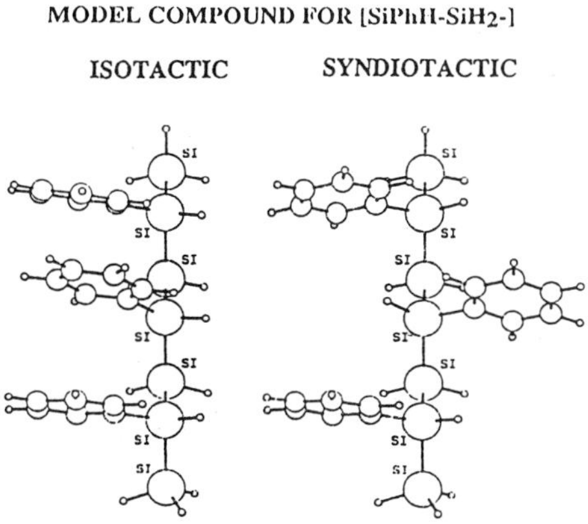

Figure 7.   [-SiPhH-SiH$_2$-] Model Compounds used for Molecular
Mechanics Study.[20]

energy. Thus both the iso- and syn-[-SiPhMe-] model compounds prefer the TT state almost exclusively.

The calculated torsion angles are as one would expect for such a highly congested molecule. Specifically, the _trans_ conformations deviate slightly (ca. 2-5°) from the idealized value of $\phi = 0°$. Again, the _gauche_ states are in all cases at least slightly less twisted (and so less compact) than the idealized $\phi = \pm120°$.

Table 7. Relative Energies E^a and Associated Torsion Angles ϕ for the Iso and Syn Isomeric Forms of the [-SiPhMe-] Model Compound

	iso		syn	
	E^a	ϕ, deg	E^a	ϕ, deg
TT	0.0	±2.2	0.0	±2.9
TG$^+$	5.5	116.3	3.8	116.3
TG$^-$	4.6	-115.6	3.9	-113.5
G$^+$G$^+$	6.2	119.8	5.6	107.3
G$^-$G$^-$	6.6	-112.2	7.5	-118.8
G$^-$G$^+$	>10	b	>10	b

^{a}In kcal mol^{-1}, taken relative to $E = 0.0$ kcal mol^{-1} for the TT ($\phi = 0°$) conformation. ^{b}No local minimum found.

For the [-SiPhH-SiH$_2$-] model compounds, the results[20] in Table 8 reveal sharp differences between the iso and syn forms in terms of conformational preferences. In particular, the iso isomer shows a strong preference for T states over G states in general and for the TT state in particular. However, this preference for T states is less exclusive compared with that described above for [-SiPhMe-]. By comparison, the [-SiPhH-SiH$_2$-] chain is much less congested sterically and hence considerably more flexible. In contrast to its iso form, the syn-[-SiPhH-SiH$_2$-] model compound shows a strong preference for G states in general and for G$^+$G$^+$ (over TT) in particular by roughly the same amount of conformational energy (ca. 2.5 kcal/mol) as the iso form prefers TT over G$^+$G$^+$ states.

The reason for this reversal in conformational preferences can be traced to the observation that both the iso and syn chains appear to prefer conformations in which the phenyl substituents are nearly overlapping. For the iso form this situation is achieved in the TT state, while for the syn form this occurs more nearly in the alternative G states. The phenyl...phenyl interplanar distance is about 3.8Å for both the iso form at TT and the syn form at G^+G^+. This distance represents a near minimum in the potential energy for $C_{Ar}\cdots C_{Ar}$ interactions. Hence, in the case of [-SiPhH-SiH$_2$-], the calculations predict for the iso form an all-<u>trans</u>, planar zigzag, conformation for the syn form sequences of G^+G^+ and TG^+ states corresponding to a helical conformation.

Table 8. Relative Energies E^a and Associated Torsion Angles ϕ for the Iso and Syn Isomeric Forms of the [-SiPhH-SiH$_2$-] Model Compound

	iso		syn	
	E^a	ϕ, deg	E^a	ϕ, deg
TT	0.0	±3.0	0.0	±3.5
TG^+	2.3	120.9	-1.4	107.9
TG^-	1.1	-116.4	-0.67	-101.6
G^+G^+	2.8	121.2	-2.3	120.6
G^-G^-	2.8	-121.95	-2.4	-119.8
G^-G^+	>10[b]	+105.2	>10[b]	+95.2

[a]In kcal mol^{-1}, taken relative to E = 0.0 kcal mol^{-1} for the TT (ϕ = 0°) conformation. [b]No local minimum found.

Considering the calculated torsion angles for [-SiPhH-SiH$_2$-], the TT state is again represented by a pair of nearby minimum energy conformations straddling ϕ = 0°. For the iso form the <u>gauche</u> minima (except of course the highly repulsive $G^{\pm}G^{\mp}$ states) are located very near the standard ϕ = ±120° values. For the syn form, $G^{\pm}G^{\pm}$ states are located very near ϕ = 120°, while $TG^{\pm}$ states are receded from this value.

<u>Calculated Geometrical Parameters</u>. Values of the MM geometry-optimized bond lengths and bond angles for each conformational state are given in Tables 9 and 10, respectively, for the [-SiPhMe-] and [-SiPhH-SiH$_2$-] model compounds. Values of the bond lengths vary only slightly with conformation for both systems. Average values taken from Tables 9 and 10 are 2.35, 1.87, and 1.49Å for the Si-Si, Si-C$_{Me}$, and Si-H bonds, respectively. For Si-C$_{Ar}$ bond lengths, the average value is 1.85Å in Table 9 and 1.88Å in Table 10. This difference is attributable to differences in the Si-related force-field parameters used for the [-SiPhMe-][12] (Table 9) and [-SiPhH-SiH$_2$-][20] (Table 10) model compounds.

Calculated bond angles in general vary with conformation more so than do bond lengths due to the "softer" nature of their deformation energy functions in the force field. This is reflected in Tables 9 and 10 where, for instance, the Si-Si-Si bond angle ranges from 111.6° to 120.2° for [-SiPhMe-]. On the other hand the values of bond angles listed for [-SiPhH-SiH$_2$-] (Table 10) are surprisingly uniform over all conformations, with deviations ranging only about 1-2°. These smaller deviations reflect the less severe steric congestion found in [-SiPhH-SiH$_2$-] relative to that in [-SiPhMe-].[20]

Table 9. Selected Values of the MM Geometry-Optimized Bond Lengths[a] and Bond Angles[b] for the Iso and Syn Forms of the [-SiPhMe-] Model Compounds

	TT		TG$^+$		TG$^-$		G$^+$G$^+$		G$^-$G$^-$	
	iso	syn	iso	syn	iso	syn	iso	syn	iso	syn
bond lengths[a]										
Si-Si	2.354	2.357	2.356	2.353	2.354	2.353	2.361	2.353	2.359	2.366
Si-C$_{Me}$	1.867	1.868	1.868	1.867	1.867	1.868	1.867	1.867	1.867	1.867
Si-C$_{Ar}$	1.851	1.850	1.849	1.850	1.851	1.849	1.852	1.850	1.852	1.853
bond angles[b]										
Si-Si-Si	113.9	111.6	117.7	116.6	116.2	116.5	119.7	120.2	120.0	116.2
Si-Si-C$_{Me}$	107.8	109.6	106.4	107.6	108.7	107.8	108.4	106.5	107.3	107.6
Si-Si-C$_{Ar}$	110.2	108.9	109.2	109.7	107.9	107.6	110.6	108.1	108.8	110.9
C$_{Me}$-Si-C$_{Ar}$	111.2	111.3	111.3	111.6	111.9	111.8	108.9	110.8	109.5	107.5

[a]In units of Angstroms. [b]In units of degrees.

Table 10. Selected Values of the MM Geometry-Optimized Bond Lengths[a] and Bond Angles[b] for the Iso and Syn Forms of the [-SiPhH-SiH$_2$-] Model Compound

	TT		TG$^+$		TG$^-$		G$^+$G$^+$		G$^-$G$^-$	
	iso	syn	iso	syn	iso	syn	iso	syn	iso	syn
bond lengths[a]										
Si-Si	2.340	2.340	2.341	2.340	2.337	2.342	2.338	2.346	2.337	2.341
Si-C$_{Me}$	1.490	1.489	1.489	1.489	1.489	1.489	1.489	1.489	1.489	1.488
Si-C$_{Ar}$	1.877	1.878	1.877	1.877	1.877	1.878	1.877	1.881	1.876	1.879
bond angles[b]										
Si-Si-Si	109.3	110.2	109.1	110.2	111.2	110.1	110.9	107.9	110.9	109.7
Si-Si-C$_{Me}$	108.6	107.9	108.4	108.4	110.1	107.6	109.3	109.5	109.8	109.3
Si-Si-C$_{Ar}$	109.0	109.3	109.4	108.2	108.4	109.4	108.8	108.4	108.4	108.2
C$_{Me}$-Si-C$_{Ar}$	109.2	109.3	109.4	108.2	108.4	109.4	108.8	108.4	108.4	108.2

[a]In units of Angstroms. [b]In units of degrees.

Chain Statistics Calculations. The structural information given in Tables 9 and 10 for the [-SiPhMe-] and [-SiPhH-SiH$_2$-] model compounds, in conjunction with the conformational energy and torsion angle data given in Tables 7 and 8, were used to calculate the characteristic ratio C_n for n = 200, 300 and 400 backbone bonds. The number 400 would correspond to a chain of molecular weight of about 48,000 for [-SiPhMe-] and 27,200 for [-SiPhH-SiH$_2$-]. The associated statistical weight parameters[23] σ, λ, ψ, ψ', ω, and ω' calculated at 25°C for each polymer are listed in Table 11.

Table 11. Values of the Statistical Weight Parameters Computed at 25°C for Iso and Syn Forms of [-SiPhMe-] and [-SiPhH-SiH$_2$-]

	[-SiPhMe-]		[-SiPhH-SiH$_2$-]	
	iso	syn	iso	syn
σ	0.00	0.00	0.17	10.10
λ	0.00	0.00	0.02	3.10
ψ	0.31	0.05	0.48	4.65
ψ'	0.03	0.00	0.05	12.58
$\omega = \omega'$	0.00	0.00	0.00	0.00

For this analysis, values of the backbone Si-Si bond length and Si-Si-Si bond angle were fixed at 2.35Å and 113.9°, respectively, for [-SiPhMe-] and at 2.34Å and 109.2°, respectively, for [-SiPhH-SiH$_2$-]. These values were obtained by taking Boltzmann-factor weighted averages of those corresponding values listed in Tables 9 and 10.

In a similar fashion, torsion angles associated with the T, G$^+$, and G$^-$ states were set at 0°, 116°, and -116° for both iso- and syn- [-SiPhMe-] and at 0°, ±116.4° and 0°, ±120.0° for iso- and syn- [-SiPhH-SiH$_2$-], respectively. These values were chosen similarly based on weighted averages taken from the results of the conformational energy calculations (Tables 7 and 8). For [-SiPhMe-], it should be noted that all T states were set at 0°, as opposed to the actual values found from the energy calculations, largely as a mathematical convenience. Preliminary calculations showed that setting all T states to, for example, ±3° had a negligible effect on the calculated C_n values.

The RIS-calculated[23] values of C_n for n = 200, 300, and 400 and the associated temperature coefficient $d(\ln C_{n=400})dT$ at T_K = 298 K for the chains considered here are listed in Table 12. The essentially rodlike configurations adopted by the [-SiPhMe-] chains, as manifested by their virtual exclusive preference for TT states, give rise to very large and nonconvergent values of C_n. In this context, these chains are more nearly rigid rods than random coils.

In contrast, the $C_{n=400}$ values for [-SiPhH-SiH$_2$-] are nearly convergent. In fact, the value $C_{n=400}$ = 18.6 for the iso form is fully converged. This value reflects a fairly extended chain but one that remains somewhat flexible conformationally owing to the inclusion of sequences such as TTTG$^+$TG$^-$T and TTTG$^+$TTG$^-$. These and related sequences would tend to alter the direction of chain propagation away from the rodlike configurations associated with [-SiPhMe-]. In contrast, the C_n values for the syn form of [-SiPhH-SiH$_2$-] are substantially higher, and at n = 400 are still not yet convergent. These results are consistent with chains having sequences almost exclusively G$^+$G$^+$G$^+$G$^+$ giving rise to an inflexible and highly extended helical configuration. (Note that the associated G$^-$G$^-$G$^-$G$^-$ sequences are energetically and hence statistically much less probable by virtue of the asymmetry of the substituents).

Calculated values of d(ln $C_{n=400}$)dT at 25°C (Table 12) are all negative in sign and uniformly small.[20] The negative sign is interpreted to mean that an increase in temperature causes conformational transitions away from the low-energy conformational states of high spatial extension to the alternative high-energy, more conformationally compact, states.[23] Thus for each polymer an increase in temperature results in a decrease in the unperturbed dimensions of the chains. The small magnitude of the d(ln $C_{n=400}$)/dT values listed in Table 12 indicates that each polymer exhibits a strong preference for a particular conformational state and hence that each is quite inflexible.

Table 12. Calculated Values of the Characteristic Ratios $C_{n=200}$, $C_{n=300}$, and $C_{n=400}$ and Associated Temperature Coefficient d(ln $C_{n=400}$)/dT at 25°C for [-SiPhMe-] and [-SiPhH-SiH$_2$-]

	$C_{n=200}$	$C_{n=300}$	$C_{n=400}$	d(ln $C_{n=400}$)/dT[a]
[-SiPhMe-]				
iso	140	209	274	-7.8×10^{-4}
syn	130	188	240	-3.4×10^{-3}
[-SiPhH-SiH$_2$-]				
iso	17.9	18.4	18.6	-4.5×10^{-3}
syn	50.4	67.0	78.8	-7.9×10^{-3}

[a] In units of degrees^{-1}.

Comparison with Experiment. These calculated C_n values are qualitatively in agreement with the preliminary experimental values of Cotts, et al.,[31] obtained from dilute solution light-scattering

measurements. They found a value $C_{n=\infty} = 64 \pm 20$ for a presumably
atactic [-SiPhMe-] sample having a weight average of 400 ± 75 backbone
chain atoms. While this experimental value is substantially smaller
than our calculated values of 240-270, the basic picture of a nearly
fully extended chain persists. By comparison, the $C_{n=\infty}$ values reported
by Cotts, et al.,[31] for a number of presumably non-stereoregular dialkyl
polysilanes fall in the 14-30 range. Given these experimental values,
the calculated values of 19 and 79 obtained, respectively, for iso- and
syn-[-SiPhH-SiH$_2$-] appear reasonable.[2,32]

Some Recent Studies: Models for the para-Alkoxy Diaryl Substituted Polysilanes

Recently Miller and Sooriyakumaran prepared and characterized the
first soluble poly(bis(p-(alkoxyphenyl)silanes).[7b] Whereas the
poly(bis(p-alkyl-phenyl)silanes) (and their _meta_ analogues) absorb near
$\lambda_{max} = 400$ nm,[33] the corresponding poly(bis(p-alkoxyphenyl)silanes)
absorb near $\lambda_{max} = 325$ nm.[7b] This effect seems to be conformational
since at elevated temperatures the 325 nm band shifts to 400 nm. It is
mysterious that an apparently subtle structural perturbation would
produce such a dramatic effect on the electronic spectra. In fact, the
electron-donating effect of the p-alkoxy group might suggest a $\lambda_{max} >$
400 nm, not less. Miller and Sooriyakumaran[7b] have proposed that
twisted (i.e., _gauche_) conformations arise in these polymers to avoid
unfavorable dipole interactions arising in the _trans_ form between
neighboring para-substituted ether groups.

Both MM and MO calculations aimed at elucidating the origins of
this unusual behavior are currently in progress.[34] Even the smallest of
the model compounds representative of these polymer chains are found to
contain 87 "heavy" (i.e., non-hydrogen) atoms and 77 hydrogen atoms.
Clearly, for such large molecular systems an _ab initio_ MO analysis would
be prohibitive. Even a MNDO-type semiempirical MO calculation imposes
impressive demands on most computer systems. Nevertheless, the unusual
phenomena observed in these polymers are almost certainly electronic in
nature, so use of MO approaches is essential for a full delineation.

Recent preliminary MM calculations[35] have contributed support for
the explanation proposed by Miller and Sooriyakumaran.[7b] For these
highly illustrative studies, model compounds were devised so as to
conserve computer resources yet preserve the essential features of the
parent polymers under study. The representative model compounds thus

selected were $CH_3[Si(CH_3)_2]_7CH_3$ for poly(bis(p-methylphenyl)silane) (PBMS) and $CH_3[Si(OCH_3)_2]_7CH_3$ for poly(bis(p-methoxyphenyl)silane) (PBMOS). The essential simplification embodied in each model compound is the elimination of the phenyl groups in the substituents. Admittedly such an abridgment is not trivial. Nevertheless our purpose was to eliminate the phenyl group serving in its role as a spacer between the chain backbone and would-be para substituent. Hence, the resulting effect will be to bring the substituent dipoles into closer proximity of the chain backbone and thus to maximize any impact their interactions might have on conformation.

For each model compound, MM conformational energies E were calculated vs. torsion rotations ϕ_1 and ϕ_2 about adjacent centrally-located Si-Si bonds. Values of E were computed for each pair of ϕ_1 and ϕ_2 values varied in increments of $10°$ over the range 0-360°. Energy minimization was carried out at each grid point on the ϕ_1, ϕ_2 surface with all internal degrees of freedom permitted to relax while holding ϕ_1 and ϕ_2 fixed. The MM2(87) program was employed,[25] and the dielectric constant was set to 3.0 to represent a nonpolar solvent.

The E vs. ϕ_1, ϕ_2 scans are plotted as contour energy maps shown in Figure 8 for $CH_3[Si(CH_3)_2]_7CH_3$ and in Figure 9 for $CH_3[Si(OCH_3)_2]_7CH_3$. For $CH_3[Si(CH_3)_2]_7CH_3$ (Figure 8) the global minimum is located at the trans, trans (TT) conformation ($\phi_1 = 180°$, $\phi_2 = 180°$). The alternative $TG^\pm$ and $G^\pm T$ states are on average 0.5 kcal mol^{-1} higher in energy than TT. Thus the $CH_3[Si(CH_3)_2]_7CH_3$ model compound is predicted to assume an all-trans conformation at room temperature. This is consistent with the all-trans conformation being attributed to the red-shifted λ_{max} in the electronic spectrum of PBMS.[7b,33]

The analogous energy scan for the $CH_3[Si(OCH_3)_2]_7CH_3$ model compound (Figure 9) exhibits some significant differences from that found above for $CH_3[Si(CH_3)_2]_7CH_3$. In this case the $TG^\pm$ and $G^\pm T$ states correspond to the minimum energy.[35] The alternative TT state is thus an average 0.4 kcal mol^{-1} higher in energy than the $TG^\pm$ and $G^\pm T$ states. Therefore a preference for gauche states over the trans state is predicted for $CH_3[Si(OCH_3)_2]_7CH_3$. Again, this result is consistent with deviations from an all-trans conformation in PBMOS as suggested by Miller and Sooriyakumaran to explain its blue-shifted λ_{max} and its thermochromism.[7b]

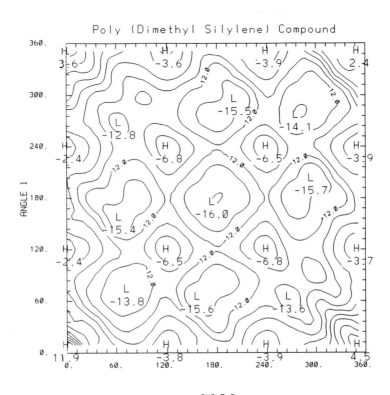

Figure 8. Conformational Energy Map for the PBMS Model Compound, as derived from Geometry-optimized Molecular Mechanics Calculations.[35] The <u>trans</u> conformation corresponds to $\phi = 180°$.

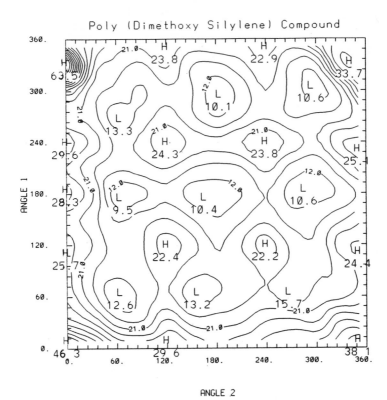

Figure 9. Conformational Energy Map for the PBMOS Model Compound, as derived from Geometry-optimized Molecular Mechanics Calculations.[35] The _trans_ conformation corresponds to $\phi = 180°$.

Molecular computer graphics studies suggest that the preference for _gauche_ states over the _trans_ state in $CH_3[Si(OCH_3)_2]_7CH_3$ is not steric in origin, thus implying that electrostatics, and in particular dipole...dipole interactions, may play an important role. These initial studies further encourage the need for more detailed theoretical analysis.[34]

ACKNOWLEDGMENT

Acknowledgment is made to the donors of The Petroleum Research Fund, administered by the American Chemical Society, for partial support of this research. The author also acknowledges Biosym Technologies, Inc. (San Diego, CA) for supplying the Insight/Discover program for molecular modeling. The author also wishes to acknowledge Mr. Qamar Abbassi for his assistance in constructing Figures 8 and 9.

REFERENCES

1.   R.J. West, J. Organomet. Chem., _300_, 327(1986), and references cited therein.
2.   L.A. Harrah and J.M. Zeigler, Macromolecules, _20_, 2039(1987).

222

3.   For recent reviews, see: (a) J.M. Zeigler, Synthetic Metals, 28, C581(1989); (b) R.D. Miller and J. Michl, Chem. Rev., 89, 1359(1989).

4.   R.D. Miller, D. Hofer, J. Rabolt, R. Sooriyakumaran, C.G. Willson, G.N. Fickes, J.E. Guillet and J. Moore, "Polymers for High Technology: Electronics and Photonics," (M.J. Bowden and S.R. Turner, Eds.), ACS Symp. Ser. 346, American Chemical Society, Washington, D.C., 1987, page 170.

5.   A number of relevant articles can be found in "Electronic and Photonic Applications of Polymers," (M.J. Bowden and S.R. Turner, Eds.), Advances in Chemistry Series 218, American Chemical Society, Washington, D.C., 1988.

6.   R.D. Miller and R. Sooriyakumaran, J. Polym. Sci., Part A, Polym. Chem., 25, 111(1987).

7.   (a) J.F. Rabolt, D. Hofer, R.D. Miller and G.N. Fickes, Macromolecules, 19, 611(1986); (b) R.D. Miller and R. Sooriyakumaran, ibid., 21, 3120(1988).

8.   H. Kuzmany, J.F. Rabolt, B.L. Farmer and R.D. Miller, J. Chem. Phys., 85, 7413(1986).

9.   R.D. Miller, B.L. Farmer, W. Fleming, R. Sooriyakumaran and J. Rabolt, J. Am. Chem. Soc., 109, 2509(1987).

10.   F.C. Schilling, A.J. Lovinger, J.M. Zeigler, D.D. Davis and F.A. Bovey, Macromolecules, 22, 3055(1989).

11.   W.J. Welsh, K. Beshak, J.L. Ackerman, J.E. Mark, L.D. David and R. West, Polym. Prepr. (Am. Chem. Soc., Div. of Polym. Chem.), 24(1), 131(1983).

12.   J.R. Damewood, Jr. and R. West, Macromolecules, 18, 159(1985).

13.   J.R. Damewood, Jr., Macromolecules, 18, 1793(1985).

14.   K. Takeda, H. Teramae and N. Matsumoto, J. Am. Chem. Soc., 108, 8186(1986).

15.   W.J. Welsh, L. Debolt and J.E. Mark, Macromolecules, 19, 2978(1986).

16.   B.L. Farmer, J.F. Rabolt and R.D. Miller, Macromolecules, 20, 1169(1987).

17.   J.V. Ortiz and J.W. Mintmire, J. Am. Chem. Soc., 110, 4522(1987).

18.   P.R. Sundararajan, Macromolecules, 21, 1256(1988).

19.   J.W. Mintmire and J.V. Ortiz, Macromolecules, 21, 1189(1988).

20.   W.J. Welsh, J.R. Damewood, Jr., and R. West, Macromolecules, 22, 2947(1989).

21.   W.J. Welsh and W.D. Johnson, Macromolecules, 23, 1881 (1990); ibid., PMSE J., 61, 222 (1989).

22.   R.W. Bigelow and K.M. McGrane, J. Polym. Sci.: Part B: Polym. Phys., 24, 1233(1986); R.W. Bigelow, Organometallics, 5, 1502(1986).

23.   P.J. Flory, "Statistical Mechanics of Chain Molecules," Interscience, New York, 1969.

24.   W.L. Jorgensen, J. Am. Chem. Soc., 103, 677(1981); ibid., J. Chem. Phys., 87, 5304(1983).

25.   MM2(87) Force Field Parameters, courtesy of Prof. N.L. Allinger, Department of Chemistry, University of Georgia, Athens, GA, 30602; U. Burkert and N.L. Allinger, "Molecular Mechanics", ACS Monograph 177, American Chemical Society, Washington, D.C., 1982.

26.   A. Abe, R.L. Jernigan and P.J. Flory, J. Am. Chem. Soc., 88, 631(1966).

27.   P.R. Sundararajan and P.J. Flory, J. Am. Chem. Soc., 96, 5025(1974).

28.   V. Crescenzi and P.J. Flory, J. Am. Chem. Soc., 86, 141(1964).

29.   M.J.S. Dewar, E.G. Zoebisch, E.F. Healy and J.J.P. Stewart, J. Am. Chem. Soc., 107, 3902(1985).

30.   M.J.S. Dewar and W. Thiel, J. Am. Chem. Soc., 99, 4899(1977).

31. P.M. Cotts, R.D. Miller, P.T. Trefonas, III, R. West and G.N. Fickes, Macromolecules, 20, 1046(1987); P.M. Cotts, Proc. ACS Div. Polym. Mater., Sci. Eng. (PMSE), 53, 336(1985).

32. L.A. Harrah and J.M. Zeigler, "Photophysics of Polymers," (C.E. Hoyle and J.M. Torkelson, Eds.), ACS Symp. Ser. 358, American Chemical Society, Washington, D.C., 1987.

33. R.D. Miller and R. Sooriyakumaran, J. Polym. Sci.: Polym. Lett. Ed., 25, 321(1987).

34. P. Ritter and W.J. Welsh, work in progress.

35. Y. Yang and W.J. Welsh, unpublished results.

ATOMIC OXYGEN RESISTANT POLY(CARBORANE SILOXANE) COATINGS

Joseph Kulig, Virginia Jefferis, David Schwamm and Morton Litt

Department of Macromolecular Science and Engineering
Case Western Reserve University
Cleveland, Ohio 44106

ABSTRACT

Atomic oxygen degradation of materials poses a severe threat to low earth orbit structures like the space station. We report here the investigations of new self healing protective coating—poly(carborane siloxanes). On exposure to atomic oxygen the upper layer of the coating oxidizes to a glassy surface which offers excellent protection to Kapton and silver substrates.

INTRODUCTION

Atomic oxygen is formed from the ultraviolet light dissociation of molecular oxygen in the upper atmosphere(1). It is the predominant neutral species in the Low Earth Orbit Environment (LEO)(2). Surface recession, changes in optical properties, and morphological changes of material by atomic oxygen were first observed on early shuttle flights. Later shuttle experiments determined the susceptibility of various materials to attack. The erosion yield or surface recession per atom of oxygen of various materials is shown in Table 1. Some materials like gold or aluminum are resistant to atomic oxygen attack. Organic materials like Kapton or graphite epoxy composites, and some metals, notably silver, degrade rapidly. These materials cannot be used on the space station without a suitable protective coating. It is projected, for example, that if unprotected, a 0.051 mm (2 mil.) thick Kapton solar array blanket would be completely eroded within six months(3).

Table 1. Erosion yield of materials exposed to atomic oxygen in LEO

material	Erosion Yield x10^{-24}cm^3/ O atom
aluminum	0.0
silver	10.5
gold	0.0
Kapton	3.0
polyethylene	3.3
epoxy	1.7
carbon	1.2

Inorganic and Metal-Containing Polymeric Materials
Edited by J. Sheats *et al.,* Plenum Press, New York, 1990

225

Our strategy is to use semiinorganic polymers which would form glassy protective coatings on exposure to atomic oxygen. Poly(carborane siloxanes), because of their high boron content, are ideal in this regard. These polymers were first developed as high temperature elastomers. They could be modified for our uses by attaching easily hydrolysable groups which cure on exposure to atmospheric moisture (RTV) to generate a crosslinked rubbery polymer. They are structurally similar to siloxanes except that an oxygen atom in the backbone is periodically replaced by a carborane functionality ($B_{10}H_{10}C_2$). The aromatic nature of the carborane unit stabilizes the silicon oxygen bond on adjacent silicon atoms(4). This enhances the thermal and oxidative resistance compared to silicones(5). On oxidation, these polymers could be anticipated to form a borosilicate glass with a net weight increase of 150% largely due to the incorporation of 15 oxygen atoms into the glass per carborane group. The incorporation of oxygen into the coating would counteract the shrinkage due to density changes as the glassy layer forms. This would alleviate stresses in the coating which could induce cracks, causing undercutting. A rubbery layer between the glass will remain because only the top layer would oxidize. This would allow the coating to heal if the glass layer were damaged and could improve the adhesion of the glass to the substrate.

EXPERIMENTAL

Materials

Silanol terminated Dexsil from Dexsil Corporation (One Hampten Park Drive Hampten, CT 06517) was used as received. 1,7–Bis(hydroxydimethyl–silyl)–m–carborane, also from Dexsil Corp., was recrystallized from high boiling petroleum ether (m.p. 98.5 − 100.2 °C, lit. 98 − 99.5 °C)(6). Methyl– triacetoxysilane from Petrarch Systems was distilled using a spinning band column (b.p. 88 − 92 °C, 1.5 torr). Toluene was dried over 5 Å molecular sieves for at least 24 hours and then distilled before using.

Equipment

Molecular weight was measured by Water's Gel Permeation Chromatograph using a differential refractometer as a detector. The solvent, tetrahydrofuran, was passed through a precolumn (PLgel 5μ, 100 Å) and two PLgel columns (5μ, 10^4 Å and 500 Å) at a flow rate of 1.0 ml/min.. The columns were calibrated with polystyrene and o–dichlorobenzene was used as a standard.

Atomic oxygen exposure was performed using a radio–frequency–generated plasma. A Structure Probe Incorporated Plasma Prep. II, 13.56 MHz RF plasma asher was used for all experiments. The plasma pressure was kept between 10 − 30 millitorr. Since air was used as the feed gas the plasma was mostly ionized nitrogen, but atomic oxygen appears to be the reacting species(3). The asher has been found to be a good test of survivability of materials even though the flux of oxygen is higher and the kinetic energy is lower than that of LEO.

Mass loss was measured using a Sartorius R160d balance, 0.00001g ± 0.00002g. The spectral reflectance measurements were performed using a Perkin Elmer U.V./visible /near–infrared spectrophotometer with an integrating sphere attachment. The chemical composition of the coatings were determined by electron spectroscopy for chemical analysis(ESCA) using a Perkin Elmer model PH1 5400 ESCA with MgKα x–rays. The spot size was 1 mm² and the angle of the analyzer was set at 45°.

Synthesis

Two types of RTV coating materials were prepared. The first was a lightly crosslinked system made by endcapping silanol terminated Dexsil with methyl–triacetoxy silane, P1. Dexsil(1.104g) and methyltriacetoxysilane(0.372g), mole ratio of 1 to 18, were refluxed in 5g of toluene for four hours; acetic acid was removed by fractional distillation. Ten milliliters of additional solvent was added and the reaction continued for two more hours. The solution of P1 was stored under dry air and refrigerated.

The second material was a more highly functionalized poly(carborane siloxane), P2. This was prepared by reacting a five mole percent excess of methyltriacetoxysilane (1.004g, 0.004563 moles) and 1,7 bis(hydroxydimethyl silyl)–m–carborane(1.128g, 0.003856 moles) in 39 g. of toluene for five hours at reflux. Acetic acid was distilled off using a fractional distillation column. The progress of the reaction was monitored by high pressure gel permeation chromatography with tetrahydrofuran as solvent. The reaction was stopped when the polystyrene equivalent weight average molecular weight of the prepolymer reached 8,000. P2 was stored like P1.

Preparation of the Coatings

The coatings were prepared under dry nitrogen in a glovebox to prevent premature curing. In the case of P1, excess solvent was first removed with a rotovapor. Dibutyl tin dioctanoate catalyst(0.0016g in toluene solution) was added to the solution which was filtered through a 0.5 micrometer filter. The P1 solution, concentration 10%, was applied onto glass slides using a casting knife. After exposure to atmospheric moisture for two days the film was cured.

Spin coating was used for preparing thin coatings on Kapton or silver substrates. In the case of P2, both 5 and 25 weight percent solutions were spin coated onto Kapton. The solution was first applied to the whole substrate surface; the excess removed by spinning. Subsequent exposure to moisture in the air for one week at room temperature cured the prepolymer to the final clear coating. The average thickness determined by weight increase was 1,100nm for the 25% solution and 100nm for the 5% solution. Since only one side was coated, for atomic oxygen exposure two samples were sandwiched together with double stick Kapton tape so only coated surfaces were exposed. For the silver specimens, an 800 Å thick layer of silver was evaporated onto one inch square quartz slides. The coating solution concentration was 1% and was spin coated like the Kapton samples, thickness of coating 700 nm. These were also cured for one week.

Exposure to Atomic Oxygen

P1 coated glass samples and P2 coated silver slides were dried under vacuum for 24 hours at 25 °C and 5 mm; the coated Kapton samples were dried for 72 hours. Drying assured that the weight loss could not be due to further loss of volatiles. Then the coatings were exposed to atomic oxygen in a 13.56 MHz RF air plasma asher at a pressure of 10 − 30 millitorr. In all experiments, whether with Kapton or silver substrates, an unprotected Kapton sample was exposed as a control to determine the atomic oxygen flux. The mass loss of the Kapton samples was periodically measured using a Sartorius R160d balance; exposure to air during weighing was less than five minutes to prevent rehydration.

RESULTS AND DISCUSSION

P1, the acetoxy end capped Dexsil polymer, reacts on exposure to moisture to form acetic acid and silanol end groups which self condense. The coatings were not transparent because some crystallization occurred before cure was completed. To correct for this, a highly functionalized coating was synthesized by reaction of methyltriacetoxy silane and 1,7 bis(hydroxydimethylsilyl)–m–carborane which produced a prepolymer with one acetoxy crosslinking site per repeat (Scheme 1).

$$AcO\left[\begin{matrix}CH_3 \\ Si-CB_{10}H_{10}C-\\ CH_3\end{matrix}\begin{matrix}CH_3 \\ Si-O-\\ CH_3\end{matrix}\begin{matrix}OAc \\ Si-O\\ CH_3\end{matrix}\right]_n-Ac \quad AcO\left[\begin{matrix}CH_3 \\ Si-C_{10}H_{10}C-\\ CH_3\end{matrix}\begin{matrix}CH_3 \\ Si-O-\\ CH_3\end{matrix}\begin{matrix}CH_3 \\ Si-O\\ CH_3\end{matrix}\right]_n-Ac$$

P2   P1

Scheme 1

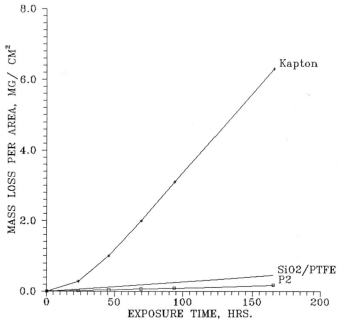

Fig. 1. Mass loss versus exposure time for unprotected and P2 protected Kapton.

The solubility of P2 and its narrow molecular weight distribution, $\overline{M}w/ \overline{M}n =$ 1.9, indicate that the steric hindrance prevented the third acetoxy group from reacting. Coatings made from this prepolymer formed clear films.

After curing, the samples were exposed to air plasma in the RF asher. Crosslinked P1 coated glass slides showed no weight loss after six day of exposure. The highly crosslinked P2 coating on Kapton was also exposed to atomic oxygen for one week, a flux of atomic oxygen equivalent to a 5 year space exposure. The coated sample showed no macroscopic change in appearance and lost very little mass (Fig. 1), whereas the 5 mil thick control became opaque and lost three quarters of its initial weight. The performance of a 8% PTFE 92% SiO_2 ion beam deposited coating is shown for comparison.

The exposed thicker P2 coating, 1,100nm, appeared uniformly waffled when observed by SEM (Fig. 2). The P1 coatings had the same appearance. The waffled morphology is similar to swelling behavior of gel films when they incorporate solvent(7). Thus the waffling confirms that the material is increasing in volume probably by incorporating oxygen. The waffling in the carborane polymers is thought to be due to the expansion of the film during initial oxidation as atomic oxygen inserts itself between silicon–carbon bonds, forming methoxy groups. The effect of coating thickness on waffling can be seen in inhomogeneous region of the films when viewed with interference microscopy. The thicker regions, shown by the greater number of Newton fringes, are heavily waffled whereas thin regions are smooth (Fig. 3). The transition from a smooth surface to a waffled region occurs between 750–1000nm thickness. This transition is also seen in the SEM where the waffling pitch changes with thickness (Fig. 4). The thinner 100nm P2 coating was smooth except where dust particles acting as stress concentrators induced cracks.

A smooth flat surface layer free from cracks was generally discernible above the waffling. When the coating was scraped after ashing and viewed edge on, a 0.1 micron thick glassy layer was seen (Fig 5). It is anticipated that the unreacted sublayer of the

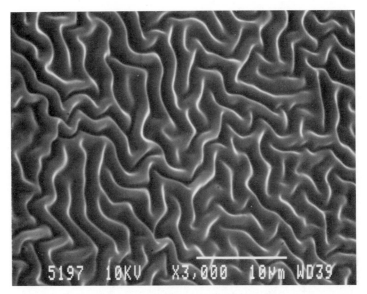

Fig. 2.   P2 coating on Kapton after one week exposure
               to atomic oxygen.

Fig. 3.   Phase contrast optical micrograph of P2 coating;
               $\lambda = 500$ nm.

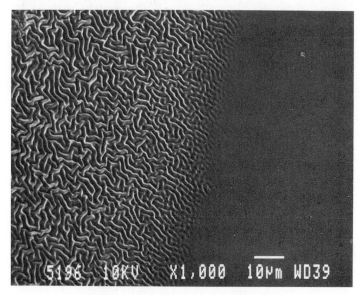

Fig. 4.   SEM photograph of P2 coating on Kapton showing
               waffling dependence on thickness.

Fig 5.   Edge view of oxidized coating.

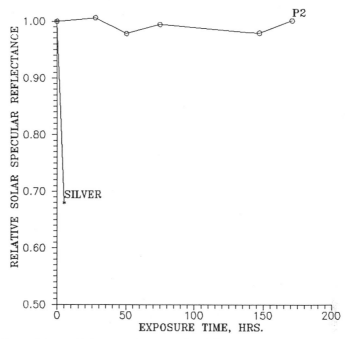

Fig. 6. Relative solar specular reflectance versus exposure to atomic oxygen for protected and unprotected silver mirrors.

coating would oxidize if the glassy region were damaged, enabling the coating to self heal. Electron Spectroscopy for Chemical Analysis, ESCA confirmed the formation of a glass. The carbon content was low as expected, but both the P1 and P2 coating surfaces were anomalously low in boron, see Table 2. However, ESCA depth profiling (to be published later) shows that the boron loss is limited to the top $300 - 400$ Å of the glassy layer. Despite the boron loss, excellent protection of silver and Kapton substrates was obtained.

Table 2. ESCA analysis of the coating surface after ashing

	Observed	Expected
P1	$Si_3C_{0.77}O_{6.4}B_{0.34}$	$Si_3O_{21}B_{10}$
P2	$Si_3C_{0.90}O_{7.5}B_{0.35}$	$Si_3O_{21}B_{10}$

Silver, because of its high solar specular reflectance; is the best candidate for mirror surfaces in the space station solar dynamic power system[8]. Silver is also the most reactive material known toward atomic oxygen. Uncoated silver rapidly degrades on exposure (Fig. 6). P2 coated silver mirrors in contrast, show only a slight loss of solar specular reflectance on exposure of up to one week. The fractional loss of the solar reflectance extrapolated to 1000 hours of ashing time (about forty years in LEO) was only 0.025.

Kapton's high specific strength and good thermal stability make it good space structural material. Space station designs envision large Kapton blankets as supports for photovoltaic arrays. Erosion of the Kapton would cause texturing of the surface from atomic oxygen attack which would interfere with the blankets' ability to act as a radiative heat sink and eventually would cause structural failure. Both P1 and P2 provided good protection for Kapton. The P2 coated Kapton substrates had very low mass loss on exposure. The ashing rate of Kapton was reduced from $1.5 \times 10^{-8} cm^2/$ sec. to $2.5 \times 10^{-10} cm^2/$ sec. This is half that observed for the best commercially available coating (3), ion beam deposited 8% PTFE and 92% SiO_2.

SUMMARY

Poly(carborane siloxane) coatings are being investigated as protective coatings against atomic oxygen for space structures in low earth orbit like the space station. These materials oxidize on exposure to atomic oxygen to form a protective glassy layer. The bulk of the coating is unaffected, while a glassy surface layer can be seen by SEM. ESCA analysis confirmed that this layer was a borosilicate glass, but the boron content was lower at the surface then expected.

The coatings provided excellent protection to various substrates. P2 protected silver mirror was estimated to lose only 2.5 % of its initial solar reflectance when extrapolated to 1000 hours ashing time. For the poly(carborane siloxane) P2 coating, the mass loss rate of Kapton was reduced to half that of the best commercial coating.

ACKNOWLEDGEMENTS

The assistance and support of NASA Lewis Research Center and particularly the help of Bruce Banks and Sharon Ruteledge is gratefully appreciated. This work was supported by the NASA Center for the Commercial Developement of Space, Case Western Reserve University.

REFERENCES

1. G. Sjolander and L. Bareiss,"Martin Marietta Atomic Oxygen Beam Facility, 18th International SAMPE Technical Conference,"(J. Hoggah, and J. Johnson, eds.), SAMPE Azusa, CA., 722 (1986)
2. L. Teichman and Stein(eds.),"NASA/SDIO Space Environmental Effects on Materials Workshop," NASA Conference Publication 3035, 1988, pages 204–208
3. B. Banks et. al. "An Evaluation of Candidate Oxidation Resistant Materials for Space Applications in the LEO", NASA Technical Memorandum 100122, 1986, page 2
4. H. Dievich,"Icosaheral Carboranes", Die Makromolekulare Chemie, 175, 425 (1974)
5. E. Peter,"Poly(dodecarborane–siloxanes), Macromolecular Science, Review Macromolecular Chemistry, C17(2), 173 (1978)
6. S. Papetti and T. Heying,"A New Series of Organoboranes. VI. the Synthesis and Reactions of Some Silyl Neocarboranes", Inorganic Chemistry, vol. 3, no. 10, 1449(1964)
7. T. Tanaka et. al. "Mechanical Instability of Gels at the Phase Transition," Nature, 325, 26(1987)
8. D. Gaulino et. al. "Oxidation Resistant Reflective Surfaces for Solar Dynamic Power Generation in Near Earth Orbit,"NASA Technical Memorandum 88865, 1986

THE SYNTHESIS OF SILICON CONTAINING POLYMERS VIA
CARBANION CHEMISTRY

James E. Loftus[1,2] and Bernard Gordon III

Department of Materials Science and Engineering, Polymer Science Program
Penn State University, University Park, PA 16802

ABSTRACT

We have synthesized some novel silarylene and silarylenesiloxane polymers via dicarbanions prepared from Lochmann's base. Lochmann's base, a powerful metalating reagent composed of equimolar amounts of n-butyl lithium and potassium t-butoxide in a hydrocarbon, has been used to dimetalate compounds such as m-xylene, 4,4'-dimethylbiphenyl, and 2,3-dimethyl-1,3-butadiene in good yields. In this work the dicarbanion of m-xylene and 4,4'-dimethylbiphenyl have been used to prepare silicon containing monomers and polymers by two different routes. The first route involves a 1:1 condensation reaction between the dicarbanion and a dichlorodiorganosilane to produce a condensation polymer. The second route involves reaction of the dicarbanion with a chlorodiorganosilane which is then converted to a bis(silanol) and then polymerized. Spectroscopic as well as thermal characterization will be presented on the polymers which have been described.

INTRODUCTION

There is a driving interest in silicon containing polymers due, in large part, to the unique properties that these materials exhibit. There are a wide variety of polymers which contain silicon atoms both in the backbone and in pendent groups. The vast majority of the silicon containing polymers which are known come in the form of silicones which are characterized by the siloxane linkage, Si-O-Si, in the backbone. There are other silicon containing polymers which are known that do not have the siloxane linkage in the backbone such as poly(silanes), characterized by the presence of a Si-Si linkage, and the poly(carbosilanes), characterized by the presence of a Si-C linkage. The work presented here is focused on silicon containing polymers with silicon-carbon and carbon-carbon backbone bonds, and with siloxane, silicon-carbon and carbon-carbon, backbone bonds.

Inorganic and Metal-Containing Polymeric Materials
Edited by J. Sheats *et al.*, Plenum Press, New York, 1990

Our group has been interested in utilizing carbanion chemistry in polymer synthesis for some time now. In this paper we present some novel chemistry to prepare polymers which contain silicon atoms in the polymer backbone. The carbanions which have been used in this study are benzylic stabilized dicarbanions which are prepared using strong base chemistry. There are essentially two polymerization schemes which were examined for preparing these polymers. The first scheme uses a polycondensation reaction between a diorganodichlorosilane and a dicarbanion. The second scheme requires the preparation of a bis(silanol), prepared from a dicarbanion, which can subsequently undergo a homocondensation reaction to form polymer. The Results and Discussion section will describe the basic synthesis of these materials and some of the physical properties associated with them.

There has been a fair amount of work done the synthesis and characterization of silarylene type polymers. Many groups have looked at the interesting chemistry involved in the preparation of these materials as well as the novel material properties which these polymers exhibit. Before discussing our work, a general overview of the synthesis and properties of some interesting silarylene and silarylenesiloxane polymers will be presented.

BACKGROUND

Silarylene Polymers. Silarylene polymers can be best described as those polymers which contain silicon atoms and phenyl units in the polymer backbone. Perhaps the best example, and the one which has been given the most scrutiny, is poly(tetramethyl-*p*-silphenylenesiloxane), (1). The major reason which has been cited for incorporating arylene units into this type of polymer has been to increase the thermal stability of the material while maintaining the low temperature flexibility which is associated with siloxane polymers.[3,4,5]

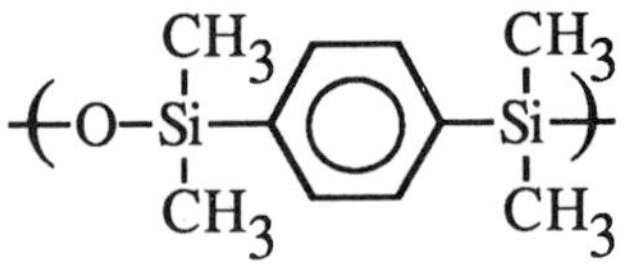

1

Poly(tetramethyl-*p*-silphenylenesiloxane) , (1), referred to as TMPS, was originally synthesized in the early 1950's, but was not routinely prepared in high molecular weight and in high yield until the work of Merker and Scott in 1964.[6] Their synthetic scheme, shown below, began with the preparation of *p*-bis(dimethylhydrogensilyl)benzene, (2). Upon reaction with a strong base and liberation of hydrogen, 2 was hydrolyzed to obtain the *p*-bis-(dimethylhydroxysilyl)benzene, (3). Careful purification of 3 and subsequent condensation in benzene with azeotropic removal of water, resulted in a tough fiber forming polymer, 1, which had a viscosity average molecular weight of several hundred thousand.

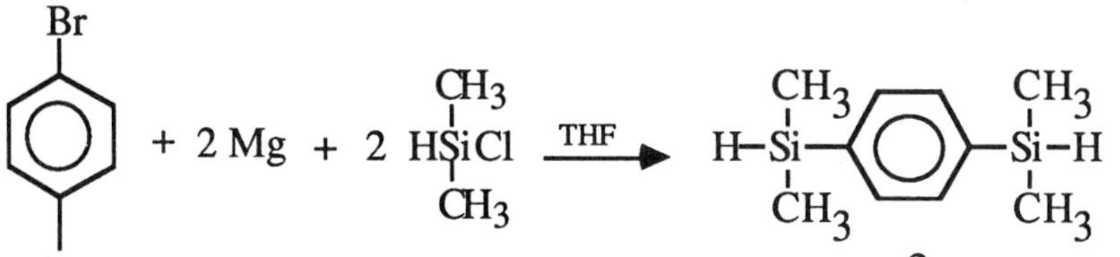

Br—C6H4—Br + 2 Mg + 2 HSi(CH3)2Cl →(THF) H-Si(CH3)2—C6H4—Si(CH3)2-H   2

2 →[1.) NaOCH2CH3 / HOCH2CH3; 2.) NaOH / H2O; 3.) KH2PO4 / H2O 0°C] HO-Si(CH3)2—C6H4—Si(CH3)2-OH   3

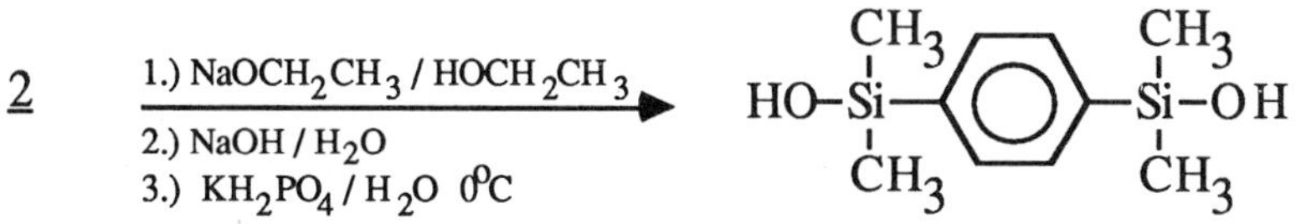

Silarylene Polymers Containing Alternating Siloxane Units. As noted in the previous section, synthesis of poly(tetramethyl-*p*-silphenylenesiloxane), 1, generated a tremendous amount of interest in the synthesis as well as the characterization of this polymer. With the introduction of almost any unique polymer system comes the inevitable increase in research activity surrounding similar polymers. Such was the case with polymers similar to 1.

Soon after Merker and Scott [6] synthesized polymer 1 in high molecular weight, they sought copolymers of 1 and dimethylsiloxane prepared by co-condensation of 3 with various molecular weight α,ω-dihydroxydimethylsiloxanes.[7] Unfortunately this method only leads to random and block copolymers of 1 and dimethylsiloxane. At that time other researchers had joined the pursuit of silarylenesiloxane polymers, but ones with more regular structures.

Since 1967 a host of different silarylenesiloxane polymers have been prepared, most of which have the general structure shown below. Breed and co-workers[7] were among the

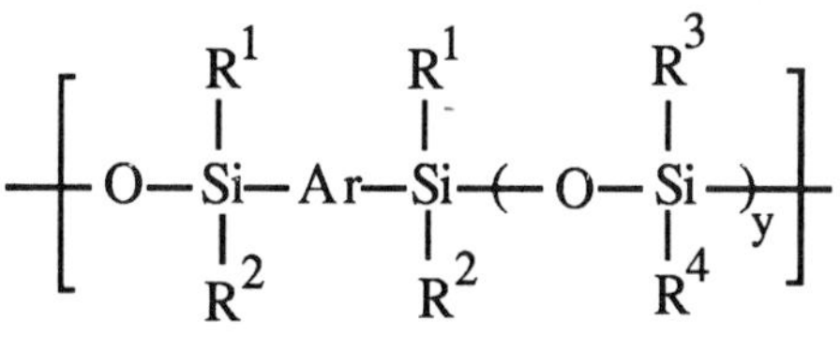

$$\left[O-\underset{R^2}{\overset{R^1}{Si}}-Ar-\underset{R^2}{\overset{R^1}{Si}}\left(O-\underset{R^4}{\overset{R^3}{Si}} \right)_y \right]$$

4

first to look at polymers with general structures like 4. They found an interesting method of preparing polymers of this nature by condensing bis(dialkylarylsilanols), similar in structure to 3, with N-methyl-siloxazanes. Cyclic siloxazanes were used because of their ease of synthesis and enhanced hydrolytic stability over simple diaminosilanes.[7] A typical polymer synthesis using these materials is shown below. Polymer <u>5</u> can be described in terms of <u>4</u>

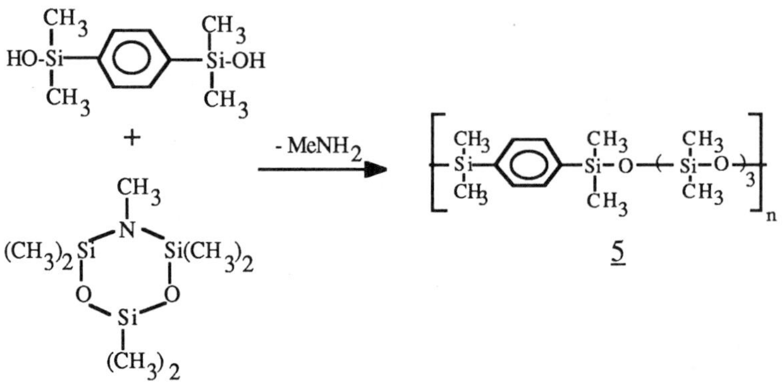

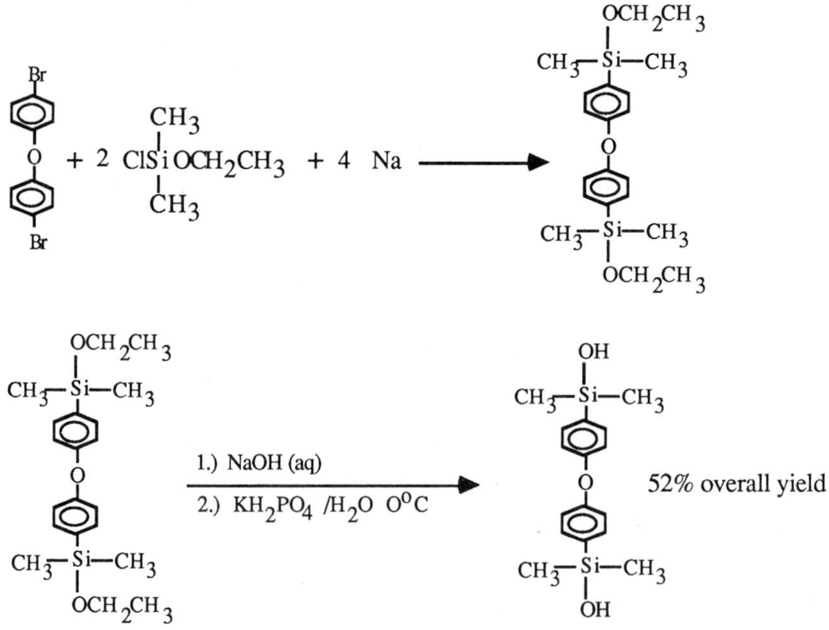

using $R^1 = R^2 = R^3 = R^4 = CH_3$, Ar = p-phenylene, and y = 3. The synthesis of the monomers is fairly straightforward. The cyclic siloxazane monomers were prepared by condensing methylamine with the corresponding α,ω-dichlorosiloxane containing the desired number of dimethylsiloxane units. The bis(arylsilanols) were prepared in an interesting manner,[7] slightly different from the method used by Merker and Scott [6] to prepare <u>3</u>. In this method, Breed and co-workers used a method whereby an aryl bromide was coupled with a chlorosilane which contained an ethoxy function. The ethoxy group was then hydrolyzed in a two step process to obtain a bis(silanol), as shown below. Polymerizations were carried out in a Wood's metal bath at 160°C by simply mixing the two monomers in correct stoichiometry and heating for eight hours. The authors report quantitative yields of the alternating polymers with no mention of any homocondensation of the bis(arylsilanol)[7] - which is moderately surprising. Two of the higher molecular weight polymers prepared had Mw's (by light scattering) of over 800,000. Some of the other polymer properties mentioned in their work are tabulated below.[7]

Table 1. Various properties of alternating silarylenesiloxane polymers having the general structure of **4**. [7]

Polymer				
Ar Group	R	y	T_g	T_{dec} N_2[a]
p,p'-diphenylether				
	R^1-R^4=CH_3	1	-	510°C
	R^1-R^4=CH_3	2	-37°C	540°C
	R^1-R^4=CH_3	3	-52°C	560°C
p-phenylene				
	R^1-R^4=CH_3	2	-62°C	510°C
	R^1-R^4=CH_3	3	-72°C	-
m-phenylene				
	R^1-R^4=CH_3	2	-75°C	480°C
	R^1,R^3,R^4=CH_3, R^2=Ph	3	-42°C	-

a.) Estimated from graphs at 10% weight loss under vacuum.

Other groups at this time were attempting similar reactions to obtain polymers having a general structure like **4**. Lai and co-workers[8] have described two different methods for the preparation of alternating silarylenesiloxane polymers. The first method involves the condensation of a bis(diorganohydroxysilphenylene) molecule with various dichlorosilanes or α,ω-dichlorosiloxanes. In all cases only low molecular weight polymers were prepared which had degrees of polymerization ranging from 3 to 50.[8] The reason for the low degree of polymerization appears to stem from the fact that HCl is given off as a condensation product. Even though the reactions are done in the presence of pyridine, an HCl acceptor, an apparent equilibrium exists between the C-Si bonds in the growing polymer chain and the HCl before the acid can be neutralized by the pyridine.[8] Later workers side-stepped this problem by using "Pike's reaction" [8] which involves the condensation of a diaminosilane and a disilanol to liberate a secondary amine by-product. According to Lai and co-workers,[8] use of this method provides high molecular weight polymers. Other workers[9] have also used this procedure to prepare alternating polymers with structures like **4** where Ar = p-phenylene, $R^1 = R^2 = R^3 = R^4 = CH_3$ or R^4 = Ph, and y = 1.

An interesting adaptation of the condensation route mentioned above, results in loss of a completely unreactive urea by-product. The condensation polymerization involves the reaction of a disilanol with a bis(ureidosilane), specifically a bis(N-phenyl-N'-tetramethylene-ureido)diorganosilane, (7), to produce N-phenyl-N'-tetramethyleneurea as the unreactive by-product. It appears that the original condensation reaction utilizing this method stemmed from the preparation of exactly alternating carborane-siloxane polymers (which will be described shortly).[10] Lenz and Dvornic were the first to apply this condensation reaction to the synthesis of exactly alternating silarylenesiloxane polymers. A general reaction scheme is shown below.

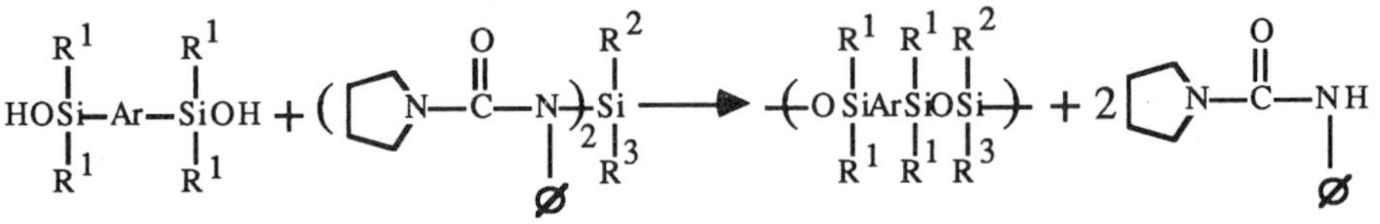

Lenz and Dvornic have done an extensive amount of work on silarylenesiloxane polymers prepared using the bis(ureidosilane) synthesis. These authors have examined in detail monomer synthesis,[11,12] structure-property relationships,[3] and mechanical properties of the crosslinked elastomers prepared from these silarylenesiloxane polymers.[4] The preparation of the bis(ureidosilane) monomer used in the condensation reactions was carried out in an interesting three-step synthesis.[11] The first step involved the preparation of a lithium-dialkylamide through reaction of a dialkylamine with butyl lithium. This was reacted in a second step with a diorganodichlorosilane to obtain a diamino substituted silane. Using a typical carbon analog reaction between an isocyanate and an amine to prepare a urea, the third step consisted of a reaction between the diaminosilane and phenylisocyanate to obtain a bis(ureidosilane) like 7. The only apparent problem which the authors encountered with this monomer was its hydrolytic instability. Reaction of 7 with one molecule of water would lead to the production of a ureidosilanol plus one molecule of the urea by-product. For a monomer which was to be used in a condensation reaction, this unsymmetrical side-product could not be tolerated.[11] The authors monitored the production of the urea by-product using ^{1}H NMR in order to ascertain the relative stabilities of these bis(ureido)silanes under various conditions, i.e., in air, vacuum, under nitrogen, and at different temperatures. The best method found for storage of the bis(ureidosilane) monomers was under nitrogen at -20°C: the monomers could be stored for up to 100 days under these conditions with production of less than 10 mole% of the urea by-product.[11]

Dvornic and Lenz [12] looked at several different polymers based on two aryl groups, these being p-phenylene and p,p'-diphenylether, where, based on structure 6, R^1=CH$_3$. The authors also used two different bis(ureidosilane) monomers, where, based on structure 7,

$R^2 = R^3 = CH_3$, and $R^2 = CH_3$, $R^3 =$ vinyl. A variety of different polymers were prepared in this study. Homopolymers of each arylene unit with both the dimethylsiloxane group and with the vinylmethylsiloxane group were synthesized. Overall, the authors prepared and examined 12 different polymers by gel permeation chromatography (GPC),[3] viscometry,[3] ^1H and ^{13}C NMR,[3] IR,[3] differential scanning calorimetry (DSC),[3,13] and thermogravimetric analysis (TGA).[14] Table 2 shows a variety of properties compiled from the above references on nine of the twelve polymers examined.

Table 2. Various properties of alternating silarylenesiloxane polymers having the general structure of $\underline{8}$. [3,13,14]

Polymer [a]		T_g	Mol. Wt.(M_n)/PD[b]	T_{dec} air [c]	T_{dec} N_2 [c]
Ar Group	% Vinyl				
p-phenylene (I)					
	0	-62°C	71,500 / 1.86	500°C	490°C
	5	-63°C	67,000 / 1.78	510°C	490°C
	100	-69°C	103,500 / 1.92	570°C	550°C
p,p'-diphenylether (II)					
	0	-24°C	100,500 / 2.13	475°C	490°C
	5	-25°C	40,000 / 1.86	-	-
	100	-34°C	143,000 / 1.63	540°C	550°C
I + II (50 mole% each)					
	0	-42°C	88,000 / 1.86	475°C	520°C
	5	-43°C	126,000 / 1.89	-	-
	100	-51°C	128,500 / 1.70	540°C	520°C

a) The polymer structure when $R^1 = R^2 = R^3 = CH_3$. The % vinyl reflects the content of vinyl groups when $R^3 = -CH=CH_2$.
b) Molecular weight of polymers precipitated in methanol, based on polystyrene standards.
c) Temperature of 10% weight loss estimated from TGA curves found in reference 14.

<u>Silarylene Polymers Containing m-Xylylene Units.</u> A majority of the silarylene type siloxane polymers which have been worked on to date contain silicon-phenyl bonds. After an extensive literature search, only one type of polymer was found which contained an arylene group that does not have silphenylene bonds in the polymer backbone. The work, done mainly by Rosenberg and Choe [15-17] of the Air Force Materials Laboratory, is based on polymers which have a general structure as shown below, referred to as poly(m-xylylene-siloxanylenes), ($\underline{9}$).

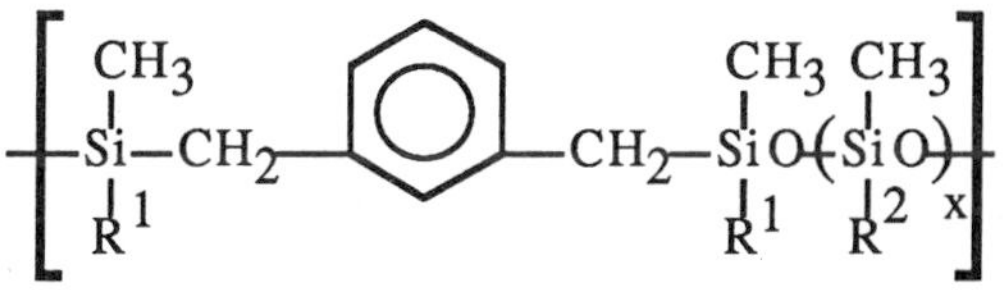

$\underline{9}$

The impetus for their research was to discover a new class of viscoelastic polymers which exhibited good thermal stability and broad use temperatures for use as fuel-resistant sealants.[16] The synthetic methods used to prepare polymers like 9 are similar to those methods which were used in studies mentioned previously. For 9 (x = 0), the bis(silanol) was prepared and then condensed using tetramethylguanidine 2-ethylhexoate as a catalyst to prepare the polymer. For 9 (x = 1) the appropriate bis(silanol) was condensed with a bis(dimethylamino)silane, and for 9 (x = 2) the appropriate bis(silanol) was condensed with a bis(dimethylamino)disiloxane. Table 3 relates some of the various properties found from polymers prepared using these methods.

Table 3. Various properties of *m*-xylylenesiloxanylene polymers having the general structure of 9. [16]

x	R^1	R^2	Mol. Wt. (M$_n$) [a]	T$_g$	T$_{dec}$ (vac)[b]
0	CH$_3$	-	38,880	-41°C	513°C
0	CF$_3$CH$_2$CH$_2$	-	-	-19°C	436°C
1	CH$_3$	CH$_3$	14,500	-62°C	537°C
1	CH$_3$	CF$_3$CH$_2$CH$_2$	33,000	-52°C	494°C
1	CF$_3$CH$_2$CH$_2$	CH$_3$	21,000	-44°C	453°C
1	CF$_3$CH$_2$CH$_2$	CF$_3$CH$_2$CH$_2$	-	-35°C	450°C
2	CH$_3$	CH$_3$	14,000	-77°C	508°C
2	CH$_3$	CF$_3$CH$_2$CH$_2$	13,000	-59°C	478°C
2	CF$_3$CH$_2$CH$_2$	CF$_3$CH$_2$CH$_2$	22,000	-38°C	446°C

a.) Molecular weight found by vapor phase or membrane osmometry.
b.) Decomposition temperature found at 10% weight loss under vacuum.

<u>Miscellaneous Silarylene Polymers.</u> Several other polymer systems have been examined which exhibit similar properties and have similar structures to those which have been discussed previously. Some of these studies have described the preparation of polymers using synthetic schemes similar to those employed for the silarylenesiloxane polymers. Others have described synthetic schemes of a completely different and fascinating nature. Since there does not appear to be any logical order in which these polymers should be discussed, the ones which have similar syntheses to those shown previously will be examined first.

The first class of polymers which will be discussed is not a true silarylene polymer, but is presented here because its synthesis has led to a number of the silarylenesiloxane polymers discussed previously. These polymers are known as poly(dodecacarborane-siloxanes), the study of which has been reviewed by Peters.[10] The carborane molecule is very interesting and becomes an entire topic in and of itself. For the sake of completeness, the three different dodecacarborane moieties examined by Peters are shown below. The *m*- and *p*-carboranes have been used almost exclusively for polymer preparation. Polymers have

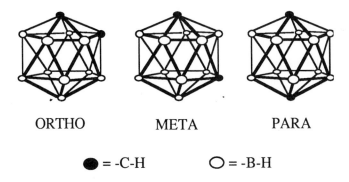

ORTHO   META   PARA

● = -C-H   ○ = -B-H

been prepared similar in structure to $\underline{5}$, where Ar is now the *m*- or *p*-dodecacarborane moiety (usually written as -CB$_{10}$H$_{10}$C-), R^1 is usually CH$_3$, R^2 is usually CH$_3$ but 1,1,1-trifluoropropyl has also been used, R^3 and R^4 can vary, usually being CH$_3$, phenyl, or 1,1,1-trifluoropropyl, and y = 0 - 5.[10]   A variety of syntheses are discussed in the review by Peters including, hydrolysis of a bis(chlorosilane) with subsequent condensation of the bis(silanol) to form polymer, condensation of a bis(silanol) with a difunctional silane (this includes the bis(ureido)silane synthesis mentioned previously), and co-condensation of two different bis(silanols).

The properties of the poly(dodecacarboranesiloxane) polymers examined in the review by Peters[10] are also quite interesting. A majority of the polymers examined exhibited crystalline melting points ranging from 40°C to 260°C depending on the polymer structure. Glass transition temperatures were also reported for most polymers, and were found to range from -88°C to 25°C. The thermal stability of these polymers, for the most part, is similar to that of the silarylenesiloxane polymers. Generally 10% weight loss occurs between 500°C and 550°C depending on the length of the siloxane segment and the nature of the silicon sidegroups.[10]

A novel class of polymers was prepared by Pittman and co-workers [18,19] who used a condensation reaction between a bis(dimethylamino)silane and a bis(silanol). The novelty of these polymers arises through the incorporation of a ferrocene unit in the polymer backbone. Ferrocene was converted to bis(dimethylaminodimethylsilyl)ferrocene and then was condensed with one of three disilanols, diphenyldihydroxysilane, 1,4-bis(hydroxy)dimethylsilyl)benzene, or 4,4'-bis(hydroxydimethylsilyl)biphenyl. Polymers which contained the ferrocene moiety exhibited slightly higher glass transition temperatures than the conventional silarylenesiloxane polymers, an increase which ranged from 20°C to 30°C depending on the final polymer structure.[18]

In the quest for polymers which exhibit exceptional high temperature properties as well as good chemical resistance, Patterson and Morris [20] of NASA experimented with some silarylenesiloxane polymers which contained perfluoroalkyl groups. The polymerization reaction employed was a typical aminosilane-silanol condensation, but the monomers employed were not so typical. Two different bis(silanol) monomers were used, 1,3-bis[p-(hydroxydimethylsilyl)phenyl]hexafluoropropane and 1,6-bis[m-(hydroxydimethylsilyl)-phenyl]dodecafluorohexane. The aminosilane which was used in the condensation reaction was 1,3-[bis(p-dimethylaminodimethylsilyl)phenyl]hexafluoropropane. These investigators have used an interesting approach in this condensation reaction, in that both the aminosilane and the bis(silanol) contained the silarylene group.

Breed and Wiley,[21] investigators on some of the original silarylenesiloxane polymers, have looked at a method for modifying heterocyclic polymers. Using some interesting chemistry, these workers were able to prepare two different types of difunctional silaryl-enesiloxane monomers as shown below. These monomers were used in polymerizations of

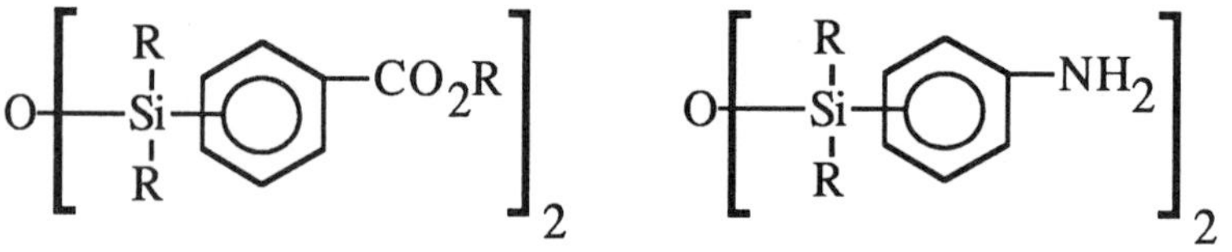

poly(benzimidazoles) and poly(benzoxazoles). The siloxane modified heterocyclic polymers exhibited exceptional solubility; they were completely soluble in N-methylpyrrolidinone, and all but one polymer in pyridine, at a concentration of 10%. The high thermal stability of these polymers was maintained in N_2, all having less than 2% weight loss after a 96 hour isothermal aging experiment at 350°C. But, because of the incorporation of the methylsiloxane linkage, the thermooxidative properties diminished, showing anywhere from a 2% to 8% weight loss in the same 350°C isothermal test in air.[21] It appears that the use of the silarylenesiloxane monomers to modify heterocyclic polymers can improve solubility, which could be important for processing, but diminishes the high temperature oxidative stability generally observed for the homoheterocyclic poly(benzimidazoles) and poly(benzoxazoles).

Two remaining polymers have been encountered in the literature which fall under the category of silarylene polymers, but these polymers have no siloxane linkages. The first polymer which will be discussed has been used by Zelei and co-workers [22] as well as Ikeda and co-workers [23] in thermal degradation studies of 1 (poly(tetramethyl-p-silphenylenesilox-ane)). The polymer used was poly(p-dimethylsilphenylene) (10) and has been prepared by a method shown below. The second polymer is poly[1-(dimethylsilyl)-4-(dimethylethyl-ene)silylbenzene]) (11) and was prepared using a well known hydrosilylation mechanism in

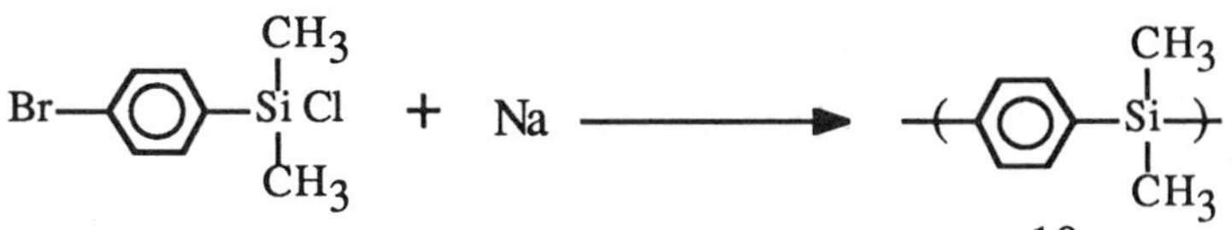

CH3
|
Br—⬡—Si Cl   +   Na   ⟶   —(⬡—Si—)—
|
CH3

CH3
|
—(⬡—Si—)—
|
CH3

10

which an Si-H adds to an olefin.[24] The synthetic scheme used for this polymer preparation is shown below. As can be seen, if the divinyl monomer is used in excess, a telechelic polymer results which contains vinyl endgroups. The molecular weight of these materials, as determined by vapor pressure osmometry, ranged from 1200 to 7800 depending on the stoichiometry used. Polymer 11 exhibited a crystalline melting point of ~ 190°C and a Tg of ~ 28°C, both, of course, depending on the molecular weight of the polymer examined.[24]

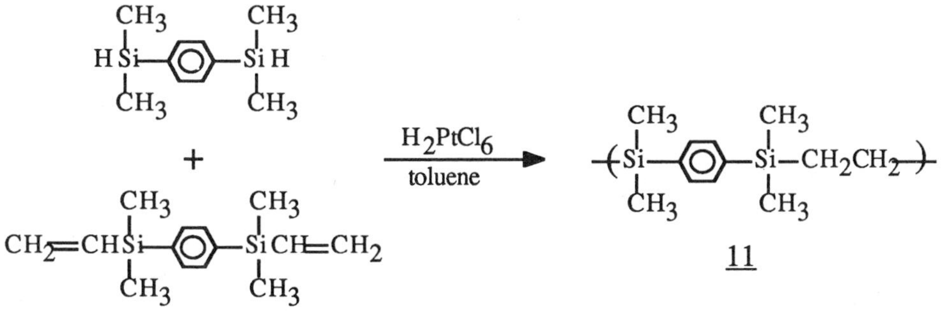

CH3   CH3
| |
H Si—⬡—Si H
| |
CH3   CH3

+

CH3   CH3
| |
CH2=CHSi—⬡—SiCH=CH2
| |
CH3   CH3

$\xrightarrow[\text{toluene}]{H_2PtCl_6}$

CH3   CH3
| |
—(Si—⬡—Si—CH2CH2—)—
| |
CH3   CH3

11

In some of the syntheses described above, carbanion or carbanion-like intermediates are used to make silicon containing monomers or polymers. Our group has done extensive work in carbanion chemistry for monomer and polymer synthesis. This work describes the use of two dicarbanion species in the preparation of several different silicon containing monomers and polymers. Some of this work closely parallels the work described above in that similar reaction schemes are utilized, but the final polymers are all different (except one). The goal of this research was to prepare different types of dicarbanions and place those moieties in a polymer backbone via Si bonds. By selecting silicon-carbon bonds and/or siloxane bonds, we were able to substantially vary the properties of the final polymers.

EXPERIMENTAL

Instrumentation. Gel permeation chromatography (GPC) was performed on a Water's 150C GPC/ALC instrument using a refractive index detector. Data reduction was done using an Apple IIe computer with software provided by Interactive Microware, Inc. Baker HPLC grade tetrahydrofuran was used as the solvent at a flow rate of 0.7 mL/min and at a temperature of 35°C. Five Water's μ-styragel columns were used with pore sizes of 10^6, 10^5, 10^4, 10^3, and 500Å. Calibration was done using polystyrene standards which ranged in molecular weight from 2,000 to 950,000.

All 200 MHz ^{1}H NMR and 360 MHz ^{1}H NMR spectra were recorded on a Bruker WP-200 and Bruker WP-360 spectrometer respectively using DISNMR software. Samples were dissolved in $CDCl_3$, and were referenced downfield to tetramethylsilane (TMS) at δ 0.00. Fourier transform infra-red spectroscopy (FTIR) was performed on a Digilab FTS-60 instrument using either KBr or NaCl windows. Typically 64 scans were taken for each sample at a resolution of 2 cm^{-1}.

Low resolution mass spectrometry was performed by the Penn State Chemistry Department on a Kratos MS 9/50. Capillary gas chromatography was performed on a Hewlett-Packard 5880A Series GC. The capillary column used was a 12 m crosslinked methylsilicone fast analysis column utilizing a flame ionization detector. Helium was used as the carrier gas at a flow rate of 2.2 mL/min under various temperature programming conditions.

Differential scanning calorimetry (DSC) was performed on a Perkin Elmer Series 7 DSC using a heating rate of 20°C/min over a temperature range of either 25°C to 300°C for higher temperature transitions or -75°C to 100°C for lower temperature transitions. Thermogravimetric analysis (TGA) was performed on a Perkin Elmer TGA-2 instrument at a heating rate of 20°C/min. Melting points were taken on a Thomas Hoover capillary melting point apparatus and are uncorrected.

Preparation of the Dianion of *m*-Xylene (12). To a suspension of 7.8 g *t*-BuOK (70 mmol) in 60 mL of hexane was added 28 mL of 2.5 M *n*-BuLi (70 mmol) under N_2. The resulting beige solution was allowed to stand for 10 min after which 2.5 mL of *m*-xylene (20 mmol) in 15 mL of hexane was slowly added. After 30 min the bright yellow anion suspension was heated to a gentle reflux and was allowed to react for 24 h. Upon cooling, the brown suspension was placed in centrifuge tubes and was centrifuged down. The hexane layer was removed, and the anions were washed once with clean hexane. The anions were then used in monomer or polymer synthesis.

Preparation of 1,3-Bis(dimethylhydrogensilylmethylene)benzene. To a N_2 purged round bottom flask in a cold bath at -40°C was added 100 mL of THF. To the THF was added 9.0 mL of chlorodimethylsilane (81 mmol). The dianion of *m*-xylene (20 mmol) was taken and the hexane layer was removed. To each centrifuge tube was added 20 mL of THF with shaking. After the anions were suspended in the THF, the contents of each one was slowly added to the round bottom flask. The reaction was allowed to proceed for 5 h after the final addition of the anion. After 5 h, the solution was washed with a saturated NaCl solution. To the organic layer which separated was added hexane or diethyl ether which forced out a second aqueous layer. The organic phase was dried over magnesium sulfate and allowed to stand for 12 h. The solution was then filtered, and the solvent was removed by rotary evaporation. The crude liquid was distilled under reduced pressure to obtain the product in a 67% yield. The b.p. was found to be 47°C at 0.3mm Hg. Mass spec. showed: M+2, 224 (4.76%); M+1, 223 (12.40%); M+, 222 (50.45%); M-163, 59 (100%). ^{1}H NMR showed: δ 7.17 (t), (J = 7.5 Hz, 1H); δ 6.83 (d), (J = 7.6 Hz, 2H); δ 6.79 (s), (1H); δ 4.02 (nonet), (J = 3.1 Hz, 2H); δ 2.17 (d), (J = 3.2 Hz, 4H); δ 0.12 (d), (J = 3.6 Hz, 12H).

<u>Preparation of 1,3-Bis(dimethylhydroxysilylmethylene)benzene.</u> To a round bottom flask was added 10 mL of ethanol and a small piece of sodium metal (~ 0.05 g). The ethanol solution was heated to a gentle reflux. After the metal had disappeared, 3.7 g of 1,3-bis(dimethylhydrogensilylmethylene)benzene (17 mmol) was added slowly with liberation of H_2. After addition was complete, the solution was allowed to react for 30 min after which it was allowed to cool. Upon cooling, the contents of the round bottom were slowly added to an Erlenmeyer flask which contained 2.5 g of NaOH, 5 mL of H_2O, and 30 mL of methanol. After addition was complete, a second solution of 35 mL of H_2O and 2.5 g of NaOH was poured into the Erlenmeyer. The solution was allowed to stir for 2 h at which time the contents were emptied into a second Erlenmeyer flask which contained 600 mL of ice water and 50 g of potassium phosphate monobasic. After addition was complete, the Erlenmeyer flask was placed in a refrigerator and allowed to sit for 24 h. The water was extracted with methylene chloride to remove organics. The crude organic phase which was collected was dried over magnesium sulfate. The solvent was removed by rotary evaporation and the crude liquid which remained was used in the polymerization without further purification. A yield was not calculated.

<u>Preparation of Poly(dimethylsiloxy-*m*-xylylene) (20).</u> To a round bottom flask fitted with a nitrogen purge was placed 4.5 g of 1,3-bis(dimethylhydroxysilylmethylene)benzene. Heat was slowly added to the reaction until the temperature reached 200°C at which point water was given off. This was allowed to stay at this temperature for 12 h before cooling. No purification steps were performed on this polymer. ^{1}H NMR showed: δ 0.03 (s), (12H); δ 2.05 (s), (4H); δ 2.32 (s), (minor); δ 6.70 (s), (1H); δ 6.75 (d), (J = 7.5 Hz, 2H); δ 7.07 (t), (J = 7.6 Hz, 1H).

<u>Preparation of Poly(dimethylsil-*m*-xylylene) (19).</u> To a round bottom flask (at temperatures ranging from -78°C to room temperature) was added 100 mL of THF. To this was added from 2.0 mL to 6.0 mL of freshly distilled dichlorodimethylsilane (16 mmol - 50 mmol). The dianion of *m*-xylene (10 mmol - 20 mmol) was rinsed and the hexane layer was removed. The anions were then suspended in THF and immediately added to the round bottom flask, either slowly or all at once. The reaction was allowed to proceed for 12 h to 24 h and then water containing a little bit of sodium bicarbonate was added. This was allowed to stand for 30 min, followed by extraction with a saturated NaCl solution. To the organic layer

which separated was added hexane or diethyl ether which forced out a second aqueous layer. The organic phase was collected and dried over magnesium sulfate for 12 h. The solution was filtered, and the solvent was removed by rotary evaporation to leave a crude viscous liquid. The liquid was dissolved in 5 mL of chloroform and was added to 200 mL of cold methanol. The tacky polymer was removed, and dried in a vacuum oven at 40°C. Molecular weights and yields varied depending on the conditions of the experiment. Typical yields ranged from 50% to 80%. ^{1}H NMR showed: δ 0.00 (s), (~6H); δ 2.11 (s), (~4H); δ 2.36 (s), (minor); δ 6.70 (s), (1H); δ 6.77 (d), (J = 7.5 Hz, 2H); δ 7.11 (t), (J = 7.5 Hz, 1H).

Preparation of Poly(diphenylsil-_m_-xylylene) (21). To a round bottom flask (at temperatures ranging from -78°C to room temperature) was added 100 mL of THF. To this was added from 5.0 mL to 11.0 mL of freshly distilled dichlorodiphenylsilane (24 mmol - 52 mmol). The dianion of _m_-xylene (20 mmol) was taken and the hexane layer was removed. The anions were suspended in THF and immediately added to the round bottom flask, either slowly or all at once. The reaction was allowed to proceed for 12 h to 24 h and then water containing a little bit of sodium bicarbonate was added. This was allowed to stand for 30 min, followed by extraction with a saturated NaCl solution. The organic phase which separated was dissolved in chloroform or methylene chloride. The organic phase was collected and dried over magnesium sulfate for 12 h. The solution was filtered, and the solvent was removed by rotary evaporation to leave a crude viscous liquid. The liquid was dissolved in 15 mL of methylene chloride and was added to 200 mL of cold methanol. The solid polymer was removed, filtered, and dried in a vacuum oven at 50°C unless otherwise noted. Molecular weights and yields varied depending on the conditions of the experiment. Typical yields ranged from 40% to 70%.

Preparation of 4,4'-Dimethylbiphenyl. In a round bottom flask equipped for stirring and reflux was placed 7.1 g of Mg powder (0.29 mol) in 75 mL of THF. To this was added 30 mL of 4-bromotoluene (0.24 mol) in 40 mL of THF in 5 mL increments and gently heated until reflux was achieved. Reflux continued even after removal of heat. After addition was complete, and upon cooling, 120 mL of benzene was added to the round bottom flask. To this solution was added 116 g of thallium bromide (0.41 mol) and was refluxed for 4 h. When cool, the solution was acidified with 100 mL of 10% HCl, filtered, the filtrate washed with benzene, and the benzene/water solution was separated. The benzene layer was collected and dried over magnesium sulfate for 12 h, filtered, and the solvent was removed by rotary evaporation. The crude product was then placed on a vacuum line and gently heated for 1 h to remove any residual 4-bromotoluene. The crude product was sublimed at 70°C and 0.05 mm Hg. The uncorrected melting point was found to be 119-120°C. The literature value is 121°C.[25] Mass spec. showed: M+2, 184, (1.04%); M+1, 183 (14.98%); M+, 182 (100%). ^{1}H NMR showed: δ 2.43 (s), (6H); δ 7.25 (d), (J = 7.8 Hz, 4H); δ 7.49 (d), (J = 8.1 Hz, 4H).

Preparation of the Dianion of 4,4'-Dimethylbiphenyl (13). To a suspension of 6.1 g t-BuOK (54 mmol) in 40 mL of hexane was added 24 mL of 2.5 M n-BuLi (60 mmol) under N$_2$. The resulting beige solution was allowed to stand for 10 min after which 3.0 g of 4,4'-dimethylbiphenyl (16 mmol) in 50 mL of hexane was slowly added. In order to get the 4,4'-dimethylbiphenyl to dissolve, the hexane solution was slightly heated with a heat gun. After 30 min the deep purple anion suspension was heated to a gentle reflux and was allowed to react for 24 h. Upon cooling, the deep purple anions were placed in centrifuge tubes and centrifuged down. The hexane layer was removed, and the anions were washed once with clean hexane. The anions were then used in monomer or polymer synthesis.

Preparation of 4,4'-Bis(dimethylhydrogensilylmethylene)biphenyl (14). To a N$_2$ purged round bottom flask in an ice bath at 0°C was added 100 mL of THF. To the THF was added 7.5 mL of chlorodimethylsilane (67 mmol). The dianion of dimethylbiphenyl (16 mmol) was taken and the hexane layer was removed. To each centrifuge tube was added 20 mL of THF with shaking. After the anions were suspended in the THF, the contents of each one was slowly added to the round bottom flask. The reaction was allowed to proceed for 1 h after the final addition of the anion. After 1 h, the solution was washed with a saturated NaCl solution. To the organic layer which separated was added hexane or diethyl ether which forced out a second aqueous layer. The organic phase was dried over magnesium sulfate and allowed to stand for 12 h. The solution was filtered, and the solvent was removed by rotary evaporation. The crude green liquid was distilled under reduced pressure to obtain 14 in a 75% yield. The b.p. was found to be 130°C at 0.03 mm Hg. Mass spec. showed: M+2, 300 (13.90%); M+1, 299 (38.66%); M+, 298 (89.58%); M-239, 59 (100%). ^{1}H NMR showed: δ 7.52 (d), (J = 8.2 Hz, 4H); δ 7.15 (d), (J = 8.2 Hz, 4H); δ 4.05 (nonet), (J = 3.6 Hz, 2H); δ 2.26 (d), (J = 3.1 Hz, 4H); δ 0.16 (d), (J = 3.6 Hz, 12H).

Preparation of 4,4'-Bis(dimethylhydroxysilylmethylene)biphenyl (15). To a round bottom flask was added 40 mL of ethanol and a small piece of sodium metal (~ 0.1 g). The ethanol solution was heated to a gentle reflux. After the metal had disappeared, a solution of 7.1 g of 42 (24 mmol) in 15 mL of ethanol was added slowly with liberation of H$_2$. After addition was complete, the solution was allowed to react for 30 min, after which it was allowed to cool. The contents of the round bottom were slowly added to an Erlenmeyer flask which contained 5.0 g of NaOH, 20 mL of H$_2$O, and 75 mL of methanol. After addition was complete, a second solution of 50 mL of H$_2$O and 5.0 g of NaOH was poured into the Erlenmeyer. The solution was allowed to stir for 2 h at which time the contents were emptied into a second Erlenmeyer flask which contained 1200 mL of ice water and 75 g of potassium phosphate monobasic. After addition was complete, the Erlenmeyer flask was placed in a refrigerator and allowed to sit for 24 h. The solution was filtered and the filtrate was collected. The water was extracted with methylene chloride to remove any remaining organics. The organic phase was dried over magnesium sulfate, filtered, and the solvent was removed by rotary evaporation. All of the crude solid which was collected was dissolved in 15 mL of chloroform and was added to an Erlenmeyer which contained 150 mL of hexane. The hexane was heated until the solid dissolved, and was allowed to cool in a refrigerator. After 24 h the yellow tinted crystals which had formed were filtered and dried. The overall yield was 68 %. ^{1}H NMR showed: δ 7.48 (d), (J = 8.0 Hz, 4H); δ 7.14 (d), (J = 8.1 Hz, 4H); δ 2.21 (s), (4H); δ 1.50 (s), (2H); δ 0.15 (s), (12H).

Preparation of Poly(4-tetramethyldisiloxymethylene-4'-methylenebiphenyl) (17). To a round bottom flask which contained 20 mL of benzene was added 2.6 g of 43 (8 mmol). To this solution was added 0.1 mL of 2-diethylaminoethanol (0.7 mmol) as catalyst. The benzene solution was slowly refluxed for 1 h while slowly allowing some to escape thus raising the temperature of the solution. After 3 h the benzene had evaporated and the temperature was at 170°C. After 12 h a very viscous, partially crosslinked polymer had formed. The soluble polymer was dissolved in 25 mL of chloroform and was precipitated in 400 mL of cold methanol. The polymer was filtered and dried in a vacuum oven. ^{1}H NMR showed: δ 0.05 (s), (12H); δ 2.13 (s), (4H); δ 7.06 (d), (J = 8.1 Hz, 4H); δ 7.45 (d), (J = 8.0 Hz, 4H).

Preparation of Poly(4-dimethylsilylmethylene-4'-methylenebiphenyl) (16). To a round bottom flask (at temperatures ranging from -78°C to room temperature) was added 100 mL of THF. To this was added from 1 to 3 equivalents of dichlorodimethylsilane. The dianion of dimethylbiphenyl was taken and the hexane layer was removed. The anions were suspended in THF and immediately added to the round bottom flask, either slowly or all at once. The reaction was allowed to proceed for 12 h to 24 h and then water containing a small amount of sodium bicarbonate was added. This was allowed to stand for 30 min, followed by extraction with a saturated NaCl solution. The organic phase which separated was dissolved in chloroform or methylene chloride. The organic phase was collected and dried over magnesium sulfate for 12 h. The solution was filtered and the solvent was removed by rotary evaporation to leave a crude product. The liquid was dissolved in 15 mL of methylene chloride and was added to 200 mL of cold methanol. The solid polymer was removed, filtered, and dried in a vacuum oven. Molecular weights and yields varied depending on the conditions of the experiment. Typical yields ranged from 50% to 85%. ^{1}H NMR showed: δ 0.03 (s), (6H); δ 2.19 (s), (4H); δ 2.42 (s), (minor); δ 7.07 (d), (J = 8.0 Hz, 4H); δ 7.50 (d), (J = 8.0 Hz, 4H).

Preparation of Poly(4-diphenylsilylmethylene-4'-methylene-biphenyl) (18). To a round bottom flask (at temperatures ranging from -78°C to room temperature) was added 100 mL of THF. To this was added from 1.5 to 2 equivalents of dichlorodiphenylsilane. The dianion of dimethylbiphenyl was taken and the hexane layer was removed. The anions were then suspended in THF and immediately added to the round bottom flask, either slowly or all at once. The reaction was allowed to proceed for 12 h to 24 h and then water containing a small amount of sodium bicarbonate was added. This was allowed to stand for 30 min, followed by extraction with a saturated NaCl solution. The organic phase which separated was dissolved in chloroform or methylene chloride. The organic phase was collected and dried over magnesium sulfate for 12 h. The solution was filtered, and the solvent was removed by rotary evaporation to leave a crude product. This was dissolved in 15 mL of methylene chloride and was added to 200 mL of cold methanol. The solid polymer was removed, filtered, and dried in a vacuum oven. Molecular weights and yields varied depending on the conditions of the experiment. Typical yields ranged from 40% to 80%.

RESULTS AND DISCUSSION

Carbanion Preparation. The preparation of delocalized carbanions has been extensively examined by our group.[26-28] We have used Lochmann's base,[29] an equimolar mixture of *n*-butyl lithium and potassium *t*-butoxide in a hydrocarbon solvent, as the exclusive metalating reagent in all our carbanion preparations. Lochmann's base has the unprecedented ability to metalate a host of allylic and benzylic moieties on a variety of molecules. The actual metalating species in this base, originally attributed to simple butyl potassium,[29] has been part of a subtle controversy for some years now. Recent investigations tend to indicate that the active species is more complicated than simple butyl potassium, probably being some type of mixed aggregate of the alkyl lithium and the potassium alkoxide.[30, 31]

Two different dicarbanions have been prepared for use in this study as illustrated below. The dicarbanion of *m*-xylene, 12, has been used previously in our work on anionic

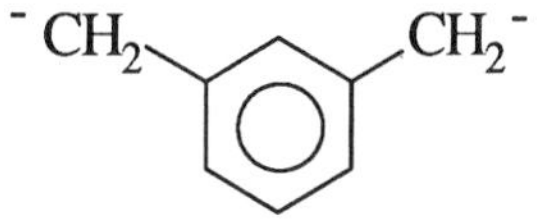

12 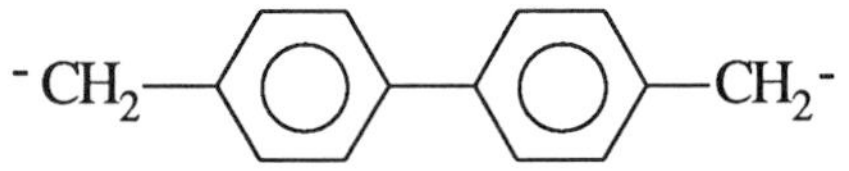   13

initiators [28] and has been prepared and characterized by others.[32] Dicarbanion 13, prepared from 4,4'-dimethylbiphenyl, has not been previously used in our work, but has recently been prepared by another group.[33] Both of these dicarbanions have been prepared using Lochmann's base in hexane and have been characterized through their respective methylation products (using either dimethyl sulfate or methyl iodide as the methylating reagent). The relative amounts of monomethylated and dimethylated adducts can be discerned from ^{1}H NMR spectra and capillary gas chromatography.

Monomer Preparation. Two monomers were prepared for this study based on dicarbanions 12 and 13. The preparation of these monomers is based on previously reported chemistry that uses a di-Grignard reagent instead of a dicarbanion as we use.[6] The basic premise behind our work was to obtain a purifiable monomer containing a bis(silane) functionality that could later be transformed into a polymerizable bis(silanol). Reaction of dicarbanion 12 or 13 with chlorodimethylsilane afforded the required functionalized molecule. The silane function was converted into an ethoxysilane through reaction with sodium ethoxide. The ethoxide was hydrolyzed to the silanol using aqueous sodium hydroxide followed by neutralization with potassium phosphate monobasic. The reaction is shown in Scheme 1 from dicarbanion preparation to silanol formation for 4,4'-dimethylbiphenyl. The bis(silanol) which is prepared from the analogous reaction with dicarbanion 12 has been prepared previously.[34] We believe that the starting material used by this group was actually α,α'-dichloro-m-xylene from which the di-Grignard was prepared.

Scheme 1. Preparation of 4,4'-bis(dimethylhydroxysilylmethylene)biphenyl.

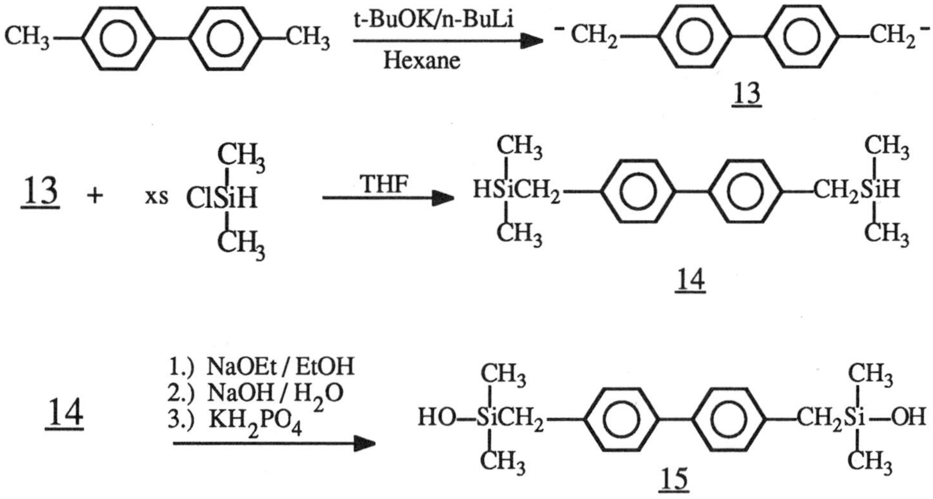

Polymer Preparation. In order to prepare silicon containing polymers which include the m-xylyl and dimethylenebiphenyl moieties, two distinct polymerization routes were chosen. The first route consists of condensation of a dichlorodialkyl(or aryl)silane with a dicarbanion, namely 12 or 13. The material which is formed consists of an all silicon-carbon and carbon-carbon backbone. The second polymerization utilizes the bis(silanol) monomers discussed in the previous section. These materials are heated and undergo a self condensation reaction to form polymers which contain silicon-carbon, silicon-oxygen, and carbon-carbon bonds in the backbone.

The first polymerization route is shown in Scheme 2 for the condensation of dicarbanion 13 with dichlorodimethylsilane to form a silarylene polymer. The polymerization which is illustrated in Scheme 2 produces only low molecular weight materials when either dicarbanion is condensed with dichlorodimethylsilane or dichlorodiphenylsilane. We had originally believed that these polymers would have high molecular weights even though the polymerization is a polycondensation reaction. Typical condensation polymerizations are difficult to drive to high molecular weights because the monomers used are usually only slightly reactive (thus the need to prepare many polymers at temperatures > 150°C). Also, some condensation polymerizations are difficult because there is an equilibrium formed between the polymer and the monomer (thus the need for high vacuum systems in many condensation polymerizations so small molecules may be removed to drive the reaction toward polymer formation). Using our system, the monomer reactivity is so high that simply titrating one monomer with another should result in better conversion and high molecular weights. Another advantage is that the condensation product is an unreactive salt molecule which does not participate in an undesirable equilibrium reaction.

Scheme 2. Condensation of dicarbanion 13 with dichlorodimethylsilane.

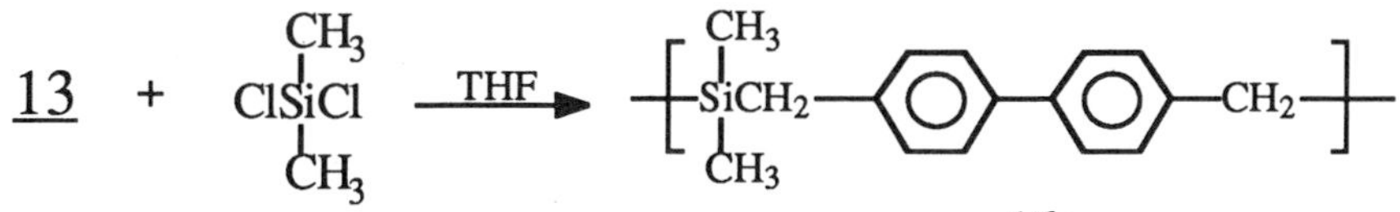

There are several reasons we can find to explain why the molecular weights of our condensation polymers are low. Firstly, because the reaction is an AA BB type polycondensation reaction, the stoichiometry of the polymerization is very critical to obtain high molecular weights. Although we are certain that the major portion (>90%) of the carbanion which is formed is the dicarbanion, some monocarbanion is present. All monocarbanionic species act to cap the chain ends and prohibit further polymerization thus limiting the overall molecular weight. Another reason for the low molecular weights appears to come from a side reaction which involves a metal-halogen exchange between the dicarbanion and the chlorosilane. Evidence for this arises in the FTIR spectra of the quenched polymers (after reaction with a small amount of water) which typically contain a band at about 2130 cm^{-1} due to an Si-H stretch. The Si-H stretch is believed to come from a water quenching of a silyl potassium moiety. This type of side reaction would result in the formation of the monocarbanion of m-xylene or dimethylbiphenyl. Another shortcoming of these highly reactive carbanions is their relative instability in the polymerization solvent, tetrahydrofuran. Although the cleavage of the THF is not a rapid reaction, enough of the dicarbanion could be quenched during the reaction to cause polymerization to be affected .

The second polymerization route involves the homopolycondensation of a bis(silanol). The polymerization is described in Scheme 3 for condensation of bis(silanol) 15. There were two major reasons to prepare the silarylenesiloxane polymers. The first and foremost reason was to obtain polymers which had higher molecular weights than the silarylene polymers described in the previous paragraph. Since we had a way of preparing and isolating pure monomer, the homocondensation polymerization of the bis(silanols) should yield much higher molecular weight materials than the polymerization shown in Scheme 2. The second reason was to examine the differences which exist between the silarylene and silarylenesiloxane polymers both spectroscopically and thermally. Although we were unable to prepare polymer 20 to any appreciable molecular weight, other workers have been able to and have characterized the material extensively.[16,17]

Scheme 3. Homocondensation of bis(silanol) 15.

Characterization of Polymers. The characterization of polymers 16 - 21 was mainly spectroscopic in nature, FTIR and ^{1}H NMR, but some thermal analyses were also performed. The thermal analyses consisted of differential scanning calorimetry, DSC, and thermogravimetric analysis, TGA. Although the thermal analyses were in no way exhaustive, a few interesting features were found in the different polymers. The molecular weights of samples were examined by GPC and are based on polystyrene standards.

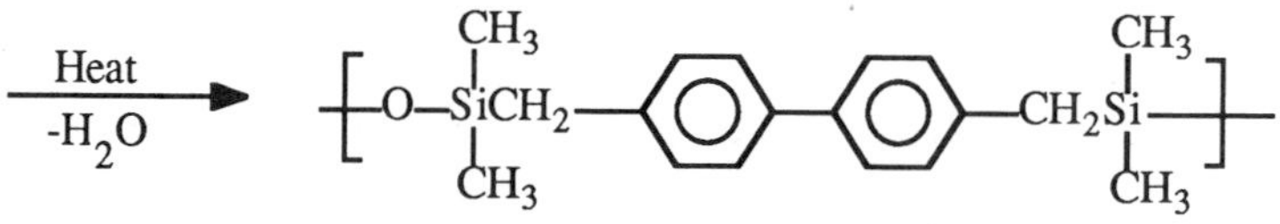

^{1}H NMR spectroscopy was perhaps the most important characterization tool we had for this work because the resonances for most of the protons in our molecules were easily assigned. In order to obtain as much information from spectroscopic analysis as possible, we prepared numerous model compounds which were subsequently examined. The results from these model compounds will not be presented here, but some of the results from the polymer spectra will be described. One of the more convenient features of working with the benzylic carbanions is the ease of characterization of the methylated products with ^{1}H NMR. For example, methylated dicarbanion 12 becomes 1,3-diethylbenzene. Ethylbenzene is a very widely used molecule for teaching on NMR spectrometers as well as for fine-tuning the instrumentation. A simple spectrum, the aliphatic region of ethylbenzene consists of a quartet for the methylene group and a triplet for the methyl group. This is much the same for the methylated benzylic carbanions. Characterization consists of examining the relative amounts

of protons from the ethyl group (triplet and quartet) as compared to the number of protons from the methyl group from the unreacted positions (a singlet).

The NMR characterization of polymers 16 - 21 is very similar, but somewhat less complicated. Since the carbanions are reacting with a chlorosilane, the benzylic methylene group will never be adjacent to another H bearing atom (unless it is a silane impurity), and will, therefore, remain a singlet. By using this characteristic along with chemical shift information found from model compounds, we have been able to examine end groups, typically methyls, and differentiate between methylene groups which are adjacent to silicon-carbon bonds or a siloxane bonds. We have also been able to use aromatic splitting in the *m*-xylylene polymers to explore substitution on the benzyl groups. A listing of the resonances as well as the integration and splitting values can be found in the experimental section.

Infra-red spectroscopy was also used as a characterization tool for polymers 16 - 21. As in the case of NMR spectroscopy, model compounds greatly helped us understand the IR results from our polymers. Also, as with NMR, the IR spectra were very simple, and absorption bands were easily assigned with very few exceptions. In two separate instances the IR results on our polymers proved very valuable. The first instance has already been alluded to, and involves the halogen-metal exchange side-reaction which was described in the previous section. Without the use of FTIR it would have been difficult to observe small amounts of Si-H with other techniques at our disposal. As a result of the IR spectra, we were easily able to detect the presence of the silane function at 2130 cm^{-1} and begin to try and understand its origin.

The second instance that FTIR helped us solve an interesting problem occurred at the beginning of this work. Initially we made attempts at preparing bis(silanols) and oligomeric α,ω-silanols through reaction of the dicarbanions with dichlorodialkylsilanes followed by hydrolysis. Synthesis and subsequent polymerization and characterization of these molecules produced IR spectra with little to no indications of siloxane linkages which typically appear as a very strong absorbance at 1060cm^{-1}. Preparation of siloxane containing model compounds as well as polymers 18 and 20 indicated that the siloxane band at 1060 cm^{-1} should have been present in the initial materials if the reaction went as we planned. Further investigation indicated that we were actually making low molecular weight polymers much like 16 and 19 which contained only very small amounts of siloxane bonds.

The thermal properties of 16 - 21 along with molecular weight results are tabulated in Table 4. Figure 1 shows the structures of all the polymers described in Table 4. The synthesis of all these materials is described in the Experimental section. Several interesting phenomenon were observed when characterizing these polymers using TGA and DSC. As can be seen, only one polymer, 17, could be made with any molecular weight, therefore, the effect of low molecular weight on physical properties should be considered.

Use of DSC allows one to obtain transition temperatures for polymers, namely glass transitions, Tg, and melting transitions, Tm. The glass transitions of all the polymers shown in Table 4 were easily observed, well pronounced transitions. (Only polymer 17 was not examined, but it is known that the Tg is < 25°C due to the physical nature of the polymer.) As can be seen the Tg's follow a fairly expected pattern. Within an arylene group, the trend is observed that the Tg decreases as one goes from an all silicon/carbon containing backbone to a siloxane containing backbone which reflects the flexibility imparted by the siloxane linkage. An increase is observed for the Tg as the size of the substituent groups on the silicon atom increase and render the chain much less flexible. Between arylene groups it is interesting to note that overall the trend is for the dimethylenebiphenyl containing polymers to have an overall higher Tg. This can be attributed to the 4-4'-biphenyl linkages in the polymer which should tend to be more rigid and rod-like than the *meta* linkages in the other polymers. With this in mind one would think that polymers 16 - 18 would be good candidates for crystallizable materials. As far as we are able to determine these materials show no crystalline melting point even after attempts at annealing the samples close to the Tg. One fascinating note concerns polymer 18. Through one rather serendipitous experiment it was found that 18 exhibited unusual liquid crystalline properties. The experiment involved a sample which was to be used for an NMR experiment. As it turned out, the sample 'gelled' in the NMR tube where, upon shaking, became liquid again. Placing some of the sample under an optical microscope with crossed polars revealed some very interesting liquid crystal-like colors. If the structure of 18 is carefully examined, the fact that it exhibits these properties is not overly surprising with the rigid structure found in the molecule. Further experimentation on this phenomenon was not carried out.

The last area which was examined was the thermal stability of these materials. Our definition of thermal stability was arbitrarily taken as that temperature where 10% weight loss was observed. As can be seen in Table 4, the thermal stability of all these materials is well over 350°C even in an air environment. The thermal stability of these materials in air is surprising seeing that the polymers contain benzylic groups. These benzylic groups should be sites for oxidation and subsequent molecular weight breakdown. Of all the polymers only 20 showed a dramatic decrease in thermal stability on going from Argon to air. The remaining polymers seem to behave similarly in air as in Argon. (Polymer 18, which actually appears to be more thermally stable in air than Argon, seems to be the only exception. This could be a phenomenon where the weight loss of the polymer is more than made up for by increasing weight from oxidation.) We will not make any claims on any of the TGA experiments, but the results are interesting. In order to characterize these materials more completely, closer examination of their thermal properties is necessary.

Table 4. Properties of silarylene and silarylenesiloxane polymers.

Polymer	GPC M_w	GPC M_n	Thermal Analysis T_g	Thermal Analysis T_{dec} (air)	Thermal Analysis T_{dec} (Argon)
16	16,100	7,500	64°C	495°C	505°C
17	170,000	25,000	< 25°C	471°C	487°C
18	8,700	2,200	110°C	540°C	487°C
19	59,000	5,800	0°C	485°C	440°C
20	29,500	7,300	-42°C	376°C	456°C
21	7,200	3,400	59°C	465°C	494°C

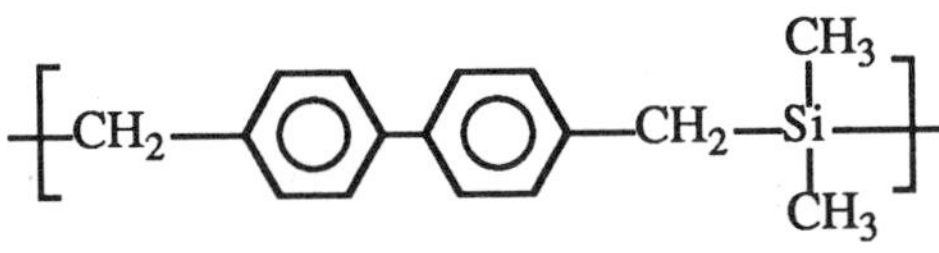

16

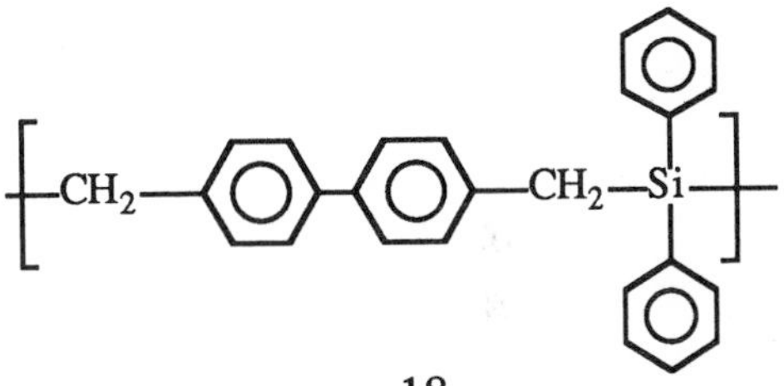

18

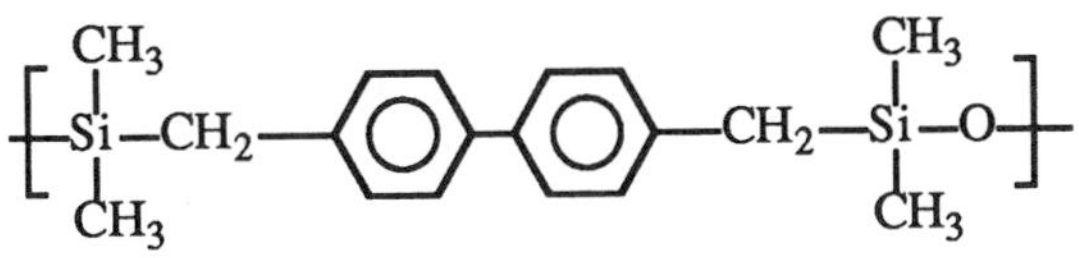

17

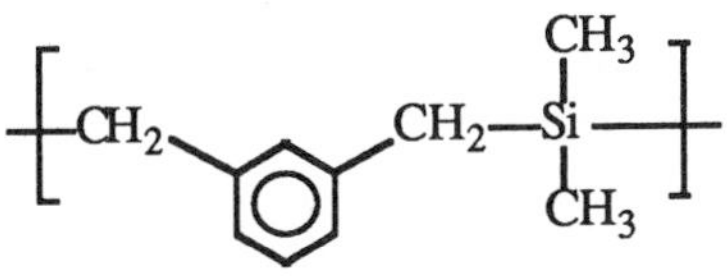

19

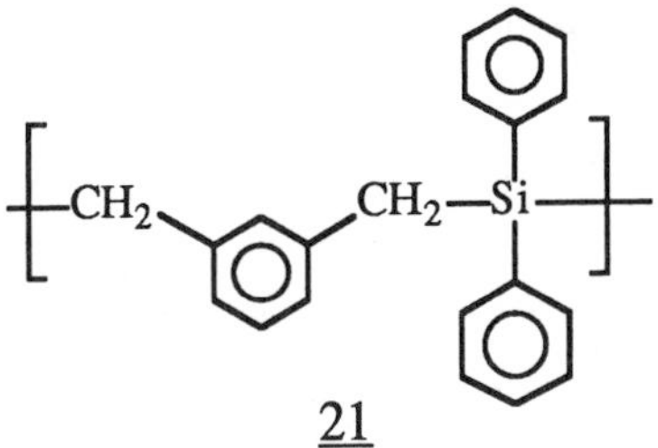

21

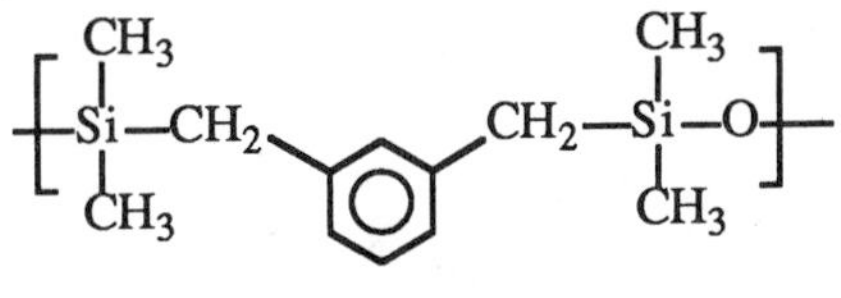

20

Figure 1. Structures of the polymers described in Table 4.

CONCLUSIONS

The use of highly reactive carbanions in polymer synthesis has proven to be an interesting subject. The work which has been described here indicates some of the potential for novel polymeric materials. Examination of the polymers prepared in this work hints at some of the endless possibilities for the preparation of similar materials. There are many other dicarbanions which can be used in analogous syntheses. One molecule which would be interesting to examine is the dicarbanion of 2,3-dimethyl-1,3-butadiene which would result in a silicon containing polymer that contains a backbone diene unit. Another area of interest which has emerged from this work is the preparation of polymers which contain other inorganic atoms in the polymer backbone such as tin. We believe materials such as these would have some very unique properties.

ACKNOWLEDGEMENTS

The authors would like to thank the National Science Foundation, Polymers Program (grant DMR 8214211) for their support. One author (J.E.L.) would also like to thank D. R. Bassett for giving him the opportunity to present this paper.

REFERENCES

1. Present address: Union Carbide Corporation, P.O. Box 8361, South Charleston, WV 25303.
2. The work presented in this paper was done at the Pennsylvania State University as part of the Ph.D. Dissertation of J. E. Loftus.
3. P. R. Dvornic and R. W. Lenz, *J. Pol. Sci. Pol. Chem.*, 20, 593 (1982)
4. M.E. Livingston, P. R. Dvornic, and R. W. Lenz, *J. Applied Pol. Sci.*, 27, 3239 (1982)
5. S. Dobos and B. Zelei, *J. Pol. Sci. Pol. Chem.*, 17, 2651 (1979)
6. R. L. Merker and M. J. Scott, *J. Pol. Sci. Part A*, 2, 15 (1964)
7. L. W. Breed, R. L. Elliott, and M. E. Whitehead, *J. Pol. Sci. Part A-1*, 5, 2745 (1967)
8. Y.-C. Lai, P. R. Dvornic, and R. W. Lenz, *J. Pol. Sci. Pol. Chem.*, 20, 2277 (1982)
9. R. E. Burks, Jr., E. R. Covington, M. V. Jackson, and J. E. Curry, *J. Pol. Sci. Pol. Chem.*, 11, 319 (1973)
10. E. N. Peters, *J. Macromol. Sci.-Rev. Macromol. Chem.*, C17, 173 (1979)
11. P. R. Dvornic and R. W. Lenz, *J. Applied Pol. Sci.*, 25, 641 (1980)
12. P. R. Dvornic and R. W. Lenz, *J. Pol. Sci. Pol. Chem.*, 20, 951 (1982)
13. N. Koide and R. W. Lenz, *J. Pol. Sci. Pol. Symp.*, 70, 91 (1983)
14. P. R. Dvornic and R. W. Lenz, *Polymer*, 24, 763 (1983)
15. R. W. Lenz and R. A. Rhein, *Polymer Preprints*, 25, 17 (1984)
16. H. Rosenberg and E. Choe in, "Organometallic Polymers," ed. C. E. Carraher, Jr., J. E. Sheats, and C. U. Pittman, Jr., Academic Press, Inc., New York, 1978, p. 239
17. I. J. Goldfarb, E. Choe, and H. Rosenberg in, "Organometallic Polymers," ed. C. E. Carraher, Jr., J. E. Sheats, and C. U. Pittman, Jr., Academic Press, Inc., New York, 1978, p. 249
18. C. U. Pittman, Jr., W. J. Patterson, and S. P. McManus, *J. Pol. Sci. Pol. Chem.*, 14, 1715 (1976)
19. W. J. Patterson, S. P. McManus, and C. U. Pittman, Jr., *J. Pol. Sci. Pol. Chem.*, 12, 837 (1974)
20. W. J. Patterson and D. E. Morris, *J. Pol. Sci. Part A-1*, 10, 169 (1972)
21. L. W. Breed and J. C. Wiley, Jr., *J. Pol. Sci. Pol. Chem.*, 14, 83 (1976)
22. B. Zelei, M. Blazso, and S. Dobos, *Eur. Pol. J.*, 17, 503 (1981)

23. M. Ikeda, T. Nakamura, Y. Nagase, K. Ikeda, and Y. Sekine, *J. Pol. Sci. Pol. Chem.*, 19, 2595 (1981)
24. X. W. He, B. Fillon, J. Herz, and J. M. Guenet, *Polymer Bulletin*, 17, 45 (1987)
25. A. McKillop, L. F. Elsom, and E. C. Taylor, *J. Am. Chem. Soc.*, 90, 2423 (1968)
26. B. Gordon, III and J. E. Loftus, *J. Org. Chem.*, 51, 1618 (1986)
27. B. Gordon, III, M. Blumenthal, A. E. Mera, and R. J. Kumpf, *J. Org. Chem.*, 50, 1540 (1985)
28. B. Gordon, III and J. E. Loftus, *Polymer Preprints*, 27, 353 (1986)
29. L. Lochmann, J. Pospisil, and D. Lim, *Tetrahedron Letters*, 2, 257 (1966)
30. J. F. McGarrity and C. A. Ogle, *J. Am. Chem. Soc.*, 107, 1805 (1985)
31. J. F. McGarrity, C. A. Ogle, Z. Brich, and H. R. Loosli, *J. Am. Chem. Soc.*, 107, 1810 (1985)
32. R. B. Bates, B. A. Hess, Jr., C. A. Ogle, and L. J. Schaad, *J. Am. Chem. Soc.*, 103, 5052 (1981)
33. R. B. Bates, F. A. Camou, V. V. Kane, P. K. Mishra, K. Suvannachut, and J. J. White, *J. Org. Chem.*, 54, 311 (1989)
34. See references in the article by H. Rosenburg and E. Choe in "Organometallic Polymers", ed. C. E. Carraher, Jr., J. E. Sheats, and C. U. Pittman, Jr., Academic Press, Inc., New York, 1978, p. 239.

REACTIONS OF ANIONIC POLY(ALKYL/ARYLPHOSPHAZENES) WITH CARBON DIOXIDE AND

FLUORINATED ALDEHYDES AND KETONES

Patty Wisian-Neilson, M. Safiqul Islam, and Tao Wang

Department of Chemistry, Southern Methodist University, Dallas, Texas 75275

ABSTRACT

Two series of new polyphosphazenes were prepared by derivatization of preformed poly(methylphenylphosphazene), $[Me(Ph)PN]_n$. This involved initial deprotonation of part of the methyl substituents with n-BuLi followed by treatment of the intermediate anion with carbon dioxide or with fluorinated aldehydes or ketones. With appropriate workup procedures, either carboxylic acid, ester, carboxylate salt, or fluorinated alcohol derivatives were obtained. These reactions and the characterization of the products are discussed in this paper. Related derivatization reactions are also discussed.

INTRODUCTION

Polyphosphazenes are inorganic polymers consisting of a backbone of alternating single and double bonds between phosphorus and nitrogen. Among polymers, polyphosphazenes are one of the most diverse classes because such a wide variety of substituents can be attached to phosphorus. The nature of these substituents greatly influences the properties, and hence, the potential applications of these polymers. With appropriate side-groups, phosphazene polymers may serve as materials that are flame retardant, flexible at low-temperatures, water repellant or water soluble, thermally stable, resistant to UV radiation and chemicals, insulators, semiconductors, or biologically inert or active. (1, 2). These materials are increasingly attracting the attention of chemists in both research and development laboratories and have been the subject of several reviews. (1, 3 - 5)

BACKGROUND

Since the first soluble polyphosphazenes were prepared in the middle of the 1960's (6), several distinct methods for synthesizing polyphosphazenes have been developed. The best studied and most commonly used method involves ring opening of the cyclic phosphazene, $[Cl_2PN]_3$, to the corresponding chlorinated polymer $[Cl_2PN]_n$, followed by replacement of the chlorines on phosphorus by nucleophilic substitution reactions. This process is the method of choice for the preparation of a very large number of polymers with amino, alkoxy, and aryloxy substituents. (3, 4, 6) A more recent variation of this process involves the ring opening of cyclic phosphazenes with both halogens and other substituents at phosphorus. (7 - 9) Alternatively, $[Cl_2PN]_n$ has been prepared by a condensation polymerization of the monomeric phosphoranimine, $Cl_2P(=O)N=PCl_3$. (10) These variations of ring opening polymerization and nucleophilic substitution processes result in polymers with substituents that are bonded to the backbone by predominantly or exclusively P-N or P-O linkages. The presence of such linkages provides pathways for the depolymerization of these polymers either chemically or thermally. Thus far, attachment of substituents to the backbone by only P-C linkages, which are relatively inert, has not been possible through the substitution process. (11)

This synthetic gap has been bridged by the development of another process that involves the condensation polymerization of suitably constructed silicon-nitrogen-phosphorus compounds. The most significant feature of this method is that it leads directly to fully P-C substituted polyphosphazenes, i.e., poly(alkyl/arylphosphazenes), $[RR'PN]_n$, where R and R' are alkyl and/or aryl groups. (5, 12, 13) The precursor N-silyl-P-trifluoroethoxyphosphoranimine reagents which serve as monomers are prepared in the reaction sequence outlined in equations 1 and 2. Thermolysis of

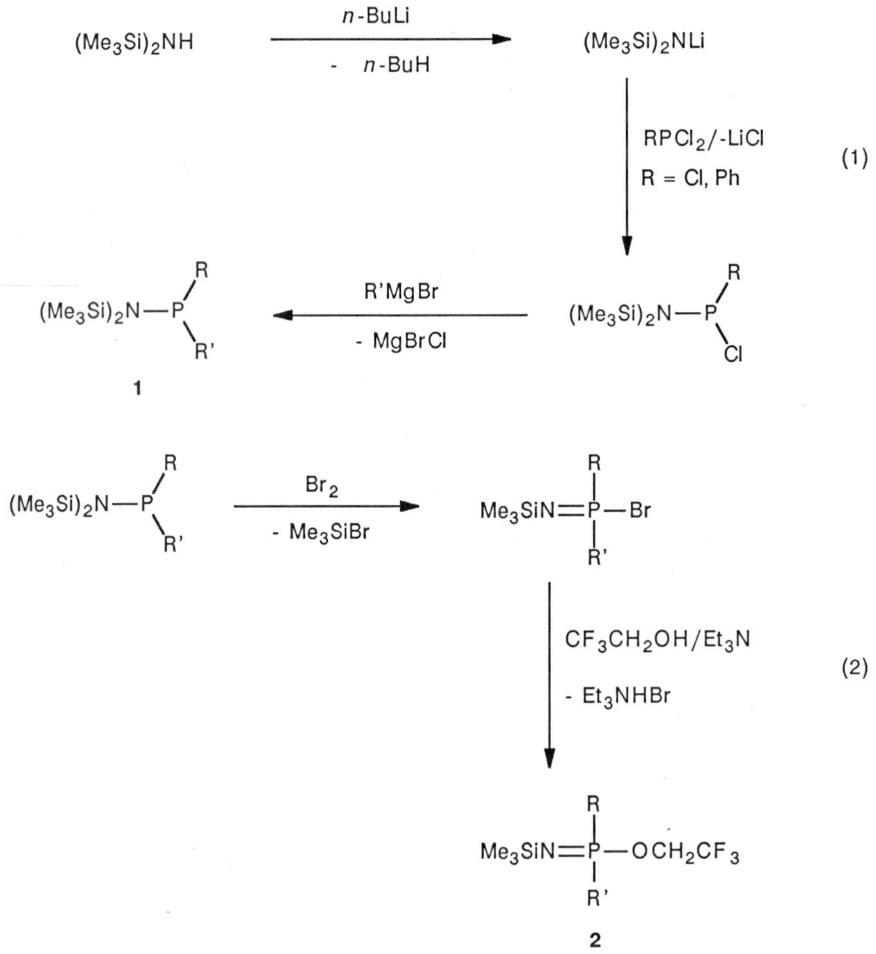

$$(Me_3Si)_2NH \xrightarrow[-\ n\text{-BuH}]{n\text{-BuLi}} (Me_3Si)_2NLi$$

$$\Big\downarrow \begin{array}{l} RPCl_2/\text{-LiCl} \\ R = Cl, Ph \end{array} \quad (1)$$

$$(Me_3Si)_2N\!-\!\overset{\displaystyle R}{\underset{\displaystyle R'}{\overset{\big|}{P}}} \xleftarrow[-\ MgBrCl]{R'MgBr} (Me_3Si)_2N\!-\!\overset{\displaystyle R}{\underset{\displaystyle Cl}{\overset{\big|}{P}}}$$

1

$$(Me_3Si)_2N\!-\!\overset{\displaystyle R}{\underset{\displaystyle R'}{\overset{\big|}{P}}} \xrightarrow[-\ Me_3SiBr]{Br_2} Me_3SiN\!=\!\overset{\displaystyle R}{\underset{\displaystyle R'}{\overset{\big|}{P}}}\!-\!Br$$

$$\Big\downarrow \begin{array}{l} CF_3CH_2OH/Et_3N \\ -\ Et_3NHBr \end{array} \quad (2)$$

$$Me_3SiN\!=\!\overset{\displaystyle R}{\underset{\displaystyle R'}{\overset{\big|}{P}}}\!-\!OCH_2CF_3$$

2

these phosphoranimines in sealed vessels at approximately 200°C for several days results in quantitative elimination of $Me_3SiOCH_2CF_3$ (eq 3) to yield the alkyl or alkyl and aryl substituted polymers. (14) With the exception of poly(diethylphosphazene), $[Et_2PN]_n$, the simple phosphazenes are soluble in several common organic solvents. The polymers, which are easily separated from the volatile silane, typically have molecular weight distributions (M_w/M_n) between 2 and 4 and M_w values in the range of 50,000 to 250,000. The glass transition temperatures for the simple poly(alkyl/arylphosphazenes) listed in equation 3 are between -46 and 37°C. Simple copolymers can be prepared by cothermolysis of different phosphoranimines.

$$Me_3SiN{=}\overset{\overset{\displaystyle R}{|}}{\underset{\underset{\displaystyle R'}{|}}{P}}{-}OCH_2CF_3 \quad \xrightarrow[- Me_3SiO\,CH_2CF_3]{160 - 190°C} \quad \left[{-}N{=}\overset{\overset{\displaystyle R}{|}}{\underset{\underset{\displaystyle R'}{|}}{P}}{-}\right]_n \qquad (3)$$

3

$$R/R' = Me/Me;\ Me/Ph;\ Et/Et;\ Et/Ph$$

DERIVATIZATION OF POLY(ALKYL/ARYLPHOSPHAZENES)

In principal, there are three general ways to vary the substituents bonded to the backbone of poly(alkyl/arylphosphazenes): (a) by introducing various groups at the phosphine precursor stage (eq 1); (b) by chemically modifying the substituents in the phosphoranimine precursors, **2**; (c) or by modifying the substituents in the preformed polymers. The first method Is limited by the availability of suitable organometallic reagents that are needed to attach the desired groups, R and R', to the phosphorus (eq 1). The second approach is more versatile and a number of potential polymer precursors have been made in this manner (eq 4). Treatment of simple P-methylphosphoranimines

$$Me_3SiN{=}\overset{\overset{\displaystyle Me}{|}}{\underset{\underset{\displaystyle Me}{|}}{P}}{-}OCH_2CF_3 \quad \xrightarrow{n\text{-}BuLi} \quad \left[Me_3SiN{=}\overset{\overset{\displaystyle Me}{|}}{\underset{\underset{\displaystyle CH_2^-}{|}}{P}}{-}OCH_2CF_3\right]$$

4

$$(4)$$

$$Me_3SiN{=}\overset{\overset{\displaystyle Me}{|}}{\underset{\underset{\displaystyle CH_2E}{|}}{P}}{-}OCH_2CF_3 \quad \xleftarrow{EX}$$

5

$$E = alkyl,\ SiMe_2R,\ CH_2CH{=}CH_2$$

with n-BuLi results in an unstable intermediate anion, **4**. This anion reacts readily with electrophiles to give new phosphoranimines, **5**, many of which contain functional groups suitable for further derivatization reactions. (15 - 17) However, many of these phosphoranimines do not undergo simple condensation polymerization. In some cases these compounds are unusually thermally stable, while in other cases complicating side reactions such as crosslinking occur. Furthermore, introduction of some groups is complicated by the reactivity of the Si-N or P-O linkages in the simple phosphoranimine. Nonetheless, these deprotonation-substitution reactions serve as models for reactions of the preformed polymers.

The preformed polymers may also be modified by various chemical reactions. There are three reactive sites in poly(methylphenylphosphazene), $[Me(Ph)PN]_n$, i.e., the lone pair on the backbone nitrogen which is susceptible to adduct formation, the phenyl group which can undergo electrophilic substitution reactions, and the methyl group with relatively acidic hydrogens that allow for deprotonation. Thus far, we have most closely examined the reactivity of the methyl group. Some simple experiments also indicate that the phenyl group is readily nitrated by sulfuric and nitric acids. (18)

The first derivatization reactions (19) of a poly(alkyl/arylphosphazene) involved the deprotonation of either 20 or 50% of the methyl groups in $[Me(Ph)PN]_n$ (eq 5). Although the intermediate anion 6 was not isolated, it reacted smoothly with various chlorosilanes (eq 6) to give poly(phosphazenes) with silylmethyl substituents. The polymers were characterized by elemental analysis and ^1H, ^{31}P, and ^{29}Si NMR spectroscopy. Both GPC analysis and intrinsic viscosity measurements demonstrated that the deprotonation-substitution process occurred with no degradation of the polymer backbone. Several of these silyl derivatives, 7, contain reactive functional groups that could serve as sites for crosslinking or additional derivatization reactions.

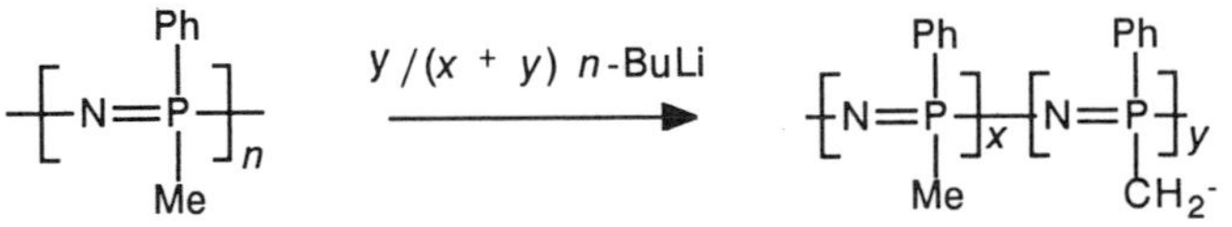

$$\qquad\qquad\qquad (5)$$

6

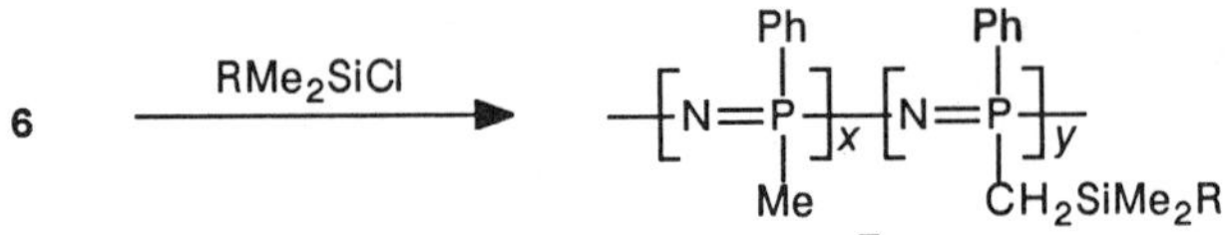

$$\qquad\qquad\qquad (6)$$

$$R = Me,\ CH=CH_2,\ (CH_2)_3CN,\ H$$

The polymeric anion intermediate is prepared by treatment of THF solutions of the polymer with the appropriate amount of n-BuLi at -78°C. Best results are obtained when this solution is allowed to stir for about one hour at this temperature before the addition of other reagents. The intermediate behaves as a simple organolithium reagent and reacts with a variety of electrophiles. Treatment with aldehydes and ketones, followed by a workup procedure that involved aqueous NH_4Cl, resulted in polyphosphazenes, 8, with alcohol functionalities attached to the backbone through two-carbon spacer groups (eq 7). This reaction sequence has been used to incorporate both additional alkyl and aryl groups as well as such interesting groups as ferrocene and thiophene. (20, 21)

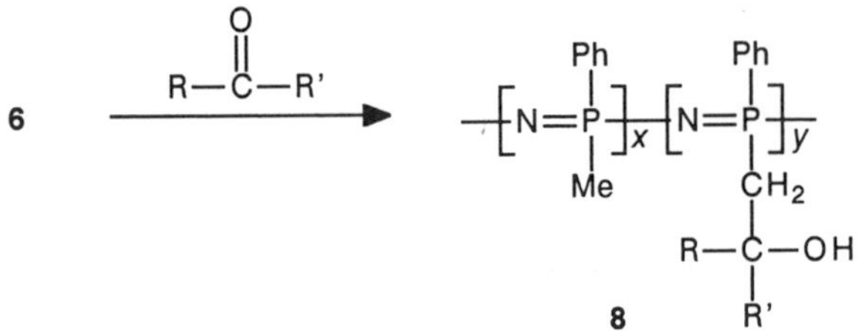

$$\qquad\qquad\qquad (7)$$

$$R = Me,\ H;\ R' = Me,\ Ph,$$

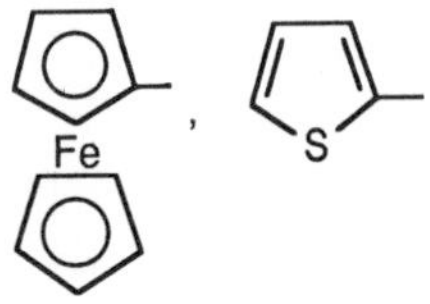

The polymer anion can also serve as sites for initiation of anionic addition polymerization and certain ring opening polymerizations. A series of graft copolymers, 9, with varying amounts of polystyrene attached to the phosphazene were made via the reaction of the anion with styrene (eq 8). (22) These reactions could be carried out with some degree of control of the length and number of grafted chains. The grafted copolymers are soluble, uncrosslinked materials that readily form transparent films. Similar reactions with other vinylic monomer systems should provide access to a variety of new inorganic-organic copolymers.

$$
6 \xrightarrow[\text{(2) H}^+]{\text{(1) CH}_2=\text{C(Ph)H}} \left[\text{N}=\underset{\underset{\text{Me}}{|}}{\overset{\overset{\text{Ph}}{|}}{\text{P}}}\right]_x \left[\text{N}=\underset{\underset{\underset{\underset{\text{Ph}}{|}}{\text{CH}_2-\text{CH}]_z\text{H}}}{|}}{\overset{\overset{\text{Ph}}{|}}{\text{P}}}\right]_y \tag{8}
$$

$$
\mathbf{9}
$$

Copolymers containing the isoelectronic Si-O backbone grafted to the P-N backbone were obtained by using the polyphosphazene anion to open the ring in hexamethylcyclotrisiloxane (23, 24). To avoid crosslinking reactions, a THF solution of the polymer anion at -78°C was treated with $[\text{Me}_2\text{SiO}]_3$ and the resulting mixture was warmed to room temperature and stirred for two hours. The reaction was quenched with Me_3SiCl to give the inorganic-inorganic copolymer system (eq 9). Differential scanning calorimetry measurements revealed two glass transition temperatures indicating the existence of two phases in the copolymers. Thermal gravimetric analysis suggested that the grafts exhibit slightly higher thermal stability than the parent polyphosphazene.

$$
6 \xrightarrow[\text{(2) Me}_3\text{SiCl}]{\text{(1) (Me}_2\text{SiO)}_3} \left[\text{N}=\underset{\underset{\text{Me}}{|}}{\overset{\overset{\text{Ph}}{|}}{\text{P}}}\right]_x \left[\text{N}=\underset{\underset{\underset{\text{CH}_2[\text{Me}_2\text{SiO}]_z\text{SiMe}_3}{|}}{|}}{\overset{\overset{\text{Ph}}{|}}{\text{P}}}\right]_y \tag{9}
$$

$$
\mathbf{10}
$$

Carboxylic Acid Derivatives

The deprotonation-substitution approach has also been used to prepare polymers with carboxylic acid and ester and carboxylated salt moieties attached to the phosphazene. (25) THF solutions of the parent polymer were treated with sufficient *n*-BuLi to facilitate deprotonation of 10, 25, or 50% of the methyl substituents. The resulting polymer anions were then treated with anhydrous carbon dioxide to produce the carboxylate salts 11 (eq 10). These salts were isolated by removal of the solvent or were used in further reactions with dilute aqueous acid solutions (eq 11) or with *p*-nitrobenzylbromide (eq 12) to give the carboxylic acids, 12, or esters, 13, respectively.

$$
6 \xrightarrow{\text{CO}_2} \left[\text{N}=\underset{\underset{\text{Me}}{|}}{\overset{\overset{\text{Ph}}{|}}{\text{P}}}\right]_x \left[\text{N}=\underset{\underset{\underset{\text{CH}_2\text{COO}^-\text{Li}^+}{|}}{|}}{\overset{\overset{\text{Ph}}{|}}{\text{P}}}\right]_y \tag{10}
$$

$$
\mathbf{11}
$$

$$
11 \xrightarrow{\text{H}^+} \left[N=\overset{\overset{\displaystyle Ph}{|}}{\underset{\underset{\displaystyle Me}{|}}{P}} \right]_x \left[N=\overset{\overset{\displaystyle Ph}{|}}{\underset{\underset{\displaystyle CH_2COOH}{|}}{P}} \right]_y \tag{11}
$$

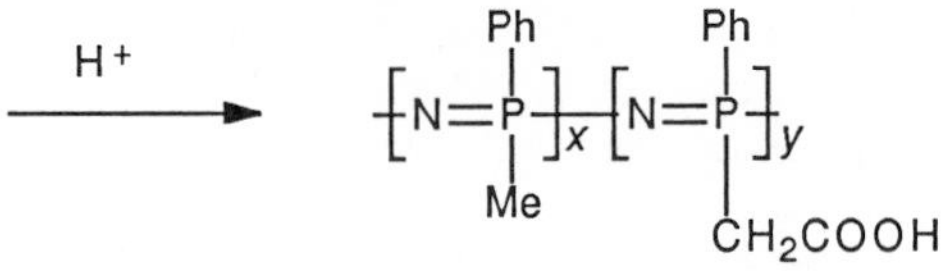

12

$$
11 \xrightarrow{\text{BrCH}_2\text{C}_6\text{H}_4\text{NO}_2} \left[N=\overset{\overset{\displaystyle Ph}{|}}{\underset{\underset{\displaystyle Me}{|}}{P}} \right]_x \left[N=\overset{\overset{\displaystyle Ph}{|}}{\underset{\underset{\displaystyle CH_2COOCH_2C_6H_4NO_2}{|}}{P}} \right]_y \tag{12}
$$

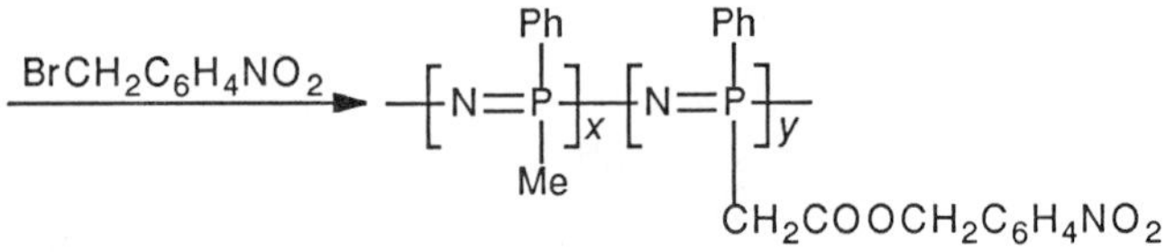

13

The solubilities of these new polymers were quite dependent on the nature of the substituents. The polymers, **11**, in which 10 and 25% of the methyl groups were converted to $CH_2COO^-Li^+$ groups (i.e., where $x = 9$, $y = 1$ and $x = 3$, $y = 1$), were insoluble in THF, hydrocarbons, and water, but soluble in 1:1 mixtures of THF and water. On the other hand, the polymer with 50% substitution (i.e., where $x = 1$) was quite water soluble and insoluble in all organic solvents. Hence, NMR spectra of the salts were obtained on swollen gels in $CDCl_3$ or in D_2O. The acid and ester derivatives **12** and **13** were soluble in THF and chlorinated hydrocarbons. These polymers were isolated and purified by precipitation of THF or CH_2Cl_2 solutions into hexanes.

All of the new derivatives were off-white colored materials that formed brittle films. They were characterized by elemental analysis and NMR spectroscopy. The salts required extended periods of drying to completely remove all solvents, after which they gave satisfactory elemental analysis. Integration of the ^1H NMR spectra indicated that the degree of substitution approached that predicted by the stoichiometry of the reaction with n-BuLi. The ^{13}C NMR spectra were straightforward.

The ^{31}P NMR spectroscopic data, on the other hand, were quite interesting. While the spectra of the salts contained only one signal (1.2 ppm), the esters exhibited two signals (1 ppm, P-Me; -5 ppm, PCH_2COOR) with intensities varying with degree of substitution. The acid derivatives exhibited very broad ^{31}P NMR spectral signals with peak maxima at 1.5 and 6.0 ppm. A third and smaller signal was observed at 4.0 ppm for the 25% substituted acid. The broad nature of these NMR signals and the relatively downfield shifts are similar to the signals in the spectra of partially protonated parent polymer, suggesting that the acid derivatives may actually autoprotonate to give a species of the type shown here. (22)

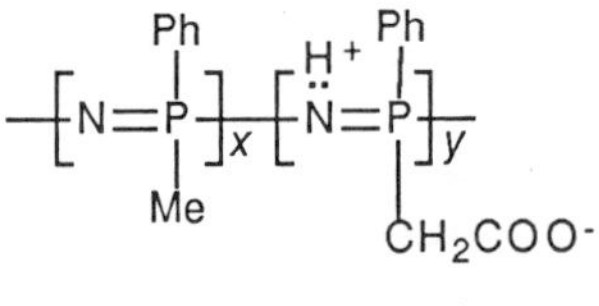

$$
\left[N=\overset{\overset{\displaystyle Ph}{|}}{\underset{\underset{\displaystyle Me}{|}}{P}} \right]_x \left[\overset{\overset{\displaystyle Ph}{|}}{\underset{\underset{\displaystyle CH_2COO^-}{|}}{\underset{\cdots}{\overset{H^+}{N}}=P}} \right]_y
$$

This is supported by IR spectra in which a broad absorption at 2650 cm^{-1} is observed, which is characteristic of P-(N-H)-P stretching (26). However, it should be noted that the carbonyl stretching frequency of 1740 cm^{-1} is more typical of the COOH moiety than of the corresponding anion COO$^-$, so the actual structure of the acid derivatives is open to debate.

As for the previous derivatives, GPC analysis of the new polymers indicated that the molecular weights were higher than that of the parent polymer and within the range of the theoretical molecular weights based on the size and number of groups attached to the final polymer. Molecular weights could not be determined for the carboxylate salts due to their insolubility in THF.

The thermal gravimetric analysis of these polymers indicated onsets of decomposition between 250 and 350°C on heating in air at a rate of 10°C per min. This is well below that of the parent polymers. Differential scanning calorimetry was used to obtain the glass transition temperatures of these polymers. For the acids, **12**, the T_g values for the 10, 25, and 50% substituted polymers were 58, 78, and 108°C, respectively. This is higher than that of the unsubstituted parent polymers and may be explained by an increased degree of hydrogen bonding. The esters, **13**, exhibited T_g values of 48, 70, and 85°C for 10, 25, and 50% substitution with this increase due to the size and number of substituents along the backbone. The thermal transitions of the salts were very broad and difficult to interpret but generally appeared between 30 and 50°C.

The salt derivatives, **11**, were readily crosslinked upon addition of the dication Cu^{+2}. Similar crosslinking to form phosphazene hydrogels has been reported previously. (27) We have not been able to prepare the corresponding acid chlorides due to severe degradation of the polymer backbone by the HCl that forms when standard procedures are used. The acid chlorides would provide ready access to a variety of carboxyl derivatives including esters. We have, however, found that the alcohol derivatives discussed above readily form related esters when treated with simple acid chlorides.

Fluorinated Alcohol Derivatives

Fluorinated polymers are of interest due to altered solubilities, thermal stability, and chemical reactivity afforded by fluoroalkyl or fluoroaryl substituents. The attachment of such groups to the phosphazene backbone by direct P-C bonds is not trivial. Thus, we have explored the possibility of attaching fluorinated substituents through reactions on the preformed polymers. Our initial work has centered on the use of relatively simple fluorinated aldehydes and ketones, R(C=O)R'. Using reaction conditions analogous to those described above for other aldehydes and ketones, several new, partially fluorinated polymers have been obtained (eq 13). Several of these new polymers have drastically different solubilities relative to the parent polymer and the other alcohol derivative. When half of the methyl groups were deprotonated ($x = y$) and treated with pentafluorobenzaldehyde, an insoluble gel formed and this remained insoluble even on quenching the reaction. The product of 10% deprotonation ($x = 9$, $y = 1$) was soluble in common solvents and the polymer from 25% deprotonation was only slightly soluble. Other aldehyde reactions displayed similar crosslinking, presumably due to complicating side reactions.

$$6 \quad \xrightarrow{\;R-\overset{\displaystyle O}{\overset{\|}{C}}-R'\;} \quad \text{(structure 14)} \tag{13}$$

	a	b	c	d
R	C_6F_5	C_6H_5	CF_3	C_6H_5
R'	H	CF_3	CF_3	CF_2CF_3

The reactions of the ketones were more straightforward and produced soluble polymers, **14b** - **d**. The thermal stability of these fluorinated alcohols is substantially lower than that of the parent polymers. The glass transition temperatures range from ca. 40 to 102°C depending on the size and number of substituents and are comparable to the simple non-fluorinated alcohols.

ACKNOWLEDGMENT

We thank the U. S. Army Research Office and The Welch Foundation for generous financial support of this project.

REFERENCES

1. *ACS Symp. Series* **1988**, *360*, Chapters 19 - 25.
2. See for example: (a) Singler, R. E.; Hagnauer, G. L.; Sicka, R. W. *ACS Symp. Series* **1984**, *260*, 143. (c) Blonsky, P. M.; Shriver, D. F.; Austin, P., Allcock, H. R. *J. Am. Chem. Soc.* **1984**, *106*, 6854. (d) Allcock, H. R. *ACS Symp. Series* **1983**, *232*, 439.
3. Allcock, H. R. *Angew Chem., Int. Ed. Engl.* **1977**, *16*, 147.
4. Allcock, H. R. *Chem. Eng. News* **1985**, *63(11)*, 22.
5. Neilson, R. H.; Wisian-Neilson, P. *Chem. Rev.* **1988**, *88*, 541.
6. Allcock, H. R.; Kugel, R. L. *Inorg. Chem.* **1966**, *5*, 1016 and 1716.
7. Allcock, H. R.; Ritchie, R. J.; Harris, P. J. *Macromolecules* **1980**, *13*, 1332.
8. Allcock, H. R.; Lavin, K. D.; *Macromolecules* **1980**, *13*, 1332.
9. Allcock, H. R.; Brennan, D. J.; Graskamp, J. M. *Macromolecules* **1988**, *21*, 1.
10. Bouchaccra, T. A.; Helioui, M.; Puskaric, E.; DeJaeger, R. J. *Chem. Res. Synop.* **1981**, 230.
11. See for example: (a) Allcock, H. R.; Evans, T. L.; Patterson, D. B. *Macromolecules* **1980**, *13*, 201. (b) Allcock, H. R. *J. Am. Chem. Soc.* **1983**, *105*, 2814.
12. Wisian-Neilson, P.; Neilson, R. H. *J. Am. Chem. Soc.* **1980**, *102*, 2848.
13. Neilson, R. H.; Hani, R.; Wisian-Neilson, P.; Meister, J. J.; Roy, A. K.; Hagnauer, G. L. *Macromolecules* **1987**, *20*, 910.
14. Wisian-Neilson, P.; Neilson, R. H. *Inorg. Synth.* **1989**, *25*, 69.
15. Wettermark, U. G.; Wisian-Neilson, P.; Scheide, G. M.; Neilson, R. H. *Organometallics* **1987**, *6*, 959.
16. Scheide, G. M.; Neilson, R. H. *Organometallics* **1989**, *8*, 1987.
17. Scheide, G. M.; Neilson, R. H. *Phosphorus, Sulfur, and Silicon* **1989**, *39*, 189.
18. Wisian-Neilson, P.; Ford, R. R.; Wood, C., unpublished results.
19. Wisian-Neilson, P.; Ford, R. R.; Roy, A. K.; Neilson, R. H. *Macromolecules* **1986**, *19*, 2089.
20. Wisian-Neilson, P.; Ford, R. R. *Organometallics* **1987**, *6*, 2258.
21. Wisian-Neilson, P.; Ford, R. R. *Macromolecules* **1989**, *22*, 72.
22. Wisian-Neilson, P.; Schaefer, M. A. *Macromolecules* **1989**, *22*, 2003.
23. Wisian-Neilson, P.; Islam, M. S.; Schaefer, M. A. *Phosphorus, Sulfur, and Silicon* **1989**, *41*, 135.
24. Wisian-Neilson, P.; Islam, M. S. *Macromolecules* **1989**, *22*, 2026.
25. Wisian-Neilson, P.; Islam, M. S.; Ganapathiappan, S.; Scott, D.L.; Raghuveer, K. S.; Ford, R. R. *Macromolecules* **1989**, *22*, 4382.
26. (a) Allcock, H. R.; Walsh, E. J. *J. Am. Chem. Soc,* **1972**, *94*, 119 and 4538. (b) Ganapathiappan, S.; Krishnamurthy, S. S. *J. Chem. Soc, Dalton Trans.* **1987**, 585.
27. Allcock, H. R.; Kwon, S. *Macromolecules* **1989**, *22*, 75.

POLYMERIC AUXIN PLANT GROWTH HORMONES BASED ON THE CONDENSATION PRODUCTS OF INDOLE-3-BUTYRIC ACID WITH BIS(CYCLOPENTADIENYL) TITANIUM IV DICHLORIDE AND DYPYRIDINE MANGANESE II DICHLORIDE

Herbert H. Stewart[a], Winn J. Soldani II,[c]
Charles E. Carraher, Jr.[b], and Lisa
Reckleben[b]

Departments of Biological Sciences[a]
and Chemistry[b]
Florida Atlantic University
Boca Raton, Florida   33431

Fancy Hibiscus[c]
1142 S.W. First Avenue
Pompano Beach, Florida   33060

ABSTRACT

   Auxin is a generic name for compounds with the ability to induce elongation in shoot cells. The major commercially employed auxin is indole-3-butyric acid, IBA. IBA was condensed with bis(cyclopentadienyl)titanium IV dichloride and dipyridine manganese II dichloride forming the corresponding poly(amine esters). These polymeric materials are more effective in promoting root formation in tests employing the hibiscus Albo Lacinatus, a commonly utilized rooting material employed in the commercial production of grafted hybrid hibiscus.

INTRODUCTION

   A living green plant -- a mature angiosperm with roots, stems and leaves and flowers, fruits and seeds -- is the stunning result of a whole series of improbable events. How from a single, fertilized egg cell, can there be the growth and differentiation and shaping of the cells, tissues and organs of an ongoing, living organism? Many of the details are not known, but investigations indicate that the principal factors that regulate plant growth and development are chemical in nature.

   These chemicals have sometimes been referred to as plant hormones. The word hormone is derived from the Greek hormaein and means "to excite"[1]. While some of these chemicals do excite or stimulate plant growth, many others exert a shaping influence through inhibitory influences. As a result, the term hormone is not a precise one for all of these plant chemicals. In addition, while many chemicals are found naturally occurring in plants(they are endogenous), literally dozens of others have been synthesized (they are exogenous to the plant) and have been found to modify plant growth. In order to make allowance for these distinctions, some plant physiologists have preferred to use the term plant growth substance. Still others prefer use of the term plant growth regulator [2]. Here, plant growth regulator (PGR) is used, recognizing that even this term has limitations, since regulation depends not only on the chemical structure of the exogeneous or endogenous chemical but also upon the nature, age, and sometimes unknown other factors of the cells of the target tissue which receive the chemical regulator, and sometimes on the complex interplay of other PGRs which may be present.

Endogeneous PGRs are produced in exceedingly minute amounts. The first PGR to be isolated and identified -- indole-3-acetic acid -- has been found in the shoot of the pineapple plant (Ananas comosus L Merr.) at a level of 6 micrograms per kilogram of plant material. According to a calculating plant physiologist, this is comparable to the weight of the proverbial needle in 20 metric tons of hay (3).

In animals, hormones are produced by specific groups of cells or glands but in plants all living cells appear to be potentially capable of PGR synthesis. As PGRs are synthesized within a cell, they may modify metabolic activities within that cell or in surrounding cells. They, also, may be transported from one part of a plant to another part of the plant by simple diffusion or by energy-consuming active transport via xylem and phloem tissue (4).

POLYMERS, PGRs AND PLANT PRODUCTION

Through the use of chemical technology, large increases have been reported in recent decades in the production of agronomic crops in the United States (5). New herbicides, fungicides, insecticides, and fertilizers have been developed and applied on a broad scale on farms and orchards. Recently, however, widespread fears have developed concerning possible air and water pollution and the contamination of fruits and vegetables with toxins. These fears have developed to the point where some consumers are reported to be deliberately looking for worm holes in apples in the supermarket as a sign of the absence of toxins.

The development and use of endogenous PGRs, which occur naturally in plants, and the careful development and use of new and existing synthetic PGRs may help to allay fears and also help to make up for possible deficiencies in crop production due to the decreased use of insecticides and fungicides in the future.

The development and use of PGRs is held to have great promise. One source goes so far as to maintain that PGRs offer plant-cultivation improvements beyond the present limits of man's imagination and that this technology may provide the type of quantum increase in plant production necessary for man's survival (6).

In order for a PGR to elicit optimum physiological response from a plant, it must satisfy at least two requirements: (1) it must be present in carefully controlled amounts (usually quite small) in a given tissue, and (2) it must be available continuously at the site of action. These two requirements may possibly be met by research on the controlled release of a polymer and a biologically-active agent, in this case a PGR.

THE FIRST PGR: AN AUXIN

Charles Darwin's inquisitiveness started a series of investigations that led to eventual isolation and identification of the first PGR (7). Why do plants turn towards the light? It was a common observation that a houseplant placed in a window would soon turn towards the light.

Darwin and his son, Francis, grew grass seedlings (Phalaris canariensis) in greenhouse flats in total darkness. Some of the greenhouse flats were momentarily exposed to the light of a candle from one side. All of the plants in those flats subsequently grew in the direction where the candle light had been given.

When Darwin carefully covered the tips of the coleoptiles of some seedlings with metal caps so that no light could get in, and then exposed them momentarily to a unilateral light source, he found no subsequent bending. Darwin hypothesized that some influence present in the tip of the plant moved down the seedling after exposure to light, causing it to bend.

What was the influence? A Dutch plant physiologist, Frits Went, cut off the tips of the coleoptiles of oat seedlings (Avena sativa) and placed them on tiny agar blocks for one hour so that the cut surfaces of the tips were placed face down and could drain into the agar. He then attached each agar block to the side of each decapitated plant, continuing to grow them in darkness. Within one hour he found the plants curving away from the side on which the agar blocks had been attached. Went concluded that a chemical was transmitted down the side of the plant from the agar block and caused the curvature. He named this chemical auxin from the Greek auxein, meaning to increase (8).

By attaching agar blocks containing auxin to the sides of decapitated plants and measuring the resulting curvature of the plants in degrees from the perpendicular, he found that, within limits, the degree of curvature was proportional to the strength of the auxin. By standardizing the procedure, a bioassay was developed by which unknown strengths of auxin could be determined based on plant curvature. Later investigators using this bioassay were able to isolate and identify this endogeneous auxin as indole-3-acetic acid or IAA (9).

IAA has an indole ring and an acetic acid side chain.

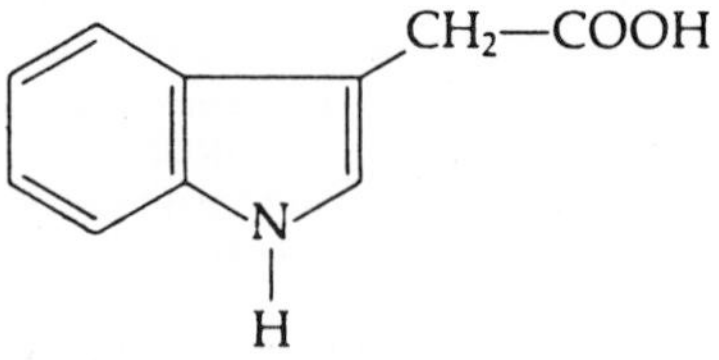

CH$_2$—COOH

indole-3-acetic acid (IAA)

Two important synthetic indoles are indole-3-butyric acid and indole-3-propionic acid. These two synthetics have the same indole ring as the naturally-occurring IAA but differ in their side chains. When longer side chains are added, the compounds appear to lose biological activity. Certain plants, however, apparently possess enzymes that shorten the side chains and transform them into biologically active ones (10).

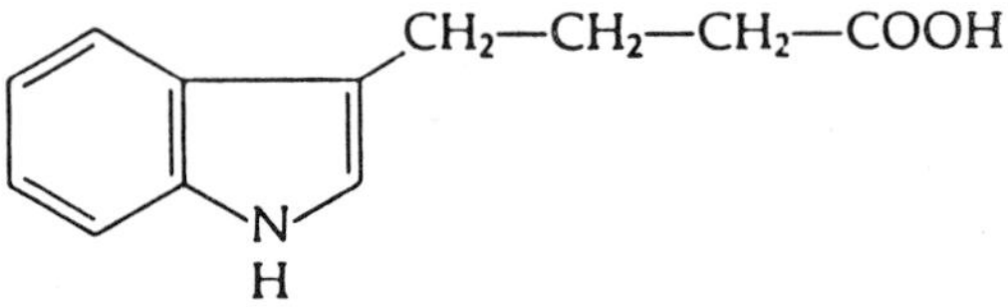

CH$_2$—CH$_2$—CH$_2$—COOH

Indole-3-butyric acid (IBA)

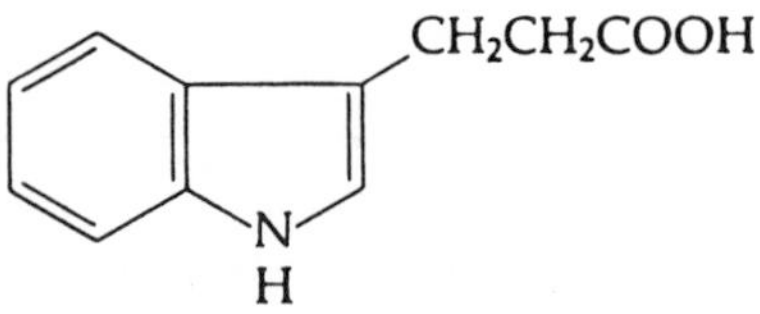

CH$_2$CH$_2$COOH

Indole-3-propionic acid (IPA)

Some biologically active compounds, such as naphthalene acetic acid (NAA), lack the indole ring but retain the acetic acid side chain present in IAA (11).

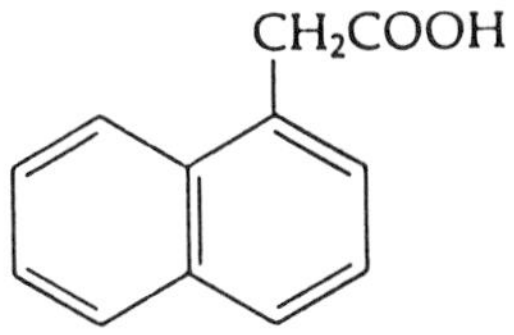

CH$_2$COOH

CH$_2$COOH

a-Naphthalene acetic acid 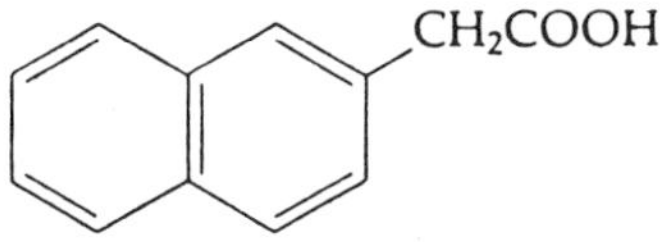   β-Naphthalene acetic acid

Another compound lacking the indole ring is 2,4-D, an auxin which becomes a potent selective weed-killer when used at high concentrations. By the substitution of the chlorine atom at various positions on the phenyl ring, other potent compounds have been developed including 2,4,5-T (12).

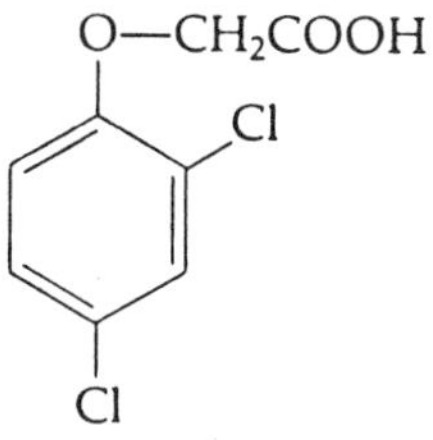

O—CH$_2$COOH

Cl

Cl

O—CH$_2$COOH

Cl

Cl

Cl

2,4-Dichlorophenoxyacetic   2,4,5-Trichlorophenoxyacetic
acid (2,4-D)   acid (2,4,5-T)

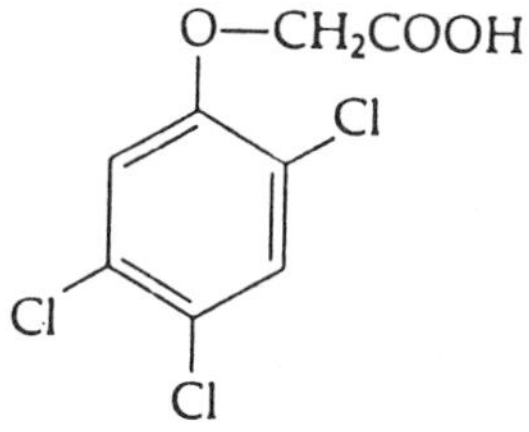

Other important members of synthetic auxins include 2,3,6-trichlorobenzoic acid and 2-methoxy-3,6-dichlorobenzoic acid in the benzoic acid group and a-naphthoxyacetic acid and 4-amino-3,5,6-trichloropicolinic acid in the naphthoxy acid group.

271

AUXINS AND CELLULAR ELONGATION

Within 10 minutes after IAA application, rapid response mechanisms result in coleoptile and stem segment cell wall deformation and elongation (13). Changes in cell metabolism appear to occur which result in the rapid pumping of protons across the cell plasma membrane and a resulting acidification of the cell wall. This acidification leads by an unknown mechanism to the hydrolysis of noncovalent cross-links between xyloglucan polymers and the cellulose microfibrils of the cell wall. The cross links are later reformed in the enlarged cell (14).

With IAA-induced cell wall loosening and an accompanying decrease in resistance to stretching and internal pressure, the cell membrane pushes outward, resulting in a decrease in turgor pressure. When the internal cell pressure becomes less positive, the water potential of the cell sap becomes more negative than in surrounding cells. Thus, water diffuses inward along the newly established energy gradient, stretching and increasing cellular volume. Reestablishment of noncovalent linkages in the cell wall leave an irreversibly enlarged cell (15).

This acid growth hypothesis is supported by empirical data including the fact that acid solutions can cause cell elongation mimicking the effects of auxin, and neutral buffers will inhibit the effect of auxin on cell elongation by preventing wall acidification.

Auxins are also involved in so-called long term responses which may be more directly gene-mediated responses. Auxins stimulate and sustain mRNA and protein synthesis to form enzymes that catalyze the production of cell wall materials, sugars, and other cellular components. Applied IAA has induced mRNA and protein synthesis in Avena coleoptiles (16), yeast cells (17), pea stem sections (18), and bean endocarp (19). Artichoke tuber disks responded to exogenously applied IAA with growth and a substantial amount of new RNA and protein synthesis. When the metabolic inhibitors actinomycin D and 8-azaguanine were added simultaneously with IAA, effects of the auxin were negated (20).

One theory holds that auxins derepress a gene, releasing a DNA template for mRNA synthesis. It is believed that at least 10 specific genes are involved in auxin action (21).

AUXINS AND APICAL DOMINANCE

Long before the discovery of auxin, it was apparent that the shaping of many species of vascular plants was determined partly by the dominance of the apical or terminal bud. In general plants that grow tall and unbranched show a strong influence on growth by the apical bud while plants that are short and shrubby show a weak influence by the apical bud. When the apical bud is removed from the plant, lateral buds immediately began active growth with the nearest lateral bud to the apex establishing dominance over the remaining buds and apparently causing them to once again become inactive.

Subsequent work with IAA indicated that this auxin which was produced at the terminal bud stimulated cellular growth in the terminal bud but as it was transported down the shoot, somehow supressed growth of the lateral buds (22). There are indications that under the influence of auxin, the cells around lateral buds produce ethylene. It is the ethylene in turn which inhibits the growth of the lateral buds (23). When the terminal bud is removed and the auxin and ethylene are no longer present, the lateral buds grow, resulting in bushy plants. The number of flowers on the plant is also increased.

AUXINS AND ROOT INITIATION

The most rapidly growing parts of the plant contain the greatest concentrations of IAA - in the tip of the coleoptile, in buds and in the tips of leaves and roots (24). The concentration of auxin drops from the tip of the coleoptile to the base. From the base of the coleoptile the concentration of auxin increases until it reaches a peak at the root tip. The amount of auxin in the shoot tip, however, is usually much greater than in the tip of the root. In fact, the concentration of auxin which promotes cellular elongation in the shoot appears to be too great for the root, and somehow results in actual inhibition of cell elongation in the root (25).

This greater sensitivity of root tissue to IAA is mirrored in the usual distribution of greater amounts of IAA in shoot tissue than in root tissue. It is also mirrored in the distribution of IAA oxidase which oxidizes IAA. There is an inverse relationship between IAA and IAA oxidase with more oxidase in regions of low growth and less oxidase in regions of high growth. In this manner, growth and shaping of the plant is determined in part by relative concentrations of IAA and IAA oxidase (26).

While the application of relatively high concentrations of IAA retard root elongation, there may be a considerable increase in the development of branch roots and root hairs (27). As a result, the practice of dipping cuttings in auxin has become commercially important in promoting the proliferation of adventitious roots. Usually, synthetic auxins, such as indole-3-butyric acid (IBA) and naphthaleneacetic acid(NAA) which are not subject to IAA oxidase, are used.

Polymer-based auxins offer promise for improved crop production in that they may be able to (1) deliver minute amounts of auxin and (2) control-release the auxin so that it is available continuously over a period of time.

OTHER EFFECTS OF AUXIN

The known biological effects of auxin are numerous. In addition to cellular elongation, apical dominance and root initiation, the more important effects include callus formation, parthenocarpy, phototropism and geotropism.

Auxins are important in cell elongation, but they also play an active role in cell division. Application of 1 per cent IAA to a detached petiole blade of bean will cause a yellow swelling where auxin is applied. This swelling is caused by rapid cell division resulting in callus tissue (28). After a time some of these cells differentiate, forming adventitious roots.

The interaction of IAA with another PGR, kinetin, which is a cytokin, has been observed by growing tobacco stem pith callus in tissue culture. Jablonski and Skoog (29) have reported that with a low cytokinin to auxin ratio in the tissue culture, the result is a mass of loosely arranged, undifferentiated cells or callus, but with a high cytokinin to auxin ratio, the result is the growth of cultured plantlets with stems and leaves.

Endogenous auxins are involved in fruit set and growth. Normally, fruit will not develop if the flower is not pollinated and fertilization occurs. In some plants, fertilization of one egg cell is sufficient but in others, such as melons and apples, several seeds must be fertilized. By treating the female flower parts with auxin it is possible to produce parthenocarpic fruit (fruit produced without fertilization) such as seedless tomatoes and cucumbers (30).

Abscission, the dropping of leaves and other plant parts, is correlated with diminished production of auxin in those parts. By treating evergreen holly (Ilex aquifolium) with auxin, leaf and berry drop during shipment is minimized. Auxins also prevents preharvest drop of citrus fruits. Large amounts of auxin, however, by causing increased amounts of ethylene, encourage fruit drop, and are used in the thinning of fruit in the production of apples and olives (31).

In an earlier discusion of the discovery of the first PGR, an auxin, it was noted that Darwin was curious about why plants turn toward the light -- the phenomenon of phototropism. Grass seedlings bent toward the light because the light side and the dark side of the shoot differed in amounts of auxin. The side of the shoot in the dark had more auxin and its cells elongated more than those on the lighted side, causing the plant to grow faster on the darker side and hence bend in the direction of the light.

Geotropism is the name given to the familiar response in plants in which roots grow down and shoots grow up. Were it not so, pity the poor farmer or gardener who would have to make sure that his seeds were always planted in the soil "right side up." A familiar explanation for roots growing down has been that they are "seeking water." It is clear that roots growing down have a certain adaptive advantage.

Starch-containing plastids called amyloplasts are contained in the central cells of the root cap where the most active growth is occurring. Amyloplasts appear to be the only cellular components with sufficient mass to be directly affected by gravity. Gravity causes the amyloplasts to fall to the lower side where they set in motion a series of reactions involving the calcium-containing component, calmodulin, and adjacent lower levels of cells so that the bottom of the root grows more slowly than the top and the root accordingly curves down. By removing the root cap and its amyloplasts, the root loses its ability to sense gravity (32). Germinating seeds have also been taken into space where there is microgravity and the roots show no downward curvature (33).

MAJOR PLANT GROWTH REGULATORS

In addition to endogenous and exogenous auxins, there are a number of other major plant growth regulators all of which have multiple functions in plants. The major PGRs are listed in Table 1. Some of the other known plant chemicals with biological activity are listed in Table 2.

HIBISCUS

For the initial rooting experiments involving the use of polymers derived from indole-3-butyric acid, cuttings from hibiscus plants were used. Specifically, a variety called Albo Lacinatus (Anderson's Crepe Pink) was used. The genus and species names are Hibiscus rosa-sinensis L. Hibiscus rosa-sinensis, or the Rose of China, has been known as the Queen Flower of the Tropics for the past 300 years (34). The particular variety described by Linnaeus in his "Species Plantarium" (1753) was probably a double red which had wide distribution throughout China, the China Sea, Indian Ocean and the Tropics. It is a member of the Malvaceae or mallow family and is considered to be a "woody plant". More recently hybridization has resulted in spectacular colors and color combinations (for instance 5th Dimension hybridized by Gordon Howard is gun metal gray and blue) and new flower shapes. Some blooms measure over ten inches across (for instance Super Star hybridized by Gordon Howard).

Since most tropical soils are deficient in plant nutrients because of high temperatures, high rainfall and leaching, many of the most desirable new varieties are grafted onto older, more established varieties (35). This type of symbiotic relationship is not unique to hibiscus and is practiced with a wide variety of fruit trees and flowers including rose bushes, citrus trees, apple trees and grape vines. Rootstocks are used which are more resistent to soil nematodes and other diseases (36). For our research Albo Lacinatus (Anderson's Crepe Pink) was employed. Anderson's Crepe Pink is widely used as a root stock in Florida and

Table 1
Major Plant Growth Regulators

Auxins

Indole-3-acetic acid; synthetic auxins including indoles, benzoic acids, naphthalene acids, chlorophenoxy acids, naphthoxy acid	apical dominance; gravitropism and phototropism; induces rooting in cuttings; inhibits abscission; stimulates ethylene synthesis; stimulates parthenocarpic fruit development

Cytokinins

N^6-adenine derivatives, phenyl urea compounds; basic effect of numerous known cytokinins may be at level of protein synthesis	apical dominance, shoot growth, fruit development delays senescence

Gibberellins

Gibberellic acid; GA_3. more than 65 gibberellins isolated from natural sources	shoot elongation; regulates seed enzyme production in cereals; stimulates flowering in long-day plants and biennials

Abscissic acid

ABA	stomatal closure; leaf and fruit abscission; dormancy

Ethylene

$CH_2 = CH_2$	fruit ripening; leaf and flower senescence; abscission

---

Australia and it roots readily in the intense South Florida sunlight. Because of its wide usage, its characteristics are well known and since it forms a good root system generally within a month to six weeks it was chosen as the representative of woody plants for the current studies.

METALS USED IN PRESENT ROOTING EXPERIMENTS

Manganese is an essential trace element and functions as an enzyme activator in cellular respiration. Specifically,

Table 2
Selected Plant Chemicals with Biological Activity

Phenolics

phenol derivatives ranging from catechol, caffeic acid, and aesculin to anthocyanidins and other complex polyphenolics	powerful fungicidal and bacteriocidal agents; may protect plant from disease and predators

Vitamins

all vitamins essential for humans found in plants and plant parts	function as cofactors in enzymic reactions in plants as well as in animals; present in minute amounts

Cyclitols

intermediate in conversion of glucose into glucuronic and galacturonic acids	found in plasma membranes conjugates with IAA to form possible storage form; found in storage tissues of seeds

Brassinolide

steroid; first isolated from pollen; has been synthesized in small amounts	enhances plant yield in radish, potato, bean and lettuce when used in small amounts

Triacontanol

straight-chain primary alcohol, 1-hydroxytriacontane; first isolated from alfalfa leaves	increases plant growth on soybean, corn, wheat, rice, tomato and carrot when applied in minute amounts

Amo 1618 and Cycocel

contain quarternary ammonium group	retards stem elongation; may accelerate flowering; leaves frequently are greener

Phosphon D

phosphonium group	effects similar to Amo 1618 and Cycocel

(continued)

Juglcne

polyphenolic naphtha-
quinone

derived from roots, bark
and fruit hulls of black
walnut, diffusate kills
other plants in vicinity;
example of bewildering
array of secondary plant
substances which kill
insects, other predators
and other intra- and
interspecific plants;
research in its infancy

Figure 1.   Picture of Albo Lacinatus bloom and leaves.

it serves as an activator for malic dehydrogenase and oxalosuccinic decarboxylase, which are enzymes in the Krebs cycle (37). Manganese also acts as an activator for the enzymes nitrite reductase and hydroxylamine reductase which are involved in nitrogen metabolism (38).

Manganese is thought to be involved in electron transfer from water to chlorophyll during the light reaction of photosynthesis (39). It may also be involved in the reduction or oxidation of indole-3-acetic acid (IAA) (40).

Manganese is found in soil in bivalent, trivalent and tetravalent forms but is available for plant uptake only as a bivalent ion in soil solution or as an exchangeable ion absorbed to soil colloids (41). Since the reduced bivalent form of manganese is taken up by the plant, poorly aerated, acid soils favor the availability of manganese. On the other hand, well-aerated alkaline soils favor the oxidation of manganese, making it unavailable for plant uptake.

Chlorotic and necrotic spots in the interveinal areas of leaves are symptoms of manganese deficiency in plants (42). In tomato leaves, the chloropasts are the first part of the plant to be affected (43).

Titanium in the form of bis(cyclopentadienyl)titanium IV dichloride, BCTD, was the other metal incorporated in the polymers and used in the initial rooting experiments. Titanium is not known to play an essential role in plant growth.

OTHER KNOWN METALS IN PLANTS

Iron has multiple functions in plants. It is necessary for electron transport in the cytochrome system in the mitochondria. It is also incorporated into ferredoxin which plays a key role in light reactions in photosynthesis (44).

Soils are usually not deficient in iron, but pH plays a major role in iron availability and plant uptake. Iron·is normally available in acid soils but is not available in neutral or alkaline soils (45). This availability can be corrected by complexing iron with a chelate such as ethylenediamine tetraacetic acid (EDTA) (46).

Leaves deficient in iron show a characteristic network of green veins with chlorotic of yellow interveinal areas (47). The younger leaves show the most chlorosis.

Copper is present in chloroplasts in plastocyanin which is essential as an electron carrier in photosynthesis (48). Copper is also essential for the functioning of phenolases, which are plastid enzymes.

Very little copper is found dissolved in soil solution (49). The divalent copper cation is adsorbed strongly to soil colloids and is relatively exchangeable (50). Copper complexed with soil organic matter is not there in the right form and may be a major reason for copper deficiency in organic soils. Also, the addition of phosphate results in the formation of insoluble copper phosphate.

Copper deficiency causes a necrosis of the tips and young leaves that proceeds along the margin of the leaf and creates a withered appearance.

Zinc may be involved in the biosynthesis of tryptophan, a precursor of indole-3-acetic acid (IAA) (51). Zinc also participates in the metabolism of the plant as an activator of several enzymes. Zinc may act as an activator for some phosphate transferring enzymes such as hexose kinase or triosephosphate dehydrogenase. Zinc deficiency results in the accumulation of soluble nitrogen compunds such as amides and amino acids (52). Apparently zinc plays an important role in protein synthesis.

Zinc in the divalent form is absorbed on soil particles in an exchangeable form. Zinc concentration in soil solution is probably quite low (53). As soil pH rises, zinc becomes less available, so that plants growing in alkaline soils frequently show zinc deficiency. The addition of calcium hydrogen phosphate creates relatively insoluble zinc phosphate (54).

Zinc deficiency is first seen as interveinal chlorosis of the older leaves starting at the tips and margins. Smaller leaves and shortened internodes result in stunted growth. Leaves are smaller, distorted in shape, and clustered on branches creating a rosette effect. Zinc deficiency also results in lower production of seeds in beans and peas and the production of citrus fruit (55).

Calcium plays multiple roles in plant physiology and a deficiency of calcium shows up early and in dramatic ways in plant appearance. Calcium is a constituent of cell walls in the form of calcium pectate. The middle lamella is composed mainly of calcium and magnesium pectates (56). Calcium also plays a role in the formation of cell membranes and lipid structures (57). Calcium in small amounts is necessary for normal mitosis and it also plays a role as an activator for numerous enzymes (58). Calcium is thought to play a role in carbohydrate translocation and somehow in the development of mitochondria (59).

A major reason for an acid soil is the predominance of exchangeable hydrogen ions and a lack of exchangeable metal cations. Low pH soils are undesirable for plant growth, partly because soluble aluminum and iron compounds become available and tie up phosphate ions. The addition of calcium to the soil creates a condition where through cation exchange Ca+2 exchanges for some of the H+ ions absorbed on clay colloids or micelles in the soil. In this manner, calcium becomes available for plant uptake and at the same time the soil pH is raised.

Calcium is relatively immobile in plants and deficiency symptoms appear first in the younger leaves. Meristematic regions of stem, leaf and root tips are affected and eventually die. Roots become short, stubby and brown (60).. Cell walls become rigid and brittle.

Magnesium is a key constituent of the chlorophyll molecule without which photosynthesis would not occur (61). Magnesium is also required as an activator of enzymes involved in carbohydrate metabolism and the synthesis of nucleic acids (DNA and RNA) from nucleotide polyphosphates (62).

Like calcium, magnesium apparently is mobile in the plant with deficiency symptoms showing up first in the basal leaves, and as the deficiency becomes more pronounced, in the younger leaves (64). Since magnesium is a constituent of chlorophyll, the most noticeable deficiency symptoms occur in interveinal chlorosis of the leaves.

Potassium is well-known for the role that it plays in the opening and closing of stomata (65). It also plays some role in respiration, photosynthesis and chlorophyll development. During the early stages of potassium deficiency an accumulation of carbohydrates is observed. This may be due to impaired protein synthesis - the carbon skeletons that would go into protein synthesis are accumulated as carbohydrates (66).

Potassium is present in the soil in a fixed form, an exchangeable form, and a soluble form on soil colloids. As soluble K+ is taken up by the plant, exchangeable K+ is released, which in turn results in fixed K+ being slowly released.

Plants with potassium deficiency show a mottled chlorosis with necrotic areas at tip and margin of leaf. In some plants the marginal regions may roll inward and the tip of the leaf curves downward (67).

Plants utilize molybdenum in minute amounts. The presence of one part per billion may eliminate molybdenum deficiency in plants. Evidence indicates that molybdenum plays a role in both nitrogen and phosphorus metabolism. Deficiency symptoms include an interveinal mottling with the leaf margins becoming brown. The leaf tissues wither leaving only the midrib and a few pieces of leaf blade and resulting in a characteristic appearance called whip tail (68).

Boron is another metal utilized in minute amounts in plants. The borate ion complexes readily with polyhydroxy compounds such as sugar and it is thought that sugar is transported across the cell membrane as a borate complex, or that the borate ion may be associated with the cell membrane where it could complex with a sugar molecule and facilitates its passage across the membrane (69). Boron also plays a role in DNA synthesis in meristems. In addition, boron has been implicated in a whole host of other processes; cellular differentiation and development, nitrogen metabolism, fertilization, active salt absorption, hormone metabolism, water relations, fat metabolism, phosphorus metabolism, and photosynthesis. Boron, however, may only affect these processes indirectly through its effect on sugar translocation (70).

A deficiency of boron results in the death of the plant shoot tip because of the role boron plays in DNA synthesis. Leaves may develop a thick coppery texture, curl and become brittle. Flowers do not form and root growth is stunted, resulting in heart rot in sugar beets and cork formation in apples (71).

GOALS OF OUR RESEARCH

The goal of our research has been to develop polymeric materials containing both auxin plant growth hormones and also metals essential for plant growth which can release these materials in controlled amounts over a prolonged time period. It would therefore be possible to stimulate plant growth for an entire season with an initial application to the soil or to the seedlings prior to planting. This paper describes preliminary experiments with the growth hormones indole 3-butyric acid and two organometallic species, bis-cyclopentadienyl titanium dichloride and dipyridyl manganese (II) dichloride. Other studies with different metals such as Fe, Co, Zn and B and different hormones will be described subsequently.

EXPERIMENTAL

The following chemicals were used without further purification: dipyridine manganese (II) dichloride, DMC, (Research Organic and Inorganic Chemical Co., Sun Valley, CA), bis-cyclopentadienyl titanium dichloride, BCTD (Aldrich, Milwaukee, WI); indole-3-butyric acid, IPA, (Aldrich) and indole-3-propronic acid, IPA, (Aldrich).

Condensation polymerizations were carried out in a one quart Kimax emulsifying jar placed on a Waring Blender (Model 1120) with a no load speed of about 18,000 rpm. An aqueous solution of IBA or IPA containing two equivalents of sodium hydroxide is added to a stirred solution of the metal-containing reactant contained in chloroform or carbon tetrachloride. The product is recovered by suction filtration as a precipitate, washed and dried.

Molecular weight measurements were taken using a Brice-Phoenix BP-3000 Universal Light Scattering Photometer. Infrared spectra were recorded employing a Mattson Alpha-Centauri FTIR employing KBr pellets. High resolution electron impact (HREI) positive ion mass spectral studies were carried out at the Midwest Center for Mass Spectrometry, Lincoln, Neb. A Kratos MS-50 Mass Spectrometer, operating at 8KV accelerating and a ten second per decade scan rate, was used for all measurements.

Rooting experiments were carried out employing sterilized coarse "Terra-Lite" perlite and Hibiscus rosa-sinensis, variety Albo Lacinatus (Anderson's Crepe Pink) wooden canes about 10 to 12 inches long and 0.25 to 0.75 inches in diameter. Fifty canes were employed for each test group. The stocks were dipped (0.037g average of talc mix or about 7×10^{-5} g of test additive for 0.2% samples) and then inserted into the perlite. The samples were watered with a fine mist spray several times a day as needed and exposed to partial to full sun.

Figure 2. Albo Lacinatus canes in rooting setting.

Figure 3. Roots on Albo Lacinatus canes.

RESULTS AND DISCUSSION

STRUCTURE

Following is a brief description of the structural evidence that is consistent with the products having a general repeat unit of form 1 where M is Ti or Mn and R is C_5H_5N or C_5H_5.

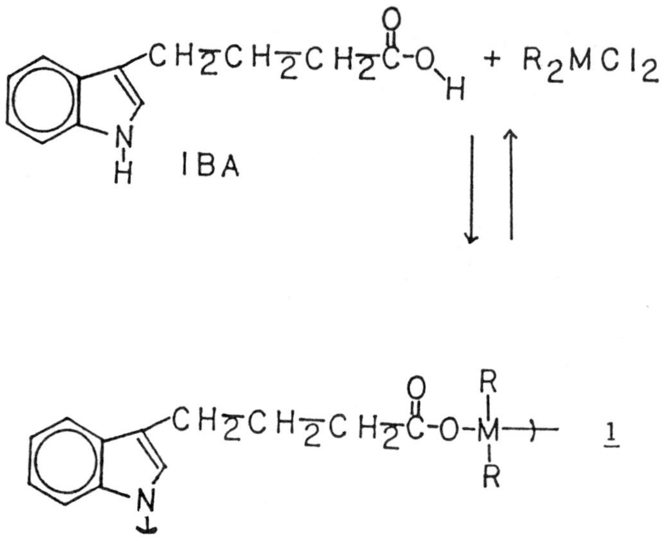

The mass spectra results are consistent with the proposed structures. For the product from BCTD and IBA there is a reasonable match with the known fragmentation pattern for Cp. There exists numerous ion fragments assignable to multiple units of the polymers. For instance the ion fragments (all ion fragments given in Daltons) at 805 corresponds to two units plus the indole ring plus a methylene minus one Cp group; 793 corresponds to two units plus an indole ring minus one Cp; 767 corresponds to two units plus an indole ring plus a propyl grouping minus two Cp units and at 743 corresponds to two units. The results of infrared spectroscopy are also consistent with the general proposed structure. Bands at about 1440, 1010 and 815 (all infrared spectral bands given in cm^{-1}) are characteristic of the pi-Cp group. The broad band in the 3400 to 3200 region is characteristic of primary and secondary amines. The band about 1650 is characteristic of the carbonyl stretching in metal esters.

For the condensation product of dipyridine manganese II dichloride, DMC, and IBA ion fragments assigned to IBA are found at 115 assigned to the indole moiety, 129 indole plus a methylene, 143 indole plus an ethylene, 157 IBA-CO_2 and 203 TBA. Ion fragments assigned as being derived from DMC are found at 133 PyMn and at 78, 51, 51 and 39 derived from Py. Ion fragments at 101, 100, 55-58, and 42 are derived

from the butyric acid moiety. Numerous higher molecular
weight ion fragments are found corresponding to one and more
units. For instance the ion fragment at 793 is assigned as
one unit plus an indole ring, plus MnPy minus CO_2; 781 one
unit plus indole plus MnPy minus a methylene minus CO_2 and
767 one unit plus indole plus MnPy minus an ethylene unit
minus CO_2.

The results of infrared spectroscopy are also consistent
with the general proposed structures. For the products from
DMC and IBA the band at about 1510 is assigned to aromatic
(pyridine) ring stretching vibrations. Bands at 1040 and
1010 are assigned to the C-H inplane deformations for
pyridine and at 1080 for the indole moiety. Bands on both
sides of 3000 are due to the C-H stretching (above to
aromatic, below to aliphatic). Bands about 660 (pyridine),
780 and 830 (indole) are assigned to C-H aromatic out of
plane deformations. A band about 3400 is characteristic of
primary and secondary amines and a band about 1580 is
characteristic of carbonyl stretching for bridged carbonyls.
Thus bands assignable to both reactants are present.

ROOTING RESULTS

Our major food, fiber, wood and ornamental plants belong
to two main classes - the gymnosperms, represented mainly by
the narrow-leaved, evergreen trees; and angiosperms,
usually broad-leaved, flowering plants. Angiosperms are
divided into two subclasses: the monocotyledons, which have
an embryo with one cotyledon, and the dicotyledons, which
have an embryo with two cotyledons. Dicots or C_3 plants
have different photosynthetic pathways as contrasted with
monocots or C_4 plants. We are utilizing a common monocot,
corn, and two dicots, bean and hibiscus.

Here we report results related to the rooting of one
variety of the hibiscus. Results appear in Tables 3-6.
There is a direct relationship between the appearance and
extent of leafing and root formation for the utilized
rooting stock Albo Lacinatus (Figure 1).

Thus in Tables 3 and 4, the incidence of leafing is
considered a direct measure of rooting progress. After ten
days leafing has begun (Table 3; Figure 2). For the most
part the higher number of stocks with leaves are those
dipped in polymer-containing talc. Stocks dipped in monomer
also show a decent increase over IBA.

Implicit in the auxin action is the idea that the auxin
causes a pH decrease in the vicinity of the cell wall.
While many auxins have acid functional groups, auxins
themselves are not believed to cause the acidic pH within
the cell wall matrix but somehow their presence "triggers"
factors that bring about this decreased pH. Even so, the
presence of protons appears to activate this response so
that compounds that release protons or cause a generation of
protons may act to trigger the "acid growth".
This may be responsible for the initial mildly positive
responses of the cuttings dipped in talc containing the
metallic monomers since the monomers produce acidic

Table 3

Table 3
Leaf formation after ten days

Test Material (% Additive)	%-Leaves	Test Material (% Additive)	%-Leaves
PMn (1%)	36	IBA (0.2%)	16
PTi (0.2%)	26	Talc	14
PMn (0.04%)	26	Nothing	12
MMn (0.2%)	22	PTi (0.04%)	8
PTi (1%)	22		
PMn (0.2%)	20		
MTi (0.2%)	20		

where P = polymer,   M = monomer

Table 4
Leaf formation after 20 days

Test Material (% Additive)	% Leaves	Test Material (% Additive)	% Leaves
PMn (1%)	100	IBA (0.2%)	80
PMn (0.2%)	100	MTi (0.2%)	80
PTi (0.2%)	98	MMn (0.2%)	78
PMn (0.04%)	96	Talc	72
PTi (0.04%)	96	Nothing	62
PTi (1%)	86		

solutions when hydrolyzed. For instance a 0.08 M solution of BCTD gives a pH of 1.7 corresponding to a $C_{p2}Ti^{+2}=C_{p2}TiOH^+$ of 3:1 (assuming these to be the most prevalent species) (16).

$$C_{p2}Ti^{++} + H_2O \longrightarrow C_{p2}TiOH^+ + H^+$$

After 20 days the results are similar except now the samples dipped with monomers are similar to IBA itself (Table 4). All of the polymer-dipped systems exhibit a higher percentage of leafing than IBA or any of the nonpolymer systems. Further, visual inspection of the abundance of leafing favors the polymer dipped samples.

By about 24 days almost all of the stocks showed some leafing. For each of the sample groupings the ten best stocks, i.e., ten stocks showing the most leafing, were removed from the perlite and the roots counted. The top achievers were the polymers and the lowest the monomers (Table 5). Again, visual observation showed that the roots associated with the polymer-dipped stocks were markedly superior with respect to size and length. Callus formation was consistently good with the polymer-dipped samples. In fact, it appears that some roots formed without the need for prior callus formation.

Table 5
Average number of roots after 24 days

Test Material (% Additive)	Average # of roots	Test Material (% Additive)	Average # of roots
PMn (0.2%)	11.1	IBA (0.2%)	6.1
PMn (1%)	10.4	Talc	5.3
PTi (1%)	9.3	Nothing	3.6
PTi (0.04%)	7.5	MMn (0.2%)	3.3
PTi (0.2%)	7.0	MTi (0.2%)	1.9

where P = polymer, M = monomer

By this time rooting is sufficient for the plants to be planted in potting soil in preparation for eventual grafting. Even so, the experiment was continued for another nine days. Again the best stocks were selected and again the samples containing the IBA-containing polymers were generally the best with regard to root length and size, average number of roots and average number of side or branch roots (Table 6; Figure 3).

Table 6
Rooting and branching after 33 days

Test Material (% Additive)	Average # of roots (Ranking)	% - with side roots (Ranking)
PTi (1%)	16.6 (1)	100 (1)
PTi (0.2%)	13.5 (3)	100 (1)
PTi (0.04%)	6.1 (8)	75 (4)
PMn (1%)	15.1 (2)	75 (4)
PMn (0.2%)	11.5 (4)	75 (4)
PMn (0.04%)	10.5 (5)	100 (1)
IBA (0.2%)	6.3 (7)	25 (7)
Talc	6.4 (6)	12.5 (9)
Nothing	1.9 (9)	0 (11)
MTi (0.2%)	1.1 (11)	25 (7)
MMn (0.2%)	1.5 (10)	12.5 (9)

In summary, it appears that the incorporation of IBA into a polymer has a positive effect on the rooting of Albo Lacinatus. Activity of the polymeric materials is probably through a controlled release mechanism.

ACKNOWLEDGEMENTS

We are pleased to acknowledge partial support from the American Chemical Society - Petroleum Research Foundation Grant # 15508-87 and the Ragland Corporation and the cooperation of the American Hibiscus Society.

REFERENCES

1. F.W. Went and K.V. Thimann, "Phytohormones," MacMillan, N.Y. (1937).

2. L.G. Nickell (ed.), "Plant Growth Regulating Chemicals," 2 vols., CRC Press, Boca Raton, FL (1983).

3. Peter H. Raven, Ray F. Evert and Susan E. Eichhorn, "Physiology of Plants," Worth, New York, (1986).

4. G. Ray Noggle and George J. Fritz, "Introductory Plant Physiology," 2nd ed., Prentice-Hall, Englewood Cliffs, N.J. (1983).

5. R.W.F. Handy, "Plant Regulation and World Agriculture" (T.K. Scott, ed.), Plenum Press, New York (1979).

6. Jocelyn M. Miller and Amar Yahiaovi, "Bioactive Polymeric Systems" (Charles G. Gebelein and Charles E. Carraher, Jr., ed.), Plenum Press, New York (1985).

7. Charles Darwin, "The Power of Movement in Plants," Murray, London (1880).

8. F.W. Went, "On Growth-accelerating Substances in the Coleoptiles of Avena sativa," Proc. Kon. Nederl. Akad. Wetensch, 35:723, Amsterdam (1926).

9. F. Kogel, A. Haagen-Smit, and H. Erxleben, "Uber ein Neues Auxin (Heteroauxin) aus Harn., XI Mitteilung," Z. Physiol. Chem. 228:90 (1934).

10. P.W. Zimmerman, A.E. Hitchcock, and F. Wilcoxon, "Several Esters as Plant Hormones," Cont. Boyce Thompson Inst. 8:105 (1936).

11. V.C. Irvine, "Studies in Growth-promoting Substances as Related to X-radiation and photoperiodism," Univ. Colo., Studies 26:69 (1938).

12. P.W. Zimmerman and A.E. Hitchcock, "Substituted Phenoxy and Benzoic Acid Growth Substances and the Relation of Structure to Physiological Activity," Contr. Boyce Thompson Inst. 12:321 (1942).

13. F.C. Stewart and H.Y. Mohan Ram, "Determining Factors in Cell Growth: Some Implications for Morphogenesis in Plants," Adv. Morphogenesis 1:189-265 (1967).

14. B.S. Valent and P. Albersheim, "The Structure of Plant Cell Walls. v. On the Binding of Xyloglucan to Cellulose Fibers," Plant Physiol. 54:105-108 (1964).

15. L.N. Vanderhoef and R.R. Dute, "Auxin-regulated Wall Loosening and Sustained Growth in Elongation," Plant Physiol. 67:146-149 (1981).

16. Y. Masuda, E. Tanimoto, and S. Wada, "Auxin-stimulated RNA Synthesis in Oat Coleoptile Cells." Physiol. Plant. 20:713 (1967).

17. C. Schimodo, Y. Masuda, and N. Yanagishima, "Nucleic Acid Metabolism of Yeast Cells," Plant Physiol. 26:189 (1967).

18. A. Theologis, T.V. Huynh, and R.W. Davis, "Rapid Induction of Specifics mRNAs by Auxin in Pea Epicotyl Tissue," Jour. Molecular Biology 183:53-68 (1985).

19. J.A. Sacher, "Senescence: Action of Auxin and Kinetics in Control of RNA and Protein Synthesis in Subcellular Fractions of Bean Endocarp," Plant Physiol. 42:1334 (1967).

20. L. Nooden, "Studies on the Role of RNA Synthesis in Auxin Induction of Cell Enlargement," Plant Physiol. 43:140 (1968).

21. T.M. Murphy and W.F. Thompson, "Molecular Plant Development," Prentice-Hall, Englewood Cliffs, N.J., 1988.

22. K.V. Thimann, "Hormone Action in the Whole Life of Plants," Univ. of Mass. Press, Amherst (1977).

23. E.C. Sisler and S.F. Yang, "Ethylene, The Gaseous Plant Hormone," Bio Science 33:233-238 (1984).

24. T.C. Moore, "Biochemistry and Physiology of Plant Hormones," Springer. Verlag., N.Y. (1979).

25. N.B. Mandava ed., "Plant Growth Substances," ACS Symposium Series III, Am. Chemical Society, Washington, D.C. (1979).

26. A.W. Galston and L.Y. Dalberg, "The Adaptive Formation and Physiological Significance of Indole-acetic acid oxidase," Am. J. Bot. 41:373 (1954).

27. H. Hartmann, A. Kofranek, V. Rubatzky and W. Flocker, "Plant Science," 2nd edition, Prentice-Hall, Englewood Cliffs, N.J. (1988).

28. W.T. Jackson, "Effects of Indoleacetic Acid on Rate of Elongation of Root Hairs on Agrostis alba L.," Phyiol. Plant., 13:36 (1960).

29. J.R. Jablonski and F. Skoog, "Cell Enlargement and Cell Division in Excised Tobacco Pith Tissue," Physiol. Plant, 7:16 (1954).

30. F.G. Gustafson, "Inducement of Fruit Development by Growth-promoting Chemicals," Proc. Natl. Acad. Sci., U.S., 22:628 (1936).

31. S.F. Yang, "Regulation of Ethylene Biosynthesis," Hort. Sci. 15:238 (1980).

32. Michael L. Evans, Randy Moore and Karl-Heinz Hasenstein, "How Roots Respond to Gravity," Scientific American 255:6 (1986).

33. Randy Moore, C.E. McClelen, W.M. Fondren, and Chia-Lien Wang, "Influence of Microgravity on Root-Cap Regeneration and the Structure of Columella Cells in Zea Mays," Am. J. Botany 74:2 (1987).

34. R.D. Dickey, "Hibiscus in Florida," Univ. of Fla. Extension Service 168A, Gainesville, FL (1962).

35. Am. Hibiscus Soc., "What Every Hibiscus Grower Should Know," Am. Hib. Soc., Sarasota, FL (1978).

36. D.L. Ingraham and L. Rabinowitz, "Hibiscus in Florida," Inst. of Food and Agr. Sci., 04-44, Gainesville, FL.

37. W.D. McElroy and A. Nason, "Mechanism of Action of Micronutrient Elements in Enzyme Systems," Ann. Rev. Plant Physiol. 5:1 (1954).

38. A. Nason, "Enzymatic Steps in the Assimilation of Nitrate and Nitrite in Fungi and Green Plants," in "Inorganic Nitrogen Metabolism," (W.D. McElroy and H.B. Glass, eds.), Johns Hopkins Press, Baltimore, MD (1956).

39. W. Wiessner, "Inorganic Micronutrients," in "Physiology and Biochemistry of Algae" (R.A. Lewin, ed.), Academic Press, N.Y. (1967).

40. P.L. Goldacre, "The Indole-3-acetic Acid Oxidase - peroxidase of Peas," in "Plant Growth Regulation" (R.M. Klein, ed.), Iowa State Univ. Press, Ames (1961).

41. G.W. Leeper, "The Forms and Reactions of Manganese in the Soil," Soil Sci. 63:79 (1947).

42. E.J. Hewitt, "Marsh Spots in Beans," Nature 155:22 (1945).

43. E.T. Eltinge, "Effects of Manganese Deficiency Upon the Histology of Lycopersicon esculentum," Plant Physiol. 16:189 (1941).

44. K. Mengel and E.A. Kirby, "Principles of Plant Nutrition," 3rd ed., Int. Pot. Inst., Bern, Switzerland, 1982.

45. R. Devlin and F. Witham, "Plant Physiology," 4th ed., W.C. Grant Press, Boston (1983).

46. Ivan Stewart and C.D. Leonard, "Chelates as Sources of Iron for Plants Growing in the Field," Science 116:564-566 (1952).

47. E.F. Wallihan, "Relation of Chlorosis to Concentration of Iron in Citrus Leaves," Am. J. Bot. 42:101 (1955).

48. L.F. Green, J.F. McCarthy and C.G. King, "Inhibition of Respiration and Photosynthesis in Chlorella pyrenoidosa by Organic Compounds that Inhibit Copper Catalysis," J. Biol. Chem. 128:447 (1939).

49. L. Wiklander, "The Soil," in "Encyclopedia of Plant Physiology," 4:118 (W. Ruhland, ed.) Springer, Berlin (1958).

50. F.I. Arnon and P.R. Stout, "The Essentiality of Certain Elements in Minute Quantity for Plants with Special Reference to Copper," Plant Physiol. 14:371-375 (1939).

51. F. Skoog, "Relationship Between Zinc and Auxin in the Growth of Higher Plants," Am. J. Bot. 27:939 (1940).

52. A.K. Pendias and H.K. Pendias (eds.), "Trace Elements in Soils and Plants," CRC Press, Boca Raton, FL.

53. C. Bould, "Mineral Nutrition of Plants in Soils," in "Plant Physiology (F.C. Steward, ed.), Academic Press, N.Y. (1963).

54. F.T. Bingham, J.P. Martin and J.A. Chastain, Effects of Phosphorus Fertilization of California Soils on Minor Element Nutrition of Citrus," Soil Sci. 86:24 (1958).

55. D.T. Clarkson and J.B. Hanson, "The Mineral Nutrition of Higher Plants," Ann Rev. Plant Physiol. 31:239-298 (1980).

56. T.A. Bennett-Clark, "Salt Accumulation and Mode of Action of Auxin: a Preliminary Hypothesis," Butterworth, London (1951).

57. E.J. Hewitt, "The Essential Nutrient Elements: Requirements and Interactions in Plants," in "Plant Physiology" (F.C. Stewart, ed.), Academic Press, N.Y. (1963).

58. E.G. Bollard and G.W. Butler, "Mineral Nutrition of Plants," Plant Physiol. 17:77-112 (1966).

59. C. Florell, "The Influence of Calcium on Root Mitochondria," Plant Phyiol. 9:236 (1956).

60. T. Wallace, "The Diagnosis of Mineral Deficiencies in Plants by Visual Symptoms. A Colour Atlas and Guide," 3rd ed., H.M. Stationary Office, London (1951).

61. E. Kessler, "On the Role of Manganese in the Oxygen-evolving System in Photosynthesis," Arch. Biochem. Biophys. 59:527 (1955).

62. P.O.P. T'so, J. Bonner and J. Vinograd, "Physical and Chemical Properties of Microsomal Particles from Pea Seedlings," Plant Physiol. Suppl. 32:x11 (1957).

63. H. Aubert, "Trace Elements in Soil," Elsevier, Amsterdam (1980).

64. C. Lyon and C.R. Garcia, "Anatomical Responses of Tomato Stems to Variations in the Macronutrient Cation Supply," Bot. Gaz. 105:441 (1944).

65. M. Fujino, "Stomatal Movement and Action Migration of Potassium" (translated), Karaku 29:660 (1959).

66. Frank B. Salisbury and Cleon W. Ross (eds.), "Plant Physiology" 3rd ed., Wadsworth, Belmont, Calif., 1985.

67. E. Epstein, "Mineral Nutrition of Plants: Principles and Perspectives," Wiley, N.Y. (1972).

68. E.J. Hewitt and T.A. Smith, "Plant Mineral Nutrition," English Univ. Press, London (1975).

69. H.G. Gauch, "Mineral Nutrition of Plants," Ann. Rev. Plant Physio.," 8:31(1957).

70. H.G. Gauch and W.M. Duggar, "The Role of Boron in the Translocation of Sucrose," Plant Physiol. 28:457 (1953).

71. H.B. Sprague, ed., "Hunger Signs in Crops," 3rd ed., McKay, N.Y. (1964).

72. F.T. McLean and B.E. Gilbert, "The Relative Aluminum Tolerance of Crop Plants," Soi. Sci. 24:163 (1927).

73. W. Stiles, "Other Elements," in "Encyclopedia of Plant Physiology," 4:599, Springer, Berlin (1958).

COMPARATIVE INFRARED AND RAMAN SPECTROSCOPY OF THE

CONDENSATION PRODUCT OF SQUARIC ACID AND

BIS(CYCLOPENTADIENYL)TITANIUM DICHLORIDE

Melanie Williams, Charles E. Carraher,Jr.,
Fernando Medina and Mary Jo Aloi

Departments of Chemistry and Physics
Florida Atlantic University
Boca Raton, FL. 33431

ABSTRACT

The product of the synthesis reported in this study appears to be the same as that obtained by Doyle and Tobias based on comparison of the infrared bands, the color, and the solubility properties.   The product is polymeric with a weight-average degree of polymerization of 136.   The infrared and Raman data supports a structure with uncoordinated carbonyls, in contrast to the polymeric derivatives of divalent metals.   The small number of bands in the infrared and Raman spectra which are attributed to the diketosquarate portion of the molecule suggests a structure with a moderate degree of symmetry.   The cyclopentadienyl rings apparently retain their C_{5v} symmetry, at least in the solid state, but it is difficult to draw conclusions regarding the symmetry of the Cp_2TiO_2 portion of the molecule based on the far-infrared data.

INTRODUCTION

Organometallic derivatives of titanium have been widely investigated, primarily as a result of their ability to act as excellent catalysts in polymerization reactions.   The interest in organotitanium derivatives was largely initiated by the discovery of the Ziegler-Natta catalysts(1,2).

A majority of the research with organometallic derivatives of titanium involves π-cyclopentadienyl(hereafter referred to as C_5H_5 or Cp, where Cp =   cyclopentadienyl) derivatives, such as the bis(cyclopentadienyl) derivatives.   These derivatives are the most stable of the organotitanium derivatives, and thus allows for their isolation.   Chatt and Shaw(3a) have suggested that the stability imparted by these strong π-bonding ligands is due to an increase in the energy gap between the d orbitals

Inorganic and Metal-Containing Polymeric Materials
Edited by J. Sheats *et al.*, Plenum Press, New York, 1990

of the metal and the carbon-metal nonbonding orbitals. This gap results in higher activation energy for bond fission. However, other authors (3b) have suggested that stability is a function of many factors, and the presence of π-bonding ligands is no guarantee of stability.

X-ray crystallographic studies have shown that the Cp rings in bis(cyclopentadienyl)metal derivatives each occupy one vertex of a tetrahedron, with the angle between Cp-M-Cp group usually being about $130°(4a)$. Thus, the geometry of these complexes is essentially that of a distorted tetrahedron.

The bond between the Cp rings and the metal consists of one sigma bond between the A bonding orbital of the Cp ligand and the metal, and two bonds involving two electron pairs from E_1 bonding orbitales of the ligand(1). This six electron bond occupies three coordination sites on the metal, having three orbitals available for bonding to other groups.

We have synthesized a wide variety of titanocene-containing polymers derived from the condensation of bis(cyclopentadienyl) titanium dichloride with diols (1), diamines (2) and salts of dicarboxylic acids (3) (for instance 5-11).

```
       Cp                      Cp                  Cp  O   O
       |                       |                   |   ||  ||
  ---(--Ti-O-R-O---)---   ---(---Ti-NH-R-NH---)---  ---(Ti-O-C-R-C-O---)---
       |                       |                   |
       Cp                      Cp                  Cp


        1                       2                    3
```

Some of the first metal complexes of squaric acid were synthesized by West and Niu(12). These complexes contained divalent transition metal ions such as Mn(II), Fe(II), Co(II), Ni(II), and trivalent metals such as Al(III), Fe(III), an Cr(III). The divalent metal complexes were initially assumed to have structure 4, but X-ray structural studies show that these metals form a three dimensional polymer network 5 with squaric acid(13). The trivalent complexes, however, are belived to have a dimeric structure 6, with uncoordinated oxygens (hereafter referred to as the diketosquarate derivatives) (12, 14, 15). In both the divalent and trivalent metal complexes, the geometry about the metal is octahedral (or approximately octahedral), which is the preferred geometry for these metal ions.

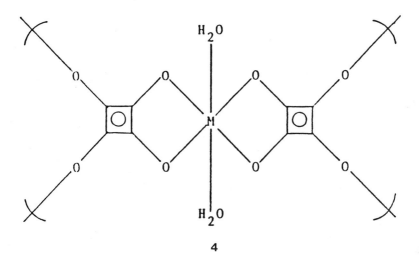

4

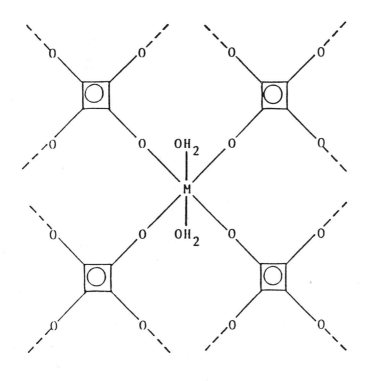

5

Since the early work of West and Niu, numerous publications have described the synthesis and properties of various metal-squarate compexes, platinum has been shown to form a mixture of salts with squaric acid (16). In addition, some trivalent main group alkyls reportedly form linear and three dimensional polymers with squaric acid (17). Complexes with the lanthanide elements (18), tantalum (19), vanadium (15), uranium (20), and titanium (20), have also been reported.

Many of the squarate complexes exhibit interesting and unusual properties. For instance, $Ni(C_4O_4)(H_2O)_2$ undergoes a low temperature phase transition to a ferromagnetically ordered state (21). Cooperative magnetic phenomena have also been observed for $Cu(C_4O_4)_3 \cdot H_2O$ and $Cu(_4O_4)(H_2O)_2$ (22-24). The magnetic phenomena observed in squarate complexes is not surprising as squaric acid itself undergoes a low temperature antiferrodistortive phase transition from a tetragonal to a monoclinic structure (25). In addition to these magnetic properties, some mixed-valence iron-squarate polymers exhibit dramatically enhanced conductivites, approximately six orders of magnitude greater than the single valence complex, as a result of the mixed valence nature of these polymers (26). The importance of thee magnetic and conductive properties is in the development of piezoelectric and semi conductor materials, respectively.

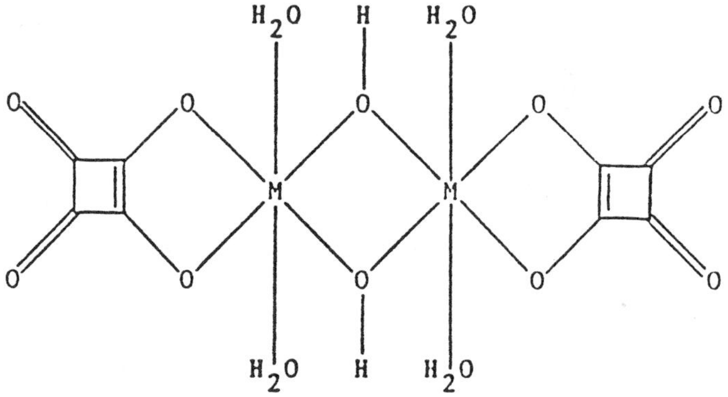

6

Few of the squarate complexes syntheseized to date contain tetravalent metals coordinated in a tetrahedral geometry. The use of bis(cyclopentadienyl)titanium(IV) dichloride (Cp_2TiCl_2 or titanocene dichloride) as the metal containing ligand provides the opportunity for obtaining such a complex.

Doyle and Tobias reported the synthesis of complex halogens products from the aqueous solution reaction of biscyclopenta-dienyltitanium IV diperchlorate with squarate acid (3,4-dihydroxy-3-cyclobutene-1,2-dione (27). Only infrared spectral data and elemental analysis results were reported. Two structures were suggested. The bidentate monomeric structure (7) and a polymeric bridging structure (8).

We recently synthesized the interfacial condensation product of biscyclopentadienyltitanium IV dichloride, Cp_2TiCl_2, and the disodium salt of squaric acid.

The products are believed to have structure (<u>8</u>) based on molecular weight measurements, infrared data, and geometric considerations.

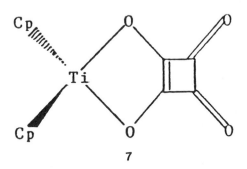

7

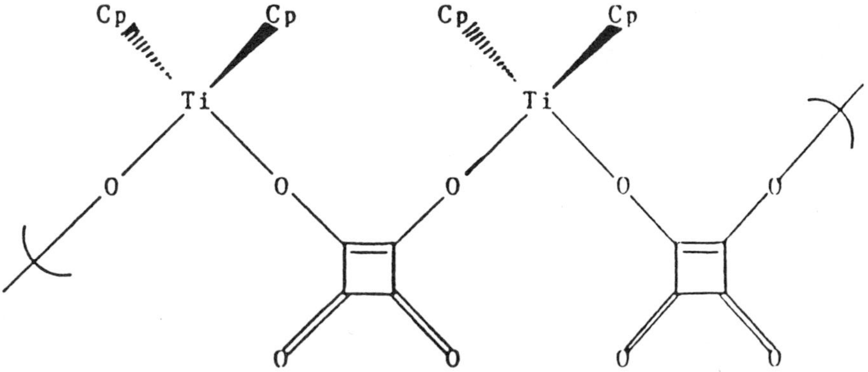

8

A detailed study of the infrared and Raman spectroscopy of the product is presented. The present study was undertaken to partially determine the influence of a polymeric nature on the vibrational properties of selected moieties.

EXPERIMENTAL

The following chemicals were utilized as received from Aldrich, Milwaukee, Wis.: Biscyclopentadienyltitanium dichloride (1271-19-8) and 3,4-dihydroxy-3-cyclobutene-1,2-dione, (99%-2892-51-5). Polycondensation was effected in a one quart Kimex emulsifying jar placed on a Waring Blendor (Model 1120) with a no-load speed of about 18,000 rpm (120 volts).

Squaric acid, neutralized with two equivalents of sodium hydroxide, was dissolved in water, and the titanocene dichloride dissolved in chloroform. The two phases were rapidly stirred for thirty seconds, at 22°C. The product was recovered as an orange-red precipitate, washed repeatedly with water and chloroform, and allowed to dry under vacuum. Molecular weight measurements were made on 0.1% solutions of the polymer in HPLC grade DMSO, employing a Brice-Phonox 3000-M Universal Light Scattering Photometer. Infrared spectrometry was carried out on KBr pellets using a Mattson Alpha-Centauri FTIR. Raman spectra were obtained with the 514.5nm line of an argon-ion laser, a Spex 1403 double monochromotor and a proton counting technique. Elemental analysis were carried out at Gailbraith Laboratories, Knoxville Tennessee. Proton NMR Spectra were recorded in d_6-DMSO, employing a GE QE300, 300 MHz FT-NMR, at room temperature.

RESULTS AND DISCUSSION

Mid-IR Region (4000-400 cm-1)

The following vibrational analysis consists mainly of an empirical evaluation, owing to the fact that many polymeric structures do not easily lend themselves to an extensive group theoretical and/or normal coordinate analysis. The interpretation of this spectrum is simplified, however, by the presence of the titanocene moiety in the polymer, as the identification of cyclopentadienyl derivatives is facilitated by characteristic bands in the infrared spectrum.

The assignment of the bands arising from the cyclopentadienyl ring vibrations can be readily accomplished by comparison with Cp_2TiCl_2 and other titanocene derivatives as the frequency of the ring vibrations are known to remain relatively constant upon substituion of the halide ligands (31-24). Examples of the ring vibrations in various Ti(III) and Ti(IV) cyclopentadienyl derivatives are given in Tables 1 and 2. The assignment of the bands arising from the cyclopentadienyl ring vibrations is based on previous studies of Cp_2TiCl_2 derivatives, as well as those for ferrocene, ruthenocene and dicyclopentadienyltin derivatives (28-33).

Previous studies on Cp_2TiCl_2 have shown that each ring can be treated on the basis of a "local" C_{5v} symmetry. Under C_{5v} symmetry, each ring gives rise to twenty-four normal vibrations, distributed as $3A_1 + A_2 + 4E_1 + 6E_2$, where the E modes are doubly degenerate. Therefore, seven normal vibrations are expected in the IR, while thirteen normal vibrations are expected in the Raman spectra (28-33). The observed bands for the ring vibrations of Cp_2TiCl_2, and the squarate polymer, and their assignments are given in Table 3.

Table 1. Infrared Frequencies of the Cp Ring Vibrations for Carboxylato Derivatives of Mono and Bis(cyclopentadienyl)titanium(III)[a,b].

	$Cp_2Ti(COOR)$ [c]		$Cp_2Ti(COOR)_2$ [d]		$CpTi(O_2CR)_2$ [e]	
	$R=CH_3$	$R=C_6H_5$	$R=CH_2CH_2$	$R=(CH=CH)$	$R=CH_3$	$R=C_6H_5$
$\upsilon 1$	3140 vw	3085 w	3085sh 3100 br	3105 m	3115 br.m	3100 w,sh
$\upsilon 2$	785 s	790 s	790 s	790 s	790 s 815 s	795 s 815 sh
$\upsilon 3$	1120 m	1120 s	1120 m	1120 m	--	--
$\upsilon 4$	--	--	--	--	--	--
$\upsilon 5$	3140 vw	3140 w	3140 w	3140w	--	--
$\upsilon 6$	1005 s	1020 s	1010 s	1010s	1015 ms	1015 ms
$\upsilon 7$	835 sh	860 s	845 sh	860 s	860 br,sh	870 sh 840 sh
$\upsilon 8$	1460 s	1445 sh	1450 s	1460 s	--	1440 sh
$\upsilon 14$	590 w	600w,sh	590 sh	595 vw	610 mw	605 w

a.Frequencies given in cm^{-1}.   b. v, very; s, strong; m, medium; w, weak; sh, shoulder; br, broad.
c-d. Taken from Reference 43.   e.Taken from Reference 46.

Table 2. Infrared Frequencies of the Cyclopentadienyl Ring Vibrations for some Bis(cyclopentadienyl)titanium(IV) Derivatives[a,b].

	$Cp_2Ti(acac)(ClO_4)$[c]	Cp_2TiF_2[d]	Cp_2TiBr_2[e]	Cp_2TiI_2[f]	$Cp_2Ti(CO)_2$[g]	$CpTiCl_3$[h]
v_1	3114m	3118sh	3110s	3104s	3030	3058
v_2	834s	820vs	819vs	818vs	794	816
v_3	--	1130vw	1130vw	1130vw	1107	1124
v_4	--	--	--	--	--	--
v_5	--	3109s	--	--	2907	--
v_6	1017m	1015s	1015s	1013s	1013	1017
v_7	865m	874m	872m 864m	866sh 858m	810	797
v_8	1437sh	1450s	1445s	1443s	1423	1435
v_{14}	--	606w	--	--	458	486

a. Frequencies given in cm^{-1}. b. v, very; m, medium; w, weak; sh, shoulder. c.Taken from Reference 50 d-f. Taken from Reference 31 g-h. Taken from Reference 28.

Table 3. Observed Infrared and Raman Frequencies for $(Cp_2Ti(C_4O_4))_n$ and Cp_2TiCl_2 Obtained in This Study-Bands Arising from Cyclopentadienyl Ring Vibrations.

υ	Description of mode	Symmetry Species	Cp_2TiCl_2 Infrared (cm-1)	Squarate Polymer Infrared (cm-1)	Squarate Polymer Raman(cm-1)
1	CH stretch	A_1	3103m	3095	*
2	CH o.p. bend	A_1	821vs	822s	825w
3	ring breathing	A_1	1130w	1149vw	1147ms
4	CH i.p. bend	A_2	1271vw	1267vw	-
5	CH stretch	E_1	-	-	-
6	CH i.p. bend	E_1	1015s	1018m	1020m
7	CH o.p. bend	E_1	871m	880m	880w
8	CC stretch	E_1	1440s	1455s	1460m
9	CH stretch	E_2	-	-	-
10	CH i.p. bend	E_2	1197vw	-	1190w
11	CH o.p. bend	E_2	1074w	1084w	1050m
12	CC stretch	E_2	1364w	1360w	1380w
13	CC i.p. bend	E_2	927w	-	-
14	CC o.p. bend	E_2	597w	595vw	605w
	$359 + 597 = 956$		957w		

* Region above 2000 not recorded.

[a] v, very; s, strong; m, medium; w, weak; o.p., out of plane; i.p., in plane.

The infrared spectra of titanocene dichloride and the squarate polymer $(Cp_2Ti(C_4O_4))_n$ are shown in Figure 1 and the Raman spectrum of $(Cp_2Ti(C_4O_4))_n$ is shown in Figure 2.

Most of the assignments for the A_1 and E_1 modes are consistent with those made previously for biscyclopentadienyl titanium (IV) derivatives (31) in similar matrices, but differs somewhat from the assignments of Balducci et al (30) for the matrix isolated spectra of the biscyclopentadienyl dihalides. Although the E_2 modes of the Cp_2TiX_2 derivatives (where X=F, Cl, Br, I), were previously assigned by Balducci, correlations can not be made with the solid state spectra. Tentative assignments for the E_2 mode have therefore been made and are given in Table 3.

The E_2 modes should be forbidden in the IR, but selection rules are not always strictly obeyed. Thus several weak infrared bands with frequencies near corresponding Raman frequencies are assigned to E_2 modes. Additionally, weak bands at 1271 cm^{-1} (Cp_2TiCl_2) and 1267 cm^{-1} (squarate polymer) are assigned to an A_2 mode. This band has been consistently observed in cyclopentadienyl derivatives even though it is strictly forbidden in both the infrared and Raman spectrum (30, 32, 33). The band is 957 cm^{-1} is believed to be a combination band.

There is little difference between the position and intensity of the Cp ring vibrations in Cp_2TiCl_2 and in the squarate polymer. This suggests that the Cp_2Ti moiety retains its original angular sandwich structure, with the $_n^5$ Cp rings tetrahedrally coordinated about the titanium atom. A change in structure would require a change in the orientation of the rings relative to one another. This orientation is the same as that found in the biscyclopentadienyltitanium (IV) dihalides, and all of the biscyclopentadienyl derivatives of titanium(1).

The remaining bands in this region of the spectrum can be assigned to the squarate portion of the molecule. These assignments are given in Table 4. A number of these assignments are facilitated by the results of previous studies on related molecules (12, 13, 34-37, 40, 43). There are to bands in the infrared spectrum above 1600 cm^{-1}, suggesting the presence of uncoordinated oxygens (or free carbonyls) (27). The band at 1792 cm^{-1} in the Raman spectrum is very strong, consistent with a symmetric stretching mode, as a totally symmetric fundamental usually results in a strong Raman line (36,40). The IR band corresponding to this fundamental appears at 1793 cm^{-1} as a moderately intense band. In contrast, the band at 1690 cm^{-1} is very weak in the Raman spectru, and appears as a strong band in the infrared spectrum at 1693 cm^{-1}, thus indicating an asymmetric stretching mode. These bands are assinged to C=O stretching modes. The symmetric stretch occurs at a higher frequency than the asymmetric stretch because in a symmetric stretch, the ring bond is forced to compress, resulting in a larger force constant, with an accompanying vibration at a higher frequency. On the other hand, the asymmetric stretch causes less strain, as the ring carbons can move to adjust to the motion, and the carbonyl carbons can keep their distance constant. The C=O symmetric stretch also occurs at a higher frequency than the asymmetric stretch in cyclobutane-1,3-dione (37), the squarate and croconate ions (C·····O) stretch (39) as well as in cyclic anhydrides and peroxides (35).

Wavelength (uM)

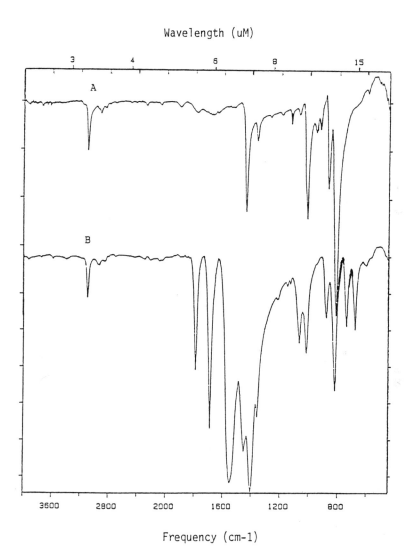

Frequency (cm-1)

Figure 1. Infrared Spectra of Titanocene Dichloride (A), and $(Cp_2Ti(C_4O_4))_n$ (B).

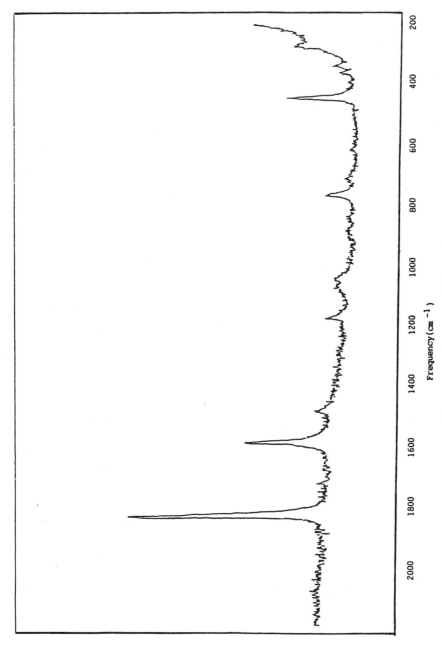

Figure 2. Raman Spectrum of $(Cp_2Ti(C_4O_4))_n$.

The band at 1555 cm^{-1} in the infrared spectrum is the second most intense band in the spectrum. Its position and intensity suggest a symmetric C=C stretch coupled with a C-O stretch. This band appears in the Raman spectrum at 1557 cm^{-1}, with a weak shoulder at 1565 cm^{-1}. A similar band has been observed in other metal containing derivatives (12, 14, 15, 17, 26) and organic derivatives of squaric acid (40), as well as many metal carboxylates (34,35). The frequency of the C=C stretch is lower than that of other cycloalkanes as a result of a longer C=C bond distance in the 4-membered rings resulting in a decreased force constant(41). For instance, an isolated double bond usually vibrates in the 1680-1620 cm^{-1} range, as do other cycloalkanes, whereas in cyclobutane this stretch is observed at 1566 cm^{-1} (42). The frequency of the C=C stretch in the polymer is grater than the corresponding frequency in squaric acid (1513 cm^{-1}). This may be a result of p-II - d-II bonding, which would increasethe electronegativity of the oxygen atom.

The most intense band in the infrared spectrum appears at 1405 cm^{-1}, but is absent in the Raman spectrum. This band is assigned to an asymmetric C-O stretch coupled to a C=C stretch. This is in agreement with the position and intensity of the C-O stretch found in organic derivatives of squaric acid (40). A band at this position is also characteristic of metal carobxylates, and usually is assigned to a symmetric O-C-O stretch (34,35). The assignment of the symmetric stretch at 1555 cm^{-1} to a higher frequency than the asymmetric stretch at 1405 cm^{-1} is again due to the ring strain effects mentioned earlier. The weak bands at 1067 cm^{-1} (IR) and 1075 cm^{-1} (Raman) are assigned to C-C stretching modes, in accordance with their intensity and position.

A characteristic vibration in all ring systems is the ring breathing mode. The band corresponding to this vibration is intense in the Raman spectrum, and is infrared active only in molecules which do not contain a center of symmetry (42).

Since nearly all of the bands in the IR and Raman spectra of the polymer are coincident, it is evident that no center of symmetry is present. A ring breathing vibration should therefore be observed in the IR spectrum. The ring breathing mode usually gives rise to a vibration in the 950-1000 cm^{-1} range in cyclobutene derivatives, but is found at considerably lower frequencies (550-750cm^{-1}) in the oxocarbons (38,39). The only band in the Raman spectrum which can be assigned to a symmetric ring breathing mode is the strong band at 755cm^{-1}. This assignment is consistent with other metal-containing derivatives of squaric acid (see table 5).

At least two C=O deformations are expected in the 550-750cm^{-1} region. However, coincidental degeneracies may occur. Additionally, since there are several bands below 700 cm^{-1}, some of these bands may be hidden. The one band which could be attributed to a C=O deformation occurs at 681 cm^{-1} (IR), and as a weak band in the Raman spectrum at 685 cm^{-1}. This band is assigned to an asymmetric C=O bending mode.

Table 5 lists the infrared and Raman frequencies (when available), for some metal-squarate complexes in which all oxygens are coordinated. In these complexes, there are no bands in the infrared spectrum above 1600 cm^{-1}. In addition, there is a very strong, broad band between 1400 cm^{-1} and 1700cm^{-1} corresponding to a combination C$\cdots\cdots$C + C$\cdots\cdots$O, and C=O stretching modes. This band is absent in the diketosquarate derivatives, and is replaced by bands corresponding to C=C, C-O, and C=O stretching modes. The position of the bands for the complexes listed in Table 5 remains fairly constant for a given vibration, which suggests similarities in the structures of these complexes.

Table 4. Observed Infrared and Raman Vibrations for the $(Cp_2Ti(C_4O_4))_n$ -Bands Arising from the Squarate Portion of the Molecule[a].

Description of Mode	Infrared(cm-1)	Raman(cm-1)
sym. C=O stretch	1793ms	1792vs
asym. C=O stretch	1693s	1690w
sym. (C=C)+(C-O) stretch	1555vs	1557s,1565sh
asym. (C-O)+ (C=C) stretch	1405vs	-----
C1-C3 stretch	1067m	1075w
sym. ring breathing	743m	755s
asym. C=O deformation	681m	685w

a. v, very; s,strong; m, medium; w, weak; br, broad; sh, shoulder; sym., symmetric;

asy., asymmetric.

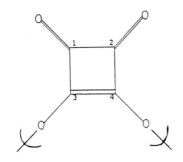

Table 5. Vibrational Frequencies of Some Squarate Complexes and the Squarate Ion. [a,b,c]

Complex	[(CH₃)₃Al]₂ C₄O₄ [d]	[(CH₃)₂Ga]₂ C₄O₄ [e]		[(CH₃)₂In]₂ C₄O₄ [f]		Divalent Metal Squarates [g]	C₄O₄²⁻ [i]	
Spectra	IR	IR	Raman	IR	Raman	IR	IR	Raman
ν(C⋯O)			1825s 1605ms 1584ms		1805s 1594m 1582m			1794w 1593s
	1587	1560s	1560s,br	1560			1530vs,br	
ν(C⋯C) +(C⋯O)						1400-1700vs,br		
ν(C⋯C)		1155s 1119m	1154s	1155sh 1118s	1154s 1122sh	1150		1123vs
	1125sh 1119ms	1099m		1098m		1105	1090s	
Ring Breathing			760m		746s			647s
δ(C⋯O)		662vs	615m 552vs		658s 647s		662vw	
δ(C⋯O)	489s,br 383m 300ms	459m 405s 272m		429m 399ms 270m			359m 259s	294

a. frequencies given in cm⁻¹. b. v, very; s, strong; m, medium; w, weak; sh; shoulder; br, broad. c. assignments of bands are taken from corresponding reference. d-f. Taken from Reference 17. g. Taken from Reference 12. i. Taken from Reference 39.

Table 6 lists the reported infrared frequencies for some diketosquarate-metal complexes which contain uncoordinated carbonyl groups, analogous to the structure proposed here for the squarate polymer. In these complexes, there are several bands above 1600 cm^{-1} attributed to C=O stretching modes. A strong band appears between 1500-1600 cm^{-1} in the spectra of all of these derivatives. This band corresponds to a C=C or C=C+C-O stretching mode. The presence of the C=O and C=C stretching modes suggests a more localized structure in these derivatives than those complexes in Table 5. Additionally, more bands appear in the infrared spectra of these derivatives than of those complexes in Table 5, which suggests that the diketosquarate complexes are less symmetric, as would be expected for D_{4h} versus C_{2v} symmetry.

Far-IR Region

The bands in the far-IR region are very difficult to assign for several reasons. First, the skeletal assignments previously reported for the monomers (titanocene dichloride and squaric acid) are conflicting and often ambiguous due to extensive coupling of these vibrations, as well as the appearance of a large number of bands in a relatively small region of the spectrum (30, 38, 44-48). Second, the symmetry of the polymer may be different from that of the monomers. Thus, direct correlations between the monomers, or derivatives, and the squarate polymer are difficult to make.

The vibrations in the far-IR region arising from skeletal modes and lattice vibrations are much more dependent on the symmetry and structure of the molecule than the vibrations in the mid-IR region. Therefore, these vibrations are influenced to a greater degree by neighboring groups, and a "characteristic" frequency may occur over a wide frequency range. In addition to these difficulties found in simpler molecules, polymers, in general, exhibit band broadening in their infrared spectra.

In spite of these difficulties, several bands in this region can be assigned by a knowledge of the expected vibrational frequency, and the effects of neighboring group substituents. Again, the analysis is based largely on the titanocene moiety. Table 7 lists the observed frequencies and assignments of the more prominent vibrational bands in the far-IR region. The infrared spectrum of this region is shown in Figure 3.

A strong band due to a Ti-O stretching mode is expected to occur in this region of the spectrum. The position of the Ti-O stretching mode has been found to vary between about 400-1000cm^{-1} in various oxygen bonded derivatives of titanium. In general, Ti-O stretching modes in compounds containing a Ti-O-Ti linkage occur at higher frequencies (700-1000cm^{-1}), whereas stretching modes for compounds containing a Ti-O-C linkage occur at lower frequencies (49). In accordance with this observation, strong bands at 434 cm^{-1} (IR) and 435 cm^{-1} (Raman), and weaker bands at 407 cm^{-1} (IR) and 406 cm^{-1} (Raman) are assigned to the symmetric and asymmetric Ti-O stretching modes, respectively.

Table 6. Vibrational Frequencies of Some Diketosquarate Derivatives and Squaric Acid. [a,b,c]

Complex Formula	$V^{III}(OH)(C_4O_4)$ $3H_2O$[c]	$Fe^{III}(OH)$ $(C_4O_4)(H_2O)$[d]	$C_4O_4H_2$[e]	$(C_4O_4)(OH)$[f] (CH_3)	(C_4O_4)[g] $(OCH_3)_2$	(C_4O_4)[h] $(O_2H_5)_2$
C=0 stretch	1801w	--	1822 1643	1818	1802	1786
C=C stretch	1615s	1620s,br		1724[i] 1587-1563	1724[i] 1587	1724[i] 1587
(C-C)+(C-O) stretch	1515s,br	1490s,br	1513	1481 1418	1481 1407	1471 1408
Other Observed Frequencies (n.a.)			1360 1318	1379 1320	1361	1370 1325
			1245		1220	1282 --
	1190					1198
	1145		1166		1143	1143
	1108	1105s				
		1085	1070	1070	1087	1075-1064
	1048w		1056	1042	1036	1020
		980				990
		--	928	930	939	--
		--	915	909-892	930	875
	875	860	856	833	833	854
				826		816
	720	750	731	791		800
	645	700	635			

a. Frequencies given in cm^{-1}. b. Assignments given are taken directly from corresponding reference. c. Taken from Ref. 15. d. Taken from Ref. 14. e. Taken from Ref. 38. f.-h. Taken from Ref. 40. n.a. These bands were not assigned with the excpetion of squaric acid. i.In this author's opinion, there is doubt about this assignment.

Table 7. Observed Vibrational Frequencies of the Squarate Polymer Below 500 cm^{-1}.

Description of Mode	Infrared(cm^{-1})	Raman(cm^{-1})
Ti-O stretch	435s	433vs
M-Cp asym. stretch	419s	------
Ti-O stretch	406s	407w
C=O deformation	389vs	390w
M-Cp sym. stretch	353m	355m
C-O deformation	329m	337m
Cp ring tilt	280w	285m
Cp ring tilt (sym)	260w	270s
C=O deformation	227 m, br	------

s, strong; m, medium; w, weak; v, very; br, broad; sym., symmetric; asym., asymmetric.

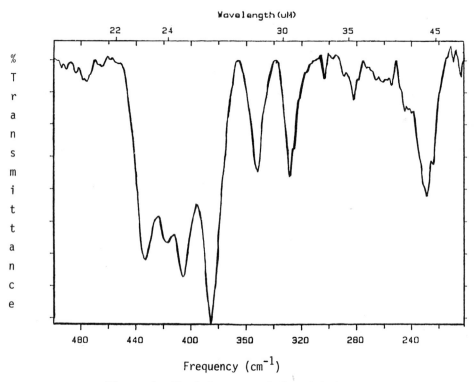

Figure 3. Far-Ir Spectrum of $(Cp_2Ti(C_2O_4))_n$.

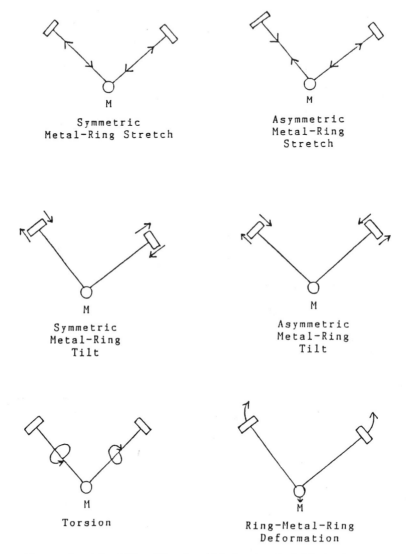

Figure 4. Metal-Ring Vibrational Modes for Cp$_2$M derivatives.

Regardless of the symmetry of the polymer, certain vibrations will arise from the Cp_2Ti moiety. The selection rules determining which symmetry species will be Raman or infrared active do, however, depend on the symmetry. The metal-ring vibrational modes which are expected to arise from the Cp_2Ti moiety are shown in Figure 4. Additional vibrarional modes corresponding to symmetry species in a given point group may be present. For instance, if the C_{2v} symmetry about the titanium is maintained, one ring tilt mode for each symmetry species is expected. Additionally, some bands may be infrared or Raman inactive. The asymmetric metal-ring stretching mode, occurring between 413 and 417 cm^{-1}, has been found to remain relatively constant in the titanocene dihalides (48). This mode is not expected to change considerably in the polymer, and has been assigned to the band at 419 cm^{-1} (IR). The symmetric M-Cp stretch is expected to occur at a lower frequency than the asymmetric stretch and is assigned to the bands at 358 cm^{-1} (IR) and 353 cm^{-1} (Raman).

The symmetric M-Cp stretching mode has been observed between 357 and 389 cm^{-1} in the metallocene dihalides (M=Ti, Zr, Hf), and shifts to lower frequencies in the order F>Cl>Br,I(48). Several ring tilting modes are expected in this spectral region, and the bands at 270 (R), 260 (IR), and 285 (R), 280 (IR) cm^{-1} are assigned to these modes, corresponding to A_1 and B_1 tilting modes, respectively, in Cp_2TiCl_2. These modes remain relatively constant for the metallocene dihalides, and are expected to occur at similar frequencies. The ring-metal-ring deformations and torsional modes are expected to occur at frequencies below 200 cm^{-1}, which are beyond the range of this study (30, 47, 40).

The remaining bands in this region of the spectrum are thought to arise from C-O or C=O bending modes. It should be stressed that the assignment of the bands in this region can only be tentative without knowledge of the crystal structure.

ACKNOWLEDGEMENTS

We acknowledge partial support from ACS-PRF 15508-87.

REFERENCES

1. P. C. Wailes, R. S. P. Coutts, and H. Weigold, "Organometallic Chemistry of Titanium, Zirconium, and Hafnium", Academic Press, New York, 1974.

2. R. Feld and P. L. Cowe, "The Organic Chemistry of Titanium", Butterworth & Co. Ltd., 1965.

3. (a) R. J. H. Clark, "The Chemistry of Titanium and Vanadium", Elsevier, New York, 1968. (b) P. S. Braterman and R. J. Cross, J. Chem. Soc., Dalton Trans., 657,(1972); D. M. P. Mingos, J. Chem. Soc., Chem. Comm., 165, (1972); W. Mowat, A. Shortland, G. Yagupsky, W. J. Hill, M. Yagupsky, and G. Wilkinson, J. Chem. Soc., Dalton Trans, 533 (1972).

4. A. Clearfield, D. K. Warner, C. H. Saldarriaga-Molina, and R. Ropal, Can. J. Chem. 53, 1622 (1975). H. Kopf and P. Kopf-Meier in "Platinum, Gold, and Other Metal Chemotherapeutic Agents", S. J. Lippard, Ed.; ACS Symposium series 209; American Chemical Society; Washington, D. C., 1983, Ch 7.

5. C. Carraher, J. Polymer Science, A-1(9), 366(1971).

6. C. Carraher, Chemtech, 744 (1972).

7. C. Carraher, "Interfacial Synthesis", Vol II, F. Millich and C. Carraher, Marcel Dekker, N.Y.,1977,Ch. 21.

8. C. Carraher and S. Bajah, Polymer(Br.), 15,9(1974).

9. C. Carraher and P. Lessek, Eur. Polymer J., 8, 1339 (1972).

10. C. Carraher, J. Chem. Ed., 46,314 (1969).

11. C. Carraher, J. Macromol. Sci-Chem., A17(8), 1293(1982).

12. R. West and H. Y. Niu, J. Am. Chem. Soc., 85, 2589 (1963).

13. M. Habenschuss and B. C. Gerstein, J. J. Chem. Phys., 61(3), 1 (1974).

14. G. Long, Inorg. Chem, 17(10), 2702 (1978).

15. S. M. Condren and H. O. McDonald, Inorg. Chem., 12, 57 (1973).

16. H. Toftlund, J. Chem. Soc., Chem. Comm., 837 (1979).

17. H. U. Schwering, H. Olapinski, E. Jungk, and J. Weidlein, J. Organomet. Chem., 76, 315 (1974).

18. E. Orebaugh and G. R. Choppin, J. Coord. Chem., 5, 123 (1976).

19. C. Sanini-Scampucci and G. Wilkinson, J. Chem. Soc., Dalton Trans., 807 (1976).

20. D. T. Ireland and H. F. Walton, Inorg. Chem., 8(4), 932 (1969).

21. B. C. Gerstein and M. Habenschuss, J. Appl. Phys., 61, 852 (1974).

22. K. T. McGregor and Z. G. Soos, Inorg. Chem., 5, 100 (1966).

23. S. de S. Barros and S. A. Friedburg, Phys. Rev., 141, 637 (1966).

24. C. G. Van Kralingen, A. C. Van Ooijen and J. Reedijk, J. Transition Met. Chem., 3, 90 (1978).

25. S. Nakashima, A. Aziza, M. LePostollec, and M. Balkanski, Phys. Stat. Sol., 94, 529 (1979).

26. J. T. Wrobleski and D. B. Brown, Inorg. Chem., 18(10), 2738 (1979).

27. G. Doyle and R. S. Tobias, Inorg. Chem., 7(12), 2484 (1968).

28. H.P. Fritz in F.G.A. Stone and R. West, Eds., Adv. Organomet, Chem., 1, 239 (1964).

29. E. Maslowsky,"Vibrational Spectra of Organometallic Chemistry", John Wiley and Sons, New York. 1973, Ch. 3.

30. G. Balducci, L. Bencivenni, G. De Rosa, R. Gigli, B. Martini and S. N. Cesaro, J. Mol. Struct., 64, 163 (1980).

31. P. M. Druce, B. M. Kingston, M. F. Lappert, T. R. Spalding and R. C. Srivastava, J. Am. Chem. Soc., (A), 2106 (1969).

32. E. R. Lippincott and R. D. Nelson, Spectrochim. Acta, 10, 307 (1958).

33. P. G. Harrison and M. A. Healy, J. Organometal. Chem., 51, 153 (1973).

34. C. N. R. Rao, "Chemical Applications of Infrared Spectroscopy", Academic Press, New York, 1963, Ch.7.

35. L. J. Bellamy, "The Infrared Spectra of Complex Molecules", 2nd Edition, Methuen, London, 1968.

36. "Oxocarbons", R. West, Ed., Academic Press, New York, 1980.

37. F. A. Miller, F. E. Kiviat, and I. Matsuhara, Spectrochim. Acta., 24A, 1523 (1968).

38. F. G. Baglin and C. B. Rose, Spectrochim. Acta., 26A, 2293 (1970).

39. M. Ito and R. West., J.Am. Chem.Soc., 85, 2580 (1975).

40. S. Cohen and S. G. Cohen, J. Am. Chem. Soc., 88 (7), 1533 (1966).

41. M. C. Tobin, "Laser Raman Spectroscopy" in Chemical Analysis Series, Vol. 35, John Wiley, New York, 1971.

42. M. Avram and M. Mateesca, "Infrared Spectroscopy", Wiley-Interscience, New York, 1972.

43. R. S. P. Coutts and P. C. Wailes, Aust. J. Chem., 20, 1579 (1967).

44. D. Gouregard and A. Novak, Sol. State Comm., 47, 453 (1978).

45. C. Puebla, J. Mol. Struct., 148, 163 (1986).

46. R. S. P. Coutts and P. C. Wailes, Aust. J. Chem., 26, 941 (1973).

47. M. Spoliti, L. Bencivenni, A. Farina, B. Martini, S. N. Cesaro, J. Mol. Struct., 65, 105 (1980).

48. E. Maslowsky and K. Nakamoto, App. Spectroscopy,
    25(2), 187 (1970).

49. "Spectroscopic Properties of Inorganic and Organo-
    metallic Compounds", The Chemical Society,
    Specialist Periodical Reports, John Wright and
    Sons, London, 1975-1982.

50. G. Doyle and S. Tobias, Inorg. Chem., 6(6), 1111
    (1967).

REACTION OF AMYLOSE WITH BISCYCLOPENTADIENYLTITANIUM DICHLORIDE

Charles E. Carraher, Jr.*, Daniel Gill*,
Yoshinobu Naoshima** and Melanie Williams*

*Department of Chemistry, Florida Atlantic
University, Boca Raton, FL 33431
**Department of Biological Chemistry
Okayama University of Science, Ridai-cho
Okayama 700, Japan

ABSTRACT

Amylose has been modified through the interfacial condensation
with biscyclopentadienyltitanium dichloride employing various mixing
procedures. The product contains largely glucose units with one
titanocene moiety connected through the ring hydroxyls. The properties
of these modified amylose derivatives and their potential applications
are discussed.

INTRODUCTION

Carbohydrates are the most abundant naturally renewing resource
on both a "molecule" and weight basis with about 400 billion tons
produced photosynthetically each year. These compounds generally
contain two or more hydroxyl groups that can undergo reactions with
metal-containing Lewis acids similar to poly(vinyl alcohol), PVA, and
diols such as 1,6-hexanediol and hydroquinone (1-2).

We have previously synthesized a number of metal-containing
products employing carbohydrates as the Lewis base (3-11). The
carbohydrates thus far investigated include simple saccharides such as
sucrose and polysaccharides such as dextran, xylan and cellulose. Here
we focus on the use of starch as the hydroxyl-containing source.

Starch occurs as microscopic granules in the roots, seeds and tubes
of plants. Corn, potatoes, rice and wheat are important commercial
sources. Over 5 billion pounds of starch are manufactured in the
U.S. each year. About half of the starch is sold as starch and
dextrin and about 1.5 billion pounds are converted into starch syrup.
Most of the starch is employed to size and stiffen weaving yarn and to
finish cloth. Starches are also used in making plywood, wallboard,
posterboard and corrugated board.

Heating starch with water causes the granules to swell and
produces a colloidal suspension from which two major components can be
isolated. The water soluble fraction, which is about 10 - 20% by weight,

Inorganic and Metal-Containing Polymeric Materials
Edited by J. Sheats *et al.*, Plenum Press, New York, 1990

is called amylose. The insoluble fraction is called amylopectin.

Amylose is a polysaccharide containing about 1000 D-glucopyranoside, hexose, units connected by an ether linkage between the C-1 site of one unit and the C-4 site of the second unit, similar to that found in maltose. Chains of D-glucose units with such alpha-glucosidic linkages tend to assume a helical arrangement. Amylopectin also has alpha-1,4-linkages but with branching occurring at intervals of 20-25 glucose units.

AMYLOSE

Prior research by our group emphasized the use of water soluble polysaccharides (3-11). Amylose is also water soluble and is commercially available in larger quantities than most other polysaccharides. Thus, amylose is a reasonable substitute to replace petroleum and coal based feedstocks.

Prior work has focused on the use of tin-containing modified polysaccharides as additives and materials that utilize the biological inhibitory properties of the organostannanes. The present study involves the use of biscyclopentadienyltitanium dichloride (titanocene dichloride) as the metal-containing moiety.

Titanocene dichloride shows solubility in organic solvents and ease of hydrolysis similar to dialkyltin dihalides but has higher thermal stability. It is widely employed as a catalyst for Ziegler Natta polymerizations and other types of homogeneous catalysis, both as a pure material and as anchored to a polymeric matrix. Thus, modified amylose derivatives may have applications as catalysts. Titanocene dichloride also has the ability to absorb ultraviolet radiation in a non-destructive manner and is frequently employed as an additive in coatings to protect against radiation damage. It may have a similar beneficial effect in materials containing amylose.

EXPERIMENTAL

Bis-cyclopentadienyltitanium dichloride (titanocene dichloride) (Aldrich, Milwaukee, WI) and amylose (Sigma Chemical Co., St. Louis, MO) were used as received. The modified amylose was prepared by the following interfacial synthetic procedure: A one-pint Kimax emulsifying jar placed on a Waring Blendor (Model 1120) with a "no load" stirring rate of 18,500 rpm was used as the reaction vessel. Two glass inlet tubes were present so that the reactants could be added while stirring was in progress. Titanocene dichloride, TDC, (0.37g, 1.5 m mole) was dissolved in 50 ml of chloroform. Amylose (0.66g, 1.0 m mole) and sodium hydroxide (0.12g, 3.0 m mole) were dissolved in 50 ml of water.

Synthesis was carried out as a function of stirring time employing three mixing procedures. The first procedure employed a layering of the

320

two phases, each containing the appropriate reactants, with stirring
occurring subsequent to the addition of the two phases (Table 1). This
procedure is analogous to a liquid bulk polymerization procedure. The
second procedure employed simultaneous addition of the two phases with
addition as stirring occurred (Table 2). This is similar to a
Y-apparatus where the reactants are added through two separate pipes with
stirring, mixing occurring as the solutions come together with the
product and solutions exiting through the third pipe. The third
procedure, and the one typically employed in most of our studies,
employed addition of one phase to the second highly stirred phase with
stirring time beginning after addition of the second phase (Table 3).
Addition typically is complete in less than 5 seconds. Reaction times
of 5 - 90 seconds were employed. The product precipitates and is
collected by suction filtration, washed repeatedly with water and dried
in vacuo at room temperature. Infrared spectra (KBr pellets) were
obtained on an Alpha Centauri FTIR. Mass spectra were obtained
employing a DuPont 491 mass spectrometer. Samples for Ti analysis were
degraded by heating with conc $HClO_4$ and Ti determined by atomic absorption spectroscopy.

RESULTS AND DISCUSSION

Synthesis

Titanocene dichloride is bright red in nonionizing solvents such as
$CHCl_3$. Hydrolysis of the Ti-Cl bonds or binding of the Cp_2Ti moiety to
alcohols, poly-ols or carboxylic acids produces a yellow or yellowish
brown color (12). The change in color from red to yellow corresponds

Table 1. Modification of amylose through reaction with titanocene
dichloride, employing layering of phases prior to rapid
stirring.

Reaction Time (Secs.)	Color Change (Secs.)	Weight Product (Grams)	Yield (%)
5	None	0.0	0
15	None	0.0	0
30	Slight	0.177	52
60	45	0.219	65
90	45	0.222	65

to completion of the reaction. Further stirring diminishes the yield
(13). This result was confirmed also for the reactions with amylose.
The yields obtained by the three different synthetic procedures are
given in Tables 1-3. The third procedure gives both the best yields
and the shortest reaction times, as determined by the observed color
change. Blank studies, in which either titanocene dichloride or amylose
is omitted, produced no product, thus confirming both materials were
incorporated.

Table 2.   Modification of amylose through reaction with
            titanocene dichloride, employing simultaneous
            addition of the two phases to the blender
            reaction jar with stirring.

Reaction Times (Secs.)	Color Change (Secs.)	Weight of Product (Grams)	Yield (%)
5	None	0.0015	4
15	None	0.206	62
30	Slight	0.22	65
60	Complete	0.21	67
90	Complete	0.24	71

Table 3.   Modification of amylose with titanocene
            dichloride by addition of one phase to stirred
            solutions of the second phase.

Reactions Times (Secs.)	Weight of Product (Grams)	Yield %	Titanium %
5	0.25	73	9
15	0.25	74	11
30	0.26	77	10
60	-	-	11
90	0.26	77	10
120	0.26	76	7

The product is a yellow or yellowish brown powder insoluble in all
common solvents.   It degrades upon prolonged stirring with NaOH solution;
hence the reaction should be terminated as soon as product formation is
complete.   It does not melt but decomposes above 250°.   Elemental analysis
shows approximately 7-11% Ti, which corresponds to 0.5-0.7 $(C_5H_5)_2$ Ti
unit per glucose subunit in the amylose.   (one $(C_5H_5)_2$ Ti unit per
glucose subunit would be 14% Ti, two would be 19% Ti and three would be
21%.   Titanocene dichloride itself is 27.1% Ti).   In our previous
studies of the reaction of titanocene dichloride with dextran
approximately 1:1 stoichiometry was also observed.   For similar
reactions with organostannane halides yields of 80% and approximately 1:1
stoichiometry was also observed.   Apparently steric factors limit the
number of titanocene units that can bind, since an excess of titanocene
is present.

STRUCTURE

It is believed that the product of the reaction between amylose and
titanocene dichloride is composed of chains containing sugar units
containing zero, one, two or three titanocene units, with an average of
of 0.5-0.7 titanocene per sugar unit (Figure 1).

Infrared spectroscopic studies of the reactants and products were
carried out employing FTIR and KBr pellets.   Bands characteristic of the
presence of both the amylose and titanocene moieties are present.   For
instance, the products exhibit infrared bands characteristic of amylose
(all bands given in cm^{-1}) at about 1650, 1480, 1440, 1360, 1275, 1240,
1160, 1100, 810 and 760 and bands characteristic of the Cp_2Ti moiety at
1405, 1030 and 855.   A band assigned to the Ti-O-C grouping is present
at 1130.   The ratios of the peak intensities of the two components also
corresponds roughly to a 1:1 stoichiometry.

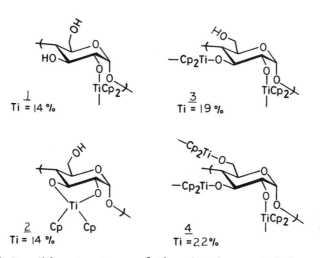

Fig. 1 Possible structures of the titanium-containing units

Mass spectral studies were also carried out using both reactants and products. Again, ion fragments characteristic of both reactants are present. Amylose alone shows only a few significant ion fragments including (all ion fragments given in m/e for individually charged species) 44 (CO_2), fragments centering about 81 (derived from the skeletal sugar ring), and about 131 (derived from the repeat unit minus CH_2OH). Products from the condensation of biscyclopentadienyl-titanium dichloride with amylose show ion fragments characteristic of the cyclopentadienyl moiety in the expected ratios (Table 4) along with ion fragments associated with the amylose at 44 and centering about 81 and 131. Ion fragments are also present about 97 (assigned to the ring plus a connective ether). There is a general absence of ions at 35/37 (Cl) and 36/38 (HCl) with the correct ratios ((3:1) corresponding to the natural abundance ratios of Cl 35 to Cl 37) consistent with the absence of Cl or Ti-Cl endgroups. This is also consistent with the yellow to yellow-brown coloration of the products.

Table 4. Ion fragments derived from cyclopentadiene.

m/e	66	65	39	38	63	67	62	61
Standard	100	47	32	8	8	7	6	5
Product	100	45	41	25*	7	14	3	3

*Present in amylose

The pKa values for the ring hydroxyls are about 10 and 12; whereas the methyl-hydroxyl has a pKa of about 14-15. In dilute sodium hydroxide solution the ring hydroxyl groups will be deprotonated, offering a highly reactive species to react with the titanocene dichloride. Thus, the preferential initial site of reaction is probably the ring-hydroxyls. As documented elsewhere, organostannanes have a high tendency to form internally cyclized products of form 2. It is possible that such cyclized products form a fraction of the titanocene-amylose products. We have previously synthesized analogous cyclic products but are unable to assign confidently specific IR bands associated with such cyclic products.

In summary, the products contain a mixture of reacted and unreacted sugar units, the reacted units containing the titanocene moiety connected through an ether linkage. Further studies of the properties and possible applications of these materials are currently in progress.

ACKNOWLEDGEMENTS

We are pleased to acknowledge partial support by the American Chemical Society - Petroleum Research Foundation Grant 19222-B7-C.

REFERENCES

1. C. Carraher and S. Bajah, Polymer, 14, 43 (1973) and 15, 9 (1974).

2. C. Carraher and J.D. Piersma, J. Macromol. Sci-Chem, A7(4), 913 (1973).

3. C. Carraher, T. Gehrke, D. Giron, D. Cerutis and H.M. Molloy, J. Macromol. Sci-Chem., A19, 1121 (1983).

4. C. Carraher, W. Burt, D. Giron, J. Schroeder, M.L. Taylor, H.M. Molloy and T. Tiernan, J. Applied Poly. Sci., 28, 1919 (1983).

5. Y. Naoshima, S. Hirono and C. Carraher, J. Polymer Sci., 2, 43 (1985).

6. C. Carraher and T. Gehrke, "Modification of Polymers," (C. Carraher and J. Moore, Eds.), Plenum, N.Y., 1983, Chpt. 18.

7. Y. Naoshima, C. Carraher, S. Hirono, T. Bekele and P. Mykytiuk, "Renewable-Resource Materials," (C. Carraher and L. Sperling, Eds.), Plenum, N.Y., 1986, pp 53-61.

8. Y. Naoshima, C. Carraher, T. Gehrke, M. Kurokawa and D. Blair, J. Macromol. Sci.-Chem., A23, 261 (1986).

9. Y. Naoshima, C. Carraher and K. Matsumoto, "Renewable-Resource Materials" (C. Carraher and L. Sperling, Eds.), Plenum, N.Y., 1986, pp 63-73.

10. Y. Naoshima, C. Carraher, S. Iwamoto and H. Shudo, Applied Organometallic Chem., 1, 245 (1987).

11. Y. Naoshima and C. Carraher, "Chemical Reactions on Polymers" (J. Benham and J. Kinstle, Eds.), ACS, Washington, D.C., 1988, Chpt. 30.

12. J. Chien, J. Phys. Chem., 67, 2477 (1963).

13. C. Carraher and R. Frary, Makromolekulare Chemie, 175, 2307 (1974).

TRIPHENYLANTIMONY-MODIFIED XYLAN

Yoshinobu Naoshima*
Charles E. Carraher, Jr.**

*Department of Biological Chemistry
Okayama University of Science, Ridai-cho
Okayama 700, Japan

**Department of Chemistry, Florida Atlantic
University, Boca Raton, FL 33431, U.S.A.

ABSTRACT

Xylan reacts with triphenylantimony dichloride
using the classical interfacial condensation
technique. The products are polyethers with Sb-O-R
moieties.

INTRODUCTION

Polysaccharides comprise the most abundant class of
naturally occurring organic material. We have focused on
the chemical modification of such hydroxyl-containing
naturally occurring materials employing the condensation
of Group IV A and B reactants with cellulose, xylan,
dextran and other saccharides (for instance 1-6).

Previously we reported the condensation of organoantimony
dihalides with salts of dicarboxylic acids (for instance
7-10) and diamines (for instance 11,12). Here we report
the initial synthesis of xylan modified organoantimony-
containing materials.

One major purpose in this research is to combine desired
features of the metal-containing reactant with the natural
occurring material to produce a product that may have
marketable properties.

Hemicelluloses form a group of polysaccharides that are
found in the primary and secondary cell walls of all land

and fresh water plants and in some seaweeds. Annual plants and woods contain about 20-35% hemicelluloses. The most abundant hemicelluloses have a backbone composed of 1-4 linked beta-D-xylopyranosyl units and are called xylans. Some xylan chains have D-glucopyranosyluronic acid units but most of the important xylans are O-acyl-4-D methylglucuronoxylans and L-arabino(4-O-methyl-D-glucurono) xylans. Most hardwood xylans have about one acid side chain for ten xylose units. The distribution and nature of the backbone and sidechain units vary with the particular plant or wood the xylan was isolated from. As with other polysaccharides, the exact nature of the chain components probably also varies as to the particular location within the plant, season of the year and particular soil, nutrient and water conditions.

Xylans are partially extractable with water from natural cell walls but typically are removed by alkaline solution extraction. To minimize contamination by lignin, the plant or woody material is usually treated with azeotropic ethanol-benzene and the lignin removed by its conversion to halocellulose. The alkaline extract can be neutralized, precipitating the more linear, less acidic xylans. The more acidic, more branched xylan is recovered as a precipitate after addition of ethanol.

Xylans, and other hemicelluloses offer many properties that are similar to exudate gums. Because of the increased price of such gums and the great abundance of hemicelluloses, there is expected to be an increased industrial market for the hemicelluloses. Xylan was employed for a variety of reasons:

- It is water soluble allowing the use of the classical interfacial, condensation system for modification.

- It is available in large quantity.

- It has only two hydroxyl groups per unit, possibly allowing the formation of non-crosslinked, modified products.

From our work and the work of many others, it is known that there is a strong tendency for xylan to form five-membered rings through reaction of the adjacent hydroxyls with organostannane dihalides. This tendency serves as the basis for a number of synthetic sequences including separation of diastereomeric diol compounds (13), selective esterification of sugars (14) and as a selective blocking agent for sugars (15-19). Since the tin-oxygen bonds have varied stabilities, the stannylene function can be employed as a protecting group for the diols or as an activating group for additional reactions. The formation of five-membered rings through reaction with organostannanes generally occurs in high yields.

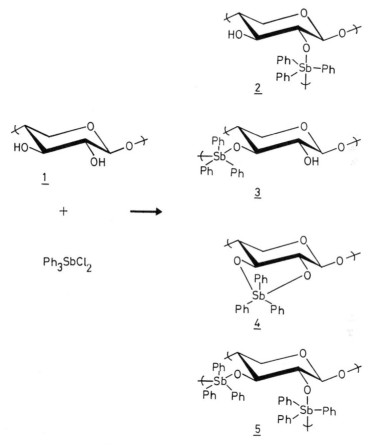

Figure 1. Possible repeat unit structures for the condensation of xylan with triphenylantimony dichloride.

Purposes for this study include the following:

First, as noted above, xylans constitute a major renewable resource which is largely untapped though they are employed in the production of furfural and as a replacement for exudate gums.

Second, it is hoped that the reaction with xylans may give products that show some solubility since they have only two hydroxyls per repeat unit. Further, the two hydroxyls are adjacent which may encourage cyclization with the dihaloantimony rather than crosslinking. Third, information gained from such studies should allow a better understanding of the reaction dependencies involved in the use of natural polyhydroxylic materials as feedstocks in the production of useful materials.

EXPERIMENTAL

Xylan (molecular weight of 23,000-25,000 from Larchmont; Sigma Chemical Co., USA), and triphenylantimony dichloride (E594-31-0) (Alfa, Danvers, MA) were used as purchased. Reactions were conducted employing a one quart Kimex emulsifying jar placed upon a Waring Blendor (Model 7011G) with a "no-load" stirring rate of 20,000 rpm. The product was collected as a precipitate employing suction. Repeated washings employing water and organic liquids assisted in the purification of the product.

Infrared spectra were obtained using Hitachi 260-10 and 270-30 spectrometers and a Digilab FTS-IMX FT-IR. The EI (70 ev) and CI (isobutane) mass spectra were obtained employing a JEOL JMS-D300 GC mass spectrophotometer connected with a JAI JHP-2 Curie Point Pyrolyzer (injector temperature-250°C; pyrolyzer temperature-590°C.). Analysis for antimony was carried out employing the usual wet analysis procedures.

Molecular weights were obtained employing light scattering photometry utilizing a Brice-Phoenix OM-3000 Universal Light Scattering Photometer.

RESULTS AND DISCUSSION

As with other polysaccharides, the exact structure of each individual chain will vary with the various chains composed of repeat units such as described in Figure 1.

Ion fragments derived from the products are consistent with a material containing moieties derived from the triphenylantimony and xylan. For instance, the product derived from triphenylantimony dichloride, TPAD, and xylan employing triethylamine, TEA, as the base gives ion fragments corresponding to all of the major ion fragments derived from xylan itself. The mass spectrum also contains ion fragments attributed to the presence of the phenyl groups with the ion fragmentation pattern in general

Table 1.   Major (>25% relative intensity) ion fragments (m/e 30-100) for the product
derived from triphenylantimony dichloride and xylan employing triethylamine as the added base.

m/e	Xylan	Product	Derivation
30	42*	176*	Xy
32		67	Ph
37		62	Ph
38		140	Ph
39	62	269	Ph,Xy
41	61	108	Xy
42	49	64	Xy
43	328	358	Xy
44	1000	1000	Xy
47		26	Ph
51		302	Ph
52		180	Ph
53		33	Ph
54		21	Ph
55	44	62	Xy
56	28	60	Xy
57	30	11	Xy
58	30	47	Xy
63		33	Ph
67		24	Ph
68		43	Ph
69	21	43	Ph
70	23	36	Xy
72	25	23	Xy
73		34	Ph
74		85	Ph
75		61	Ph
76		73	Ph
77		625	Ph
78	31	856	Ph,Xy
79	20	67	Ph,Xy
82		22	Ph
86	28	13	Xy
91	23	27	Xy
95		111	Xy
96		125	Ph

*=Relative ion intensities

agreement with that found for benzene itself (Table 1).
All ion fragments are cited in units of m/e or Daltons. Ion
fragments are also found that contain the metal, antimony.
The percentage natural abundance for antimony 121 is 57.25%
and for 122 is 42.75% or about a 6:4 ratio or 1.34:1. Ion
fragments are found at 198 and 200 in the approximate ratio
of 1.37:1 assigned to SbO; and at 352 and 354 in the ratio
of 1.34 assigned to SbO_3. The absence of ion fragments
associated with triethylamine is consistent with its
absence either as an impurity or through reaction with the
antimony dichloride.

While there are ion fragments at 35-38 that could be
attributed to Cl(35), HCl(36), Cl(37) and HCl(38) they are
not present in ratios consistent with the natural ion
abundance of chlorine (i.e., 35:37 is about 3:1). This is
consistent with the absence of large amounts of SbCl which
is in turn consistent with the presence of the antimony
moiety as a crosslinker, or more expected within a cyclic
moiety similar to that found for analogous products derived
from dihalostannanes.

The infrared spectra are also consistent with a product
containing moieties derived from both reactants. All of
the spectra are essentially identical. The spectra show
a broad based centering about 3400 (all bands are given in
cm^{-1}) attributed to the presence of unreacted hydroxyl
groups. Bands on either side of 3000 are consistent with
the presence of aromatic (SbPh) and aliphatic (Xylan)
C-H groups. A sharp band at 1440 is characteristic of the
Sb-Ph stretch. Bands assigned to the formation of the
ether linkage are found. Bands characteristic of the
symmetric C-O stretch are present at 1063 and 997. A band
characteristic to the formation of the Sb-O linkage is
assigned as appearing at 765. Bands at 738 and 690 are
characteristic of C-H out-of-plane deformation for
monosubstituted benzene. Products derived from the use of
triethylamine alone as the added base had a spectrum where
the bands associated with the triphenylantimony moiety were
diminished. Even so, mass spectra of these products
exhibit ion fragments expected from the presence of the
triphenylantimony moiety.

The absence of bands about 2700-2825cm^{-1} is consistent
with the absence of triethylamine and with the lack of ion
fragments associated with triethylamine.

Control reactions where one of the reactants was omitted
gave no precipitate consistent with the product, which is
collected as a precipitate, containing moieties derived
from both reactants. Finally, elemental analysis for
antimony is consistent with the presence of antimony.
Product yields are generally low (Table 3). By comparison,
reaction with dibutyltin dichloride under similar
conditions gives yields in the range of 40 to 60%. For
Tributyltin chloride the yields are within the range of 20
to 70% and for triphenyltin chloride the yields range from

Table 2. Fragmentation pattern for ion fragments derived from the phenyl group.

m/e	78	51	52	50	39	77	79	76	38	74	37
Standard	100	21	20	18	14	14	7	6	6	5	5
Product	100	32	21	28	31	73	8	8	16	10	7

5% to 70%. Too few analogous interfacial reactions employing hydroxyl-containing reactants with organoantimony dihalides have been studied to determine if they are intrinsically less reactive in comparison to organotin halides or if other factors are responsible for the low yields.

Table 3. Results as a function of the molar ratio of reactants.

TPAD (mmole)	Xylan (mmole)	Yield (g)	Yield (%)
1	2	-	-
2	2	0.09	9
3	2	0.20	20
4	2	0.19	19
6	2	0.24	24

Reaction conditions: Xylan and sodium hydroxide (0.5 mmole) in 50 ml water added to rapidly (18,000 rpm no load) stirred solutions of TPAD in 50 ml of CCl_4 with a 60 second stirring time at or about $25°C$.

REFERENCES

1. Y. Naoshima, S. Hirono and C. Carraher, J. Polymer Sci., 2, (1) 43(1985)

2. C. Carraher, T. Gehrke, D. Giron, D. Cerutis and H. M. Molloy, J. Macromol. Sci-Chem., A19, 1121(1983)

3. Y. Naoshima, C. Carraher and G. Hess, Polymeric Materials, 49, 215(1983).

4. C. Carraher and T. Gehrke, "Polymer Applications of Renewable-Research Materials" (Eds. C. Carraher and L. Sperling), Plenum Press, N.Y., 1983, Chpt. 4.

5. C. Carraher, P. Mykytiuk, H. Blaxall, D. Cerutis, R. Linville, D. Giron, T. Tiernan and S. Coldiron, Organic Coatings and Plastics Chemistry, 45, 564(1981).

6. C. Carraher, W. Burt, D. Giron, J. Schroeder, M. L. Taylor, H. M. Molloy and T. Tiernan, J. Applied Poly. Sci., 28, 1919(1983).

7. C. Carraher, H. Blaxall, J. Schroeder and W. Venable, Organic Coatings and Plastics Chem., 39, 549 (1976).

8. J. Sheats, H. Blaxall and C. Carraher, Polymer P. (ACS), 16(1), 655(1975).

9. C. Carraher, W. Venable, H. Blaxall and J. Sheats, J. Macromol. Sci.-Chem., A14C4), 571(1980).

10. C. Carraher and H. Blaxall, Angew.Makromol. Chemie, 83, 37(1979).

11. C. Carraher, M. Nass, D. Giron and D. R. Cerutis, J. Macromol. Sci.-Chem, A19(8&9), 1101(1983).

12. C. Carraher and M. Naas, "Crown Ethers and Phase Transfer Catalysis in Polymer Science," (Editors-L. Mathias and C. Carraher), Plenum, N.Y., Chpt. 7,1984.

13. C. Anchisi, A. Maccione, A. M. Maccioni and G. Podda, Gazzetta Chimica Italiana, 113, 73(1983).

14. R. M. Munavu and H. H. Szmant, J. Org. Chem., 41,1832(1976).

15. Y. Tsuda, M. E. Haque and K. Yoshimoto, Chem. Pharm.Bull., 31(5), 1612 (1983).

16. T. Ogawa and M. Matsui, Carbohydrate Res., 56,C1(1977).

17. D. Wagner, J. Verheyden and J. G. Moffatt, J. Org. Chem., 39, 24(1974).

18. M. Nashed and L. Anderson, Tetrahedron Letters, 39, 350(1976).

19. M. Nashed, Carbohydrate Res. 60, 200(1978).

20. L. Mathias and C. Carraher,"Crown Ethers and Phase Transfer Catalysis in Polymer Science", Plenum, N.Y. 1984.

PLATINUM II POLYAMINES: DETERMINATION OF SIZE, DEGRADATION AND BIOLOGICAL ACTIVITY

Deborah W. Siegmann, Dora Brenner, Alex Colvin, Brian S. Polner, Rickey E. Strother, and Charles E. Carraher, Jr.

Department of Chemistry
Florida Atlantic University
Boca Raton, Florida 33431

ABSTRACT

Platinum (II) polyamines have been synthesized which are polymeric analogues of the cancer drug, cis-DDP. Some of these polymers display biological activity, and so are potential chemotherapeutic agents, while others of these polymers are biologically inactive. Several properties of the various polymers were explored in an attempt to define the essential characteristics of a polymer that are responsible for biological activity. The molecular weights of the polymers were measured using light scattering photometry and Sephacryl column chromatography. While no correlation is seen between the size of a polymer and its biological activity, it was found that many of the polymer preparations contain two distinct size classes of polymer chains, suggesting that polymer synthesis can occur by two different mechanisms. The platinum polyamines were also found to degrade in dimethyl sulfoxide in a manner consistent with random scission. In water, behavior of the polymers is more complicated as polymer degradation competes with chain elongation. Again, however, the extent of polymer breakdown does not correlate with biological activity, so the biological activity of a platinum polyamine must be controlled by other, more complex factors.

INTRODUCTION

Cis-diamminedichloroplatinum (II) (also known as cis-DDP or cisplatin) is one of the major chemotherapeutic drugs currently in use. Cis-DDP produces numerous biological effects, including the induction of filamentous growth in bacteria (1,2), the induction of phage growth in lysogenic strains of bacteria (3), the inactivation of viruses (4), and the production of mutations and tumors (5-8). The most important effect of the compound is its anti-tumor activity (9), and cis-DDP is now used to treat several types of cancers, including testicular and ovarian carcinomas (10-12).

Inorganic and Metal-Containing Polymeric Materials
Edited by J. Sheats *et al.*, Plenum Press, New York, 1990

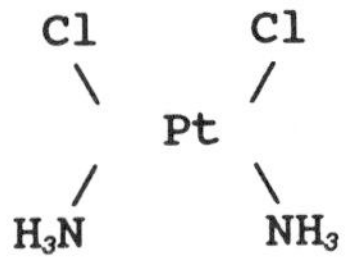

```
          Cl      Cl
           \     /
            Pt
           /     \
        H₃N       NH₃
```

Figure 1. Cis-DDP

The chemistry and mechanism of activity of cis-DDP have been
studied extensively. Cis-DDP (Figure 1) is a coordination
complex where the platinum is in the +2 oxidation state and the
ligands are in a square planar arrangement. The chloride ligands
are labile and can be easily replaced by a number of other
groups, including aquo and hydroxide ligands (13). In fact, such
substitution appears to be an important step in the mechanism of
cis-DDP's anti-tumor activity (14). In aqueous solutions with a
relatively high chloride concentration, such as plasma, cis-DDP
remains largely in its original form. However, when the chloride
concentration is low, such as in the cell cytoplasm, most of the
cis-DDP is converted into aquated species (15). The aquated
compounds can then lose their labile aquo ligands as the platinum
forms complexes with various cellular molecules to produce
biological effects. Evidence that the aquated species are
essential in producing cis-DDP's activity comes from several
sources. The aquated species are more reactive and more numerous
inside the cell than is the parent compound (15-18). In
addition, the rate of platinum interaction with cellular
macromolecules such as DNA depends upon the rate at which
equilibrium is established between cis-DDP and the aquated
species. Analogous compounds in which the amine ligands are
replaced with ligands that increase the rate of aquation show an
increased rate of platinum interaction with DNA (19).

    Cis-DDP via its aquated species can react with a variety of
cellular molecules including DNA, RNA, and protein. Of these,
DNA is the important target in producing the biological effects
of cis-DDP. The filamentous growth seen in bacteria treated with
cis-DDP indicates that DNA synthesis is selectively inhibited
(1,2). The same is true in eukaryotic cells where cis-DDP
interferes with DNA replication but does not significantly impair
RNA transcription or protein synthesis (20-25). Furthermore,
since the effective concentration of cis-DDP is very small, the
only cellular macromolecule that is likely to react with the drug
to an appreciable extent is DNA because of its large size (24-
26). The nucleophilic bases of DNA provide excellent binding
sites for platinum (27-30), with guanine reacting most rapidly
(28). Through interaction with the bases, cis-DDP can damage the
DNA in several different ways, producing interstrand DNA
crosslinks, intrastrand DNA crosslinks and DNA-protein
crosslinks. Interstrand crosslinks can form between the amino
groups of adenines on opposing strands in A-T rich regions of the
DNA (31,32). Amino groups of guanines and cytosines in opposing
strands can also be crosslinked by cis-DDP (33,34). DNA-protein
crosslinks can form between the DNA and non-histone chromosomal
proteins (35-37), with the platinum having a high affinity for
sulfur atoms within proteins (38-40). However, interstrand DNA
crosslinks and DNA-protein crosslinks account for only a very
small portion of the total DNA-platinum adducts formed by cis-DDP
(41,42). Because of this, these two types of lesions are
probably not important in producing cis-DDP's biological effects

(43-45), although some studies indicate a correlation between interstrand crosslinks and cytotoxicity (45-47). The more numerous intrastrand crosslinks form between adjacent guanines and between a neighboring adenine and guanine in the same DNA strand (48-51). The prevalence of these lesions as well as the correlation between the extent of intrastrand crosslinks and the biological effectiveness of cis-DDP indicate that the intrastrand crosslinks may be the crucial lesions, and that the mechanism of cis-DDP's anti-tumor activity is likely to involve this specific type of DNA crosslink (48-53).

The DNA crosslinks formed by cis-DDP interfere with DNA replication and eventually cause cell death. The activity of DNA polymerases is impaired by DNA-platinum adducts (54-56). Furthermore, although the cell is capable of repairing DNA-platinum adducts (50,57-59), the intrastrand DNA crosslinks produced by cis-DDP do not cause large distortions of the double helix and so may not be easily recognized and repaired (51,60,61). Despite cellular repair mechanisms, a sufficient concentration of cis-DDP will inhibit replication and prevent cell division (20-23,62-64). In some cells, cis-DDP appears to also impair the cell's ability to transcribe genes needed for mitosis (63,64), and even cells which manage to divide initially after cis-DDP treatment often do not display long-term survival (22,23).

The characteristics of cis-DDP that are important for its anti-tumor activity have been studied by testing the biological activity of closely related compounds. Platinum complexes similar to cis-DDP but containing only one labile ligand are inactive, implying that bifunctional attachment of the platinum to target molecules is essential while monofunctional attachment is ineffective (65,66). Platinum complexes which are charged are inactive, probably because such compounds have difficulty crossing the cell membrane (67,68). The nature of the ligands also influences activity. If the amine ligands of cis-DDP are replaced with other relatively non-labile ligands, the resulting compounds often retain biological activity, and a number of active compounds containing various types of amines are known (69,70). The labile ligands are more important, for if the chloride ligands of cis-DDP are replaced with ligands that are substantially less labile, the resulting compound is inactive. Substitution of the chloride ligands by ligands that are significantly more labile generates compounds that are highly toxic but which show little anti-tumor activity (71). Another important feature of cis-DDP is its configuration. Trans-DDP is relatively inactive against tumors, although it does produce toxic effects (67,72,73). Several differences exist between the reactivities of cis-DDP and trans-DDP. Trans-DDP forms different types of interstrand DNA crosslinks and intrastrand DNA crosslinks compared to cis-DDP which may produce different biological effects (14,24,43). The major intrastrand lesion produced by trans-DDP occurs between two guanines separated by an intervening base (54,74). This DNA-platinum adduct produces a major distortion in the double helix and is likely to be more easily recognized by the cellular DNA repair machinery (51,75). In fact, DNA-platinum lesions created by trans-DDP are removed more rapidly than those created by cis-DDP (36,41,76), and this difference may be largely responsible for the difference in the biological activities of the two compounds. The trans isomer also undergoes aquation at a faster rate than the cis isomer, and this may also contribute to the variation in their biological effects (70,77).

Although cis-DDP is an effective chemotherapeutic drug and much is known about its mechanism of action, there are problems associated with its use since cis-DDP causes numerous toxic side effects. Among the side effects are nephrotoxicity, gastro-intestinal disturbances, and hearing loss (78-84). The most serious of these side effects is damage to the kidney. Much of the administered cis-DDP is filtered out of the body within a few hours, exposing the kidney to a high concentration of platinum (85,86). Reducing the renal toxicity associated with cis-DDP can be accomplished in several different ways. Diuretics and prehydration are employed to somewhat reduce the toxic effects (87). There is also evidence that cis-DDP's toxic effects can be decreased if substances such as bismuth (88), selenium (89), or glutathione (90) are administered along with cis-DDP. There are a number of other drugs, including nifedipine (91), ara-C (92), etoposide (93) and 5-fluorouracil (94), that act synergistically to increase the anti-tumor effect of cis-DDP and so may permit lower effective doses of cis-DDP. Different methods of administering cis-DDP using liposomes (95-97), microcapsules (98,99), or cis-DDP complexed with other compounds (100-102) have all been tried with some success. However, nephrotoxicity remains a major limitation in the use of cis-DDP as a chemotherapeutic agent.

A second problem associated with cis-DDP treatment is the intrinsic or acquired resistance of some tumor cells to the drug. Tumors originally sensitive to cis-DDP can become resistant after several courses of chemotherapy (103). The mechanisms of cellular resistance to cis-DDP appear to be varied, and include reduced uptake of the drug by the cell (104,105), increased repair of DNA-platinum lesions (106-108), and increased cellular levels of metallothioneins (109-111) and glutathione (112). In an attempt to find platinum-containing drugs which show a broader spectrum of anti-tumor activity while producing fewer side effects, researchers have created numerous compounds which are analogues of cis-DDP (69,70). Some second generation platinum drugs such as carboplatin, iproplatin, and tetraplatin (113-115) are proving to be useful in chemotherapy.

Another approach to developing better platinum drugs is to create polymeric analogues of cis-DDP. Polymers in which platinum atoms are linked by diamines can be synthesized (Figure 2). A variety of diamines can be used to generate a family of such polymers, each with a different diamine component (116,117). Such platinum polyamines should show the same type of biological activity as cis-DDP since they possess the same crucial structural features including cis configuration and relatively labile chloride ligands. Furthermore, these polymers have several possible advantages over the monomeric cis-DDP compound. First, polymers should exhibit less movement through biological membranes compared with the monomer (118). This may reduce kidney damage which is the major toxic side effect of cis-DDP. Because cis-DDP is rapidly removed from the body (85,86), the kidneys are exposed to a high concentration of platinum. Polymers should be eliminated more slowly, reducing the exposure of the kidneys, increasing the concentration of the compound that is retained by the body, and permitting lower effective doses of the drug to be used. Second, platinum polyamines should show

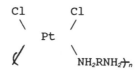

Figure 2. Platinum Polyamines

reduced rates of aquation compared to cis-DDP (119). In aqueous
solution, cis-DDP forms hydrolysis products with a reaction half-
life of nine hours at room temperature (120) or 2.4 hours at 37°C
(121). Aquated species formed outside the cell produce toxic
effects but display little anti-tumor activity (9,122). The
platinum polyamines should produce such hydrolysis products to a
lesser extent, reducing toxic side effects and allowing more of
the active compound to reach the cells. Third, polymers of this
type often degrade slowly in aqueous solution (123-125). The
platinum polyamines could break down to produce smaller, possibly
more biologically active molecules over time, acting as timed-
release drugs. Fourth, because each polymer molecule contains
many platinum atoms, the polymers might bind cooperatively to
DNA, forming more effective crosslinks than a comparable dose of
cis-DDP and producing greater anti-tumor activity.

Which of these four potential advantages, if any, actually
occurs with the platinum polymers is currently being studied.
Approximately two dozen platinum polyamines have been
synthesized, and some show biological activity while others do
not (126-128). Several of the polymers' properties are being
explored and correlated with polymer activity in an effort to
discover why certain polymers are active while others are
inactive. Such factors as the nature of the diamine component,
the extent to which a polymer is transported into the cell, the
rate of aquation, etc., could all influence the biological
activity of a polymer. Two other properties that could control
polymer activity are the size of a polymer and the tendency of a
polymer to degrade, and these will be the subject of discussion
here.

METHODS

The synthesis of the platinum polyamines has previously been
described in detail (116,129). The polymers are synthesized by
mixing equimolar amounts of a diamine and potassium tetrachloroplatinate, both in aqueous solution. After stirring for 6-168
hours at 25°-60°C (depending upon the diamine), the polymer is
isolated as a yellow to brown precipitate. Elemental analysis
and mass spectroscopy give results consistent with the predicted
structure. The cis form is the expected product because of the
trans effect which operates with platinum compounds (130,131),
and in fact the infrared and ultraviolet spectra of the polymers
are consistent with the cis configuration (132,133). Different
batches of the same polymer show no significant differences
during spectral analysis. The solid polymers are stable at room
temperature, relatively soluble in dimethyl sulfoxide (2-20
mg/ml) and slightly soluble in water (10-100 µg/ml).

The procedures for testing the biological activity of the
polymers have also been previously described (126,127). Balb/3T3
cells, either normal or transformed by the Moloney murine sarcoma

virus, are plated out in cell culture dishes at an appropriate
concentration. The following day, a polymer (initially dissolved
in dimethyl sulfoxide and then diluted to the desired
concentration with cell culture medium) is added to the cells.
The ability of the polymer to kill cells or slow cell growth is
monitored for the next three days by counting the number of
viable cells per dish and comparing the results to identical
dishes containing the same amount of dimethyl sulfoxide without
polymer.

The molecular weights of many of the polymers were
determined by light scattering photometry (134,135). In this
technique, several polymer solutions of varying concentration are
normally used to find the molecular weight. However, many of the
platinum polyamines are not sufficiently soluble to make a series
of solutions of adequate concentration. For this reason,
apparent molecular weights are measured using a 0.1% polymer
solution in dimethyl sulfoxide. While the molecular weights
obtained in this way are not absolute values, it is possible to
compare the relative sizes of different polymers.

For other polymers, the molecular weights were determined
using gel filtration on Sephacryl S-200 (Pharmacia). Sephacryl
is supplied as a suspension in water. The water is decanted off
and replaced with dimethyl sulfoxide. The dimethyl sulfoxide is
likewise replaced several times and the suspension is then
thoroughly de-gassed. A column (2.0 x 20 cm) is poured and
equilibrated with several column volumes of dimethyl sulfoxide.
Polymer solutions (1.0 ml of 0.02% - 0.10% in dimethyl sulfoxide)
are run on the column. Fractions of 0.5 ml are collected and
elution volumes (Ve) are determined by measuring the absorbances
of the fractions at an appropriate wavelength. The void volume
(Vo) of the column is found using blue dextran. Polymers whose
molecular weights are already known from light scattering
photometry are used to construct a calibration curve where Ve/Vo
is plotted against log molecular weight. The molecular weights
of other polymers are found by comparing their elution volumes to
the calibration curve.

To monitor the degradation of the polymers, a 0.1% polymer
solution in dimethyl sulfoxide or in dimethyl sulfoxide/water is
prepared, and the molecular weight over a period of days is
measured by light scattering photometry. For following polymer
degradation in water, a 0.1% polymer suspension in water is
prepared using a polymer of known molecular weight. After
stirring for a given length of time, the suspension is filtered
and the solid is allowed to air dry for 72 hours. The solid is
then placed in a vacuum desiccator until it reaches constant
weight. The apparent molecular weight of the solid is determined
by light scattering photometry using a 0.1% solution in dimethyl
sulfoxide.

RESULTS AND DISCUSSION

Polymer Synthesis

Platinum polyamines can be synthesized using a wide variety
of diamines, including pyridine derivatives, pyrimidines, amino
acids, and more complicated compounds such as methotrexate.
Table 1 lists the polymers used in this study, including the

diamine component of each polymer. Structures of the diamines are given in Figures 3 and 4.

Biological Activity

The biological activity of the platinum polyamines was tested in cell culture using two cell lines derived from mouse fibroblasts. Balb/3T3 cells resemble normal cells in their growth properties; m-MSV Balb/3T3 cells have been transformed and resemble cancer cells in their growth properties.

The effect of a polymer on cells was monitored by measuring cell growth in the presence of the polymer over several days and comparing it to control cells growing without polymer. Typical growth curves are shown in Figure 5 which demonstrates the effect of polymer 20 on both cell types. Cell growth is substantially reduced in the presence of the polymer, with both types of cells responding to the polymer to about equal extents.

Many of the polymers tested gave similar growth curves, although the polymer concentration required to produce an effect varied from compound to compound. Other polymers had no effect even at a relatively high concentration of 50. μg/ml, which is the limit of solubility for most of these polymers in cell culture medium. Table 2 summarizes the biological activities of the polymers. As the results show, there is wide variation in the activities of the different polymers with the effective concentrations spanning several orders of magnitude. Some of the polymers are extremely active while others are relatively inactive (136).

Table 1. Platinum Polyamines

Polymer Number	Diamine Component
8	2-Nitrophenylenediamine
10	2,5-Diaminopyridine
11	4,6-Diamino-2-mercaptopyrimidine
12	2,4-Diamino-6-hydroxypyrimidine
13[a]	Methotrexate
14	4,4'-Diamino-diphenylsulfone
15	2,6-Diamino-3-nitrosopyridine
16	Methotrexate
17	2-Chloro-p-phenylenediamine
18	Folic acid
19	Tryptophan
20	Tetramisole
21	Tilorone
22	Tilorone analogue 11,567
23	Acridine yellow
24	6,9-Diamino-2-ethoxyacridine
25	Proflavine
26	Histidine
27	Thiamine
28	Euchrysine
29	2,6-Hexanediamine

[a] All polymers were made with K_2PtCl_4 except 13 which was made with K_2PtI_4.

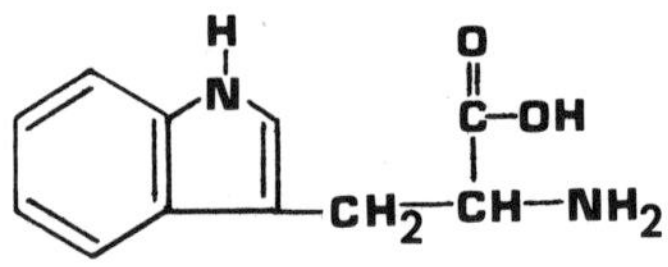

Figure 3. Diamine Structures

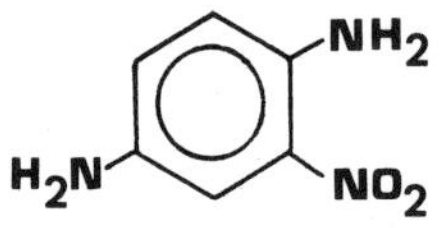

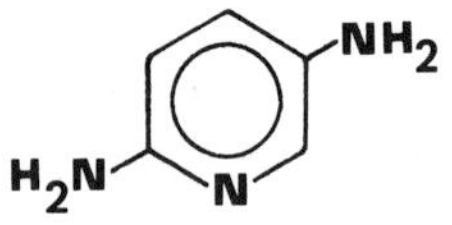

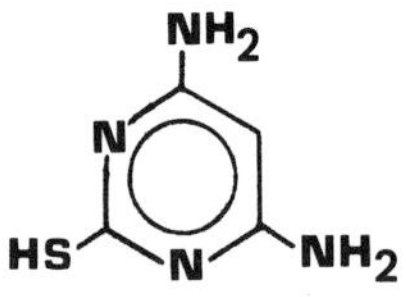

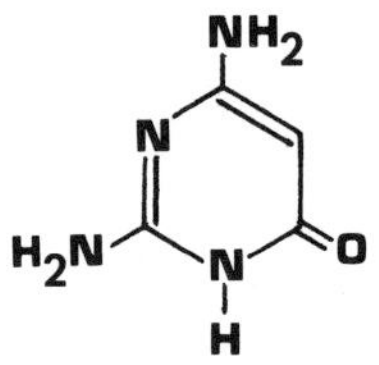

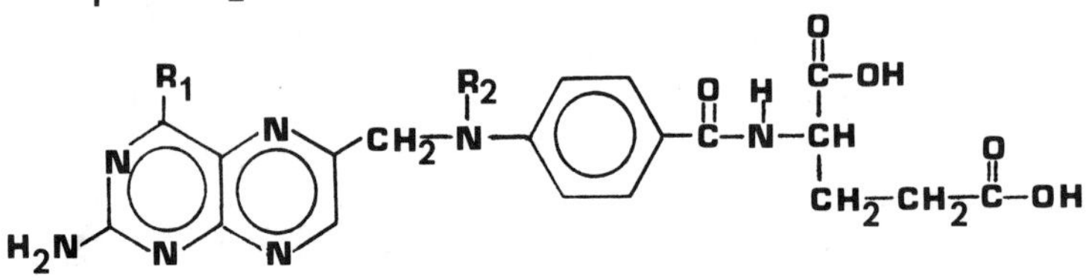

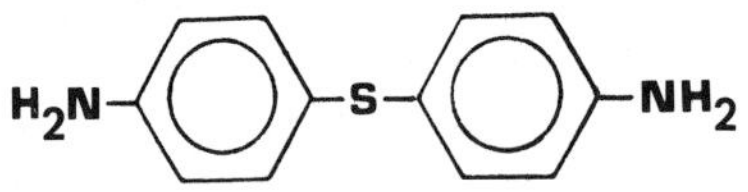

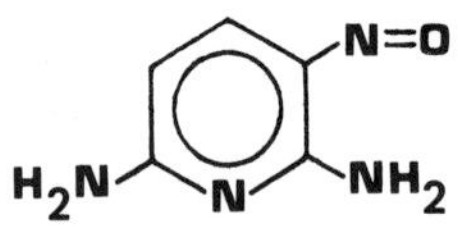

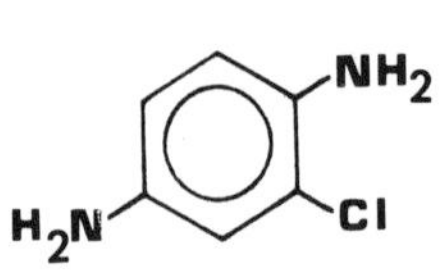

20

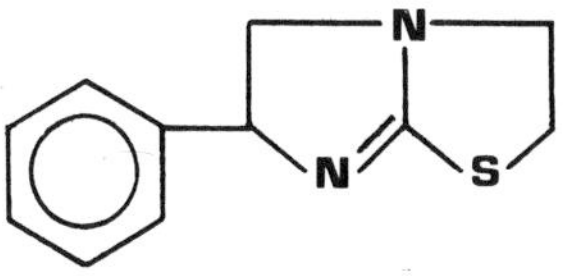

26

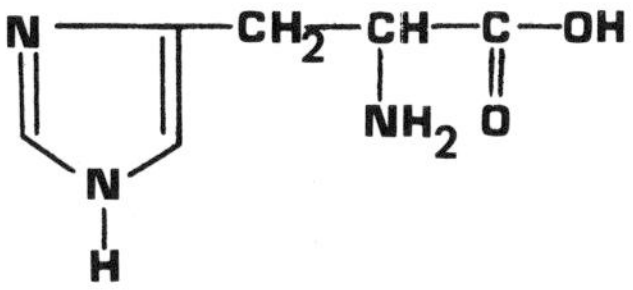

21

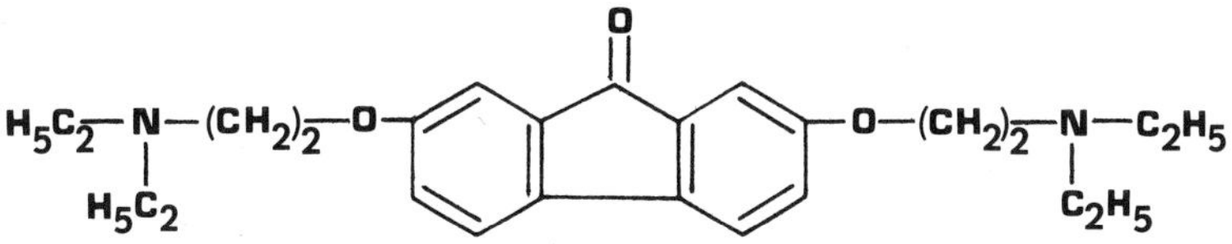

22

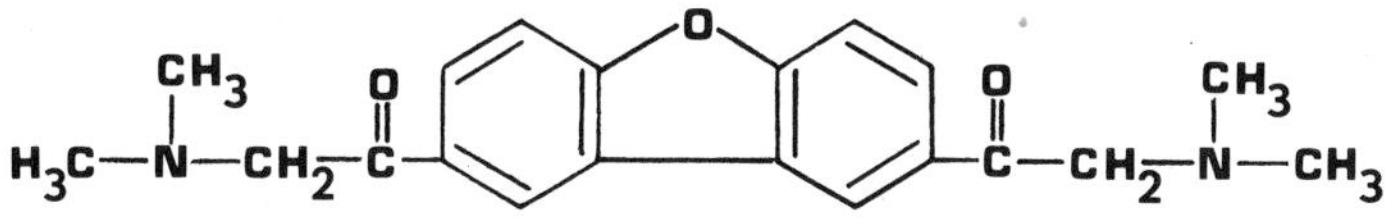

23; R$_1$=CH$_3$, R$_2$=NH$_2$
28; R$_1$=H, R$_2$=N(CH$_3$)$_2$

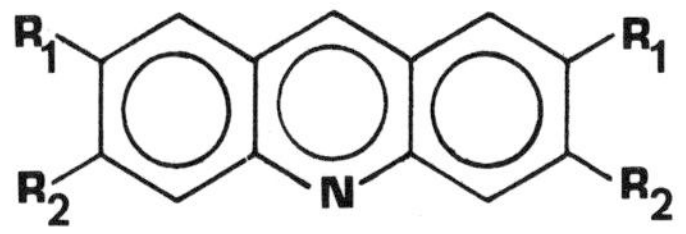

24; R$_1$=NH$_2$, R$_2$=OCH$_2$CH$_3$, R$_3$=H
25; R$_1$=R$_2$=H, R$_3$=NH$_2$

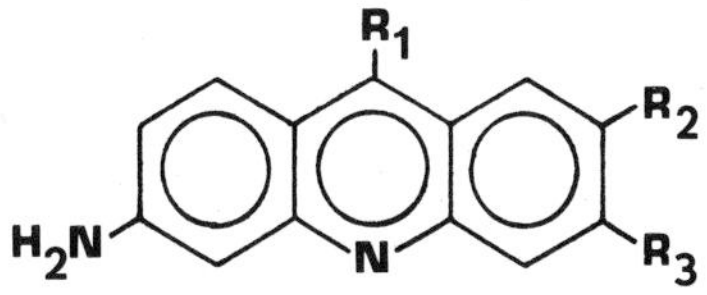

27

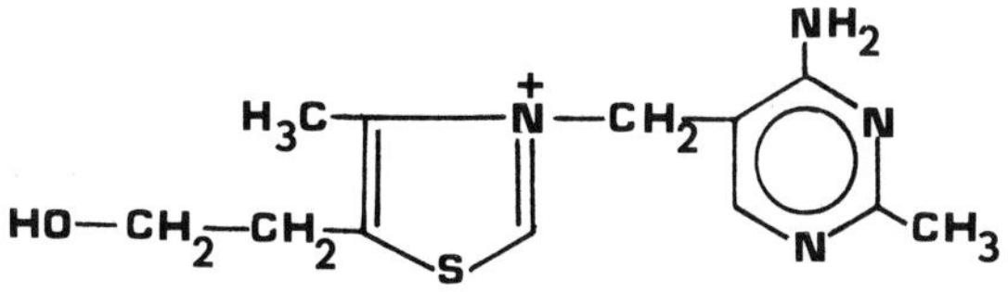

Figure 4.   Diamine Structures

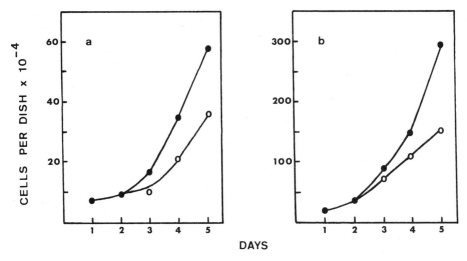

Figure 5. The effect of polymer 20 (tetramisole) on (a) normal Balb/3T3 cells and (b) transformed m-MSV-Balb/3T3 cells. Cells were plated out in dishes at the concentration shown on day 1. On day 2, polymer was added to a final concentration of 20. µg/ml. For several additional days, the number of cells per dish was counted. Cells treated with the polymer (o-o) were compared to untreated controls (●-●).

Table 2. Biological Activities of the Polymers

Polymer Number	Concentration Needed to Inhibit Cell Growth by 50%
8	5.0 µg/ml
10	no effect at 50. µg/ml
11	50. µg/ml
12	no effect at 50. µg/ml
13	0.03-0.06 µg/ml
14	no effect at 50. µg/ml
15	no effect at 50. µg/ml
16	0.03-0.06 µg/ml
17	2.0-5.0 µg/ml
18	50. µg/ml
19	no effect at 50. µg/ml
20	20. µg/ml
21	0.50-1.0 µg/ml
22	10. µg/ml
23	0.40 µg/ml
24	1.5 µg/ml
25	0.20 µg/ml
26	no effect at 25 µg/ml
27	no effect at 25 µg/ml
28	0.25 µg/ml
cis-DDP (fresh)	0.10 µg/ml
cis-DDP (stored)	10. µg/ml
trans-DDP (fresh)	25 µg/ml
trans-DDP (stored)	50. µg/ml

Although the polymers vary in effectiveness, the active polymers do share some important properties. At concentrations 2-10 fold higher than those shown in Table 2, the polymers will actively kill cells instead of merely slowing cell growth. Again, both types of cells respond similarly to higher polymer concentrations (126,127,136) Furthermore, the ability of the polymers to affect cells depends upon the growth state of the cells. Normal cells, unlike transformed cells, can enter a quiescent state where they maintain themselves but do not replicate DNA or divide. The platinum polymers display biological activity against normal growing cells and against transformed cells as described, but the polymers do not affect normal quiescent cells. Quiescent cells are not harmed when exposed to polymer concentrations that would kill actively growing cells. These studies on quiescent cells were performed with polymers 13, 16, and 17; the other polymers are likely to behave in the same way.

The properties of the biologically active polymers are similar to those displayed by cis-DDP. Like the polymers, cis-DDP kills cells and/or inhibits cell growth depending upon the concentration of the drug (22,23,63,64), and growth curves like those shown for the polymers are seen when the Balb/3T3 cell lines are exposed to cis-DDP. In addition, quiescent cells treated with cis-DDP can repair platinum lesions in their DNA and are not immediately affected by the drug as are actively growing cells (59), showing that the effectiveness of cis-DDP is dependent upon the growth state of the cells as is true for the polymers. The similar characteristics of the polymers and cis-DDP are consistent with the idea that the polymers have the same mechanism of action as cis-DDP. The same basic mode of activity for the two types of compounds is predicted from their analogous structures, and the susceptibility of actively growing cells but not quiescent cells to both the polymers and cis-DDP strongly supports the hypothesis that both types of compounds work primarily through damaging DNA. However, other factors may also play a role in the biological properties of the polymers. The resistance of quiescent cells to the compounds may be partially due to the decreased rates of RNA synthesis, protein synthesis, and transport across the cell membrane that are often seen in quiescent cells (137). In addition, the diamine component of a polymer may contribute to its biological activity. This is especially likely with those polymers whose diamine component is, by itself, a biologically active compound such as methotrexate or acridine (138,139). Thus, further study on the transport of the polymers into the cell and their effect on DNA, RNA, and protein synthesis is needed before definite conclusions can be drawn about the polymers' mechanism of action and similarity to cis-DDP. It is evident, though, that the biologically active polymers are potential chemotherapeutic agents, and preliminary results in tumor-bearing mice do show that some of the polymers have good anti-tumor activity (140).

Although the platinum polymers share many properties with cis-DDP, one important difference between the two types of compounds has been uncovered and is indicated in Table 2. The biological activity of cis-DDP depends upon the age of the solution. A freshly prepared solution of cis-DDP in either dimethyl sulfoxide or cell culture medium is 100-fold more active than a solution which has been stored for several months. This loss of activity is presumably due to the reaction of cis-DDP with the solvent. In aqueous solution, cis-DDP is converted into aquated species which probably cannot enter the cell easily because they carry a charge, and such hydrolysis products are formed in a matter of hours (120,121). In dimethyl sulfoxide, solvolysis also occurs with dimethyl sulfoxide substituting for a chloride ligand. This compound is also charged, and the anti-tumor activity of the solution decreases with time (141,142).

The inactivation of cis-DDP in dimethyl sulfoxide was further explored using the Balb/3T3 cell lines. When normal Balb/3T3 cells were exposed for three days to a 1.3 μg/ml cis-DDP solution, freshly prepared in dimethyl sulfoxide, nearly all the cells were killed. When the experiment was repeated using the same cis-DDP solution stored for one week at room temperature, the cells grew at about 90% the rate of control cells, showing that the cis-DDP solution was practically inactive. If the solution was stored for one week at 4°C, loss of activity was still substantial with cell growth inhibited by only 33% compared to controls. When stored for one week at -20°C, the cis-DDP remained almost as active as the original solution and killed a large proportion of the cells. Trans-DDP also showed loss of activity upon storage although the decrease was not nearly so severe as with cis-DDP. This is probably due to the fact that trans-DDP reacts more rapidly with water or dimethyl sulfoxide than does the cis compound (70,77,142), so a freshly prepared solution of trans-DDP still contains a large proportion of solvolysis products.

This phenomenon of inactivation of platinum compounds in dimethyl sulfoxide is important in relation to the platinum polymers. Because of the low solubility of the polymers in water, stock solutions of the polymers are routinely made in dimethyl sulfoxide. Depending upon the polymer, it may take several minutes to several days to dissolve the compound in dimethyl sulfoxide. When the polymers are tested in animals, again dimethyl sulfoxide is used as a solvent in order to produce polymer solutions of adequate concentration. If the polymers are inactivated by dimethyl sulfoxide as is cis-DDP, then this must be taken into account when interpreting the biological activity of a polymer and would indicate that dimethyl sulfoxide is not an appropriate solvent for these compounds. To test the effect of dimethyl sulfoxide on the polymers, solutions of polymers 16, 20, 23, and 25 were prepared using this solvent. The biological activity of freshly prepared solutions was compared with the activity of the same solutions stored for one week at room temperature. No difference was seen in the activity of the solutions upon storage. For all the polymers tested, there was no decrease in biological effectiveness over time when the compounds were dissolved in dimethyl sulfoxide. This result contrasts sharply with the dramatic inactivation of cis-DDP under the same conditions. In a further experiment, the activity of polymer solutions in dimethyl sulfoxide stored at 4°C did not change during a six month period. Thus solvolysis of the polymers in dimethyl sulfoxide does not occur or occurs at a very slow rate, indicating that this solvent is suitable for the polymers. In addition, the ability of the polymers to retain biological activity in dimethyl sulfoxide is a major difference in chemical reactivity compared to cis-DDP, and suggests the existence of other differences in the behavior of the platinum polymers and monomeric cis-DDP.

The biological activities of the polymers listed in Table 2 show great variation. Some polymers have a lower effective concentration compared to cis-DDP, some have about the same activity as cis-DDP, some are less active, and some do not display any biological activity at concentrations as high as 50. μg/ml. In order to create a polymer which has the best potential as a chemotherapeutic drug, it is necessary to understand what factors are responsible for determining the extent of a polymer's biological activity. One obvious factor is the diamine component which is different in each polymer. However, the diamine component does not determine a polymer's activity in any simple fashion. The biological activities of the diamines themselves have been measured, and there is no direct relationship between the activity of the diamine and the activity of a polymer containing that diamine. In some cases a diamine is more active than its polymer; in some cases the diamine is less active than its polymer. There are numerous other factors which might influence the biological activity of a polymer, such as the rate at which the polymer is transported into the cell, the rate at which it is aquated within the cell, the rate at which it reacts with cellular macromolecules, etc. Two other possible factors which could determine biological activity are the size of the polymer and its ability to degrade into smaller molecules. To study these two factors, the molecular weight of the polymers and the tendency of the polymers to break down in solution were measured.

Size of the Polymers

The molecular weights of the platinum polymers could influence their biological activity in two different ways. First, smaller polymers might be able to enter the cell more easily than larger polymers. If smaller polymers can pass through the cell membrane at a faster rate or to a greater extent than can larger polymers, then smaller polymers might have greater biological activity. Second, larger polymers contain more platinum atoms per molecule than do smaller polymers. This might result in larger polymers causing more lesions in the DNA due to cooperative binding of the many platinum atoms of the polymer molecules along the DNA. If cooperative binding occurs, then larger polymers might have greater biological activity.

Apparent molecular weights were measured by light scattering photometry using 0.1% polymer solutions. Table 3 summarizes the results for a number of the polymers. A wide variation in molecular weight is seen, ranging from 7100 daltons for polymer 13 to 584,000 daltons for polymer 14 (136). At least part of this variation is probably due to the different diamine components which produce polymers of different sizes. Polymer 19 (with tryptophan as the diamine component) has a molecular weight of only 960 and so is really only a dimer.

Table 3 also shows that different batches of the same polymer can have different molecular weights, depending upon the temperature and reaction time used to synthesize a particular batch. For instance, polymer 11-batch 2 was prepared using a 24 hour reaction time and has a larger molecular weight than polymer 11-batch 3 which was prepared using a 6 hour reaction time. Polymer 13-batch 1 was prepared using a 5 hour reaction time at 50°-60°C and it has a smaller size than polymer 13-batch 2 which

Table 3. Molecular Weights from Light Scattering Photometry

Polymer Number	Molecular Weight	Degree of Polymerization
11-batch 1	18,200	49
11-batch 2	125,000	340
11-batch 3	25,000	68
13-batch 1	7,100	11
13-batch 2	18,700	28
14-batch 1	13,100	29
14-batch 2	584,000	1,292
14-batch 3	191,000	423
16-batch 1	95,000	140
17-batch 1	24,800	67
18-batch 1	133,500	189
19-batch 1	960	2
20-batch 1	135,000	287
21-batch 1	13,000	19
22-batch 1	110,000	162

ᵃAll values given in daltons.

was made using a 24 hour reaction time at room temperature. For polymer 14, a reaction temperature of 50°-60°C (batch 2) produced a larger polymer compared to running the reaction at room temperature (batch 3). The influence of reaction time and temperature on the molecular weight of the resulting polymer appears to vary with the diamine component and it will require further study in order to determine what general principles apply. However, two conclusions can be drawn regarding the relationship of polymer size to biological activity. First, different batches of the same polymer with different molecular weights display essentially the same biological activity. Different batches of polymer 13 are all extremely active and have very similar effective concentrations. Different batches of polymer 11 are all relatively inactive. Therefore, the biological activity of a particular polymer does not appear to vary significantly with the chain length of the polymer molecules. Second, there is no direct correlation between molecular weight and biological activity when different polymers are compared. Polymer 13 is small and polymers 16 and 22 are large, yet all three polymers are active. Some batches of polymer 11 are small while polymer 18 is large, but both polymers are relatively inactive. Thus size is not the primary factor determining the biological activity of the platinum polymers.

Light scattering photometry has certain disadvantages when used to measure the molecular weights of the platinum polymers. Some of the polymers produce solutions which are too dark to give accurate transmittance readings and so cannot be used with this technique. Furthermore, light scattering photometry gives a simple weight average molecular weight but does not indicate the range of molecular weights existing in the polymer sample. For these reasons, a second method for determining size was used. Gel filtration on Sephacryl S-200 was employed. Polymers whose molecular weights had been measured by light scattering photometry were run on a Sephacryl column and the elution volumes found. A calibration curve of Ve/Vo vs. log molecular weight was constructed. The elution volumes of other polymers were then compared to the calibration curve and their molecular weights determined. It should be noted that the molecular weights found in this manner are not absolute values since they depend upon apparent molecular weights from light scattering photometry. In addition, the gel filtration technique is not ideal for the platinum polymers. Some polymers tend to bind to Sephacryl and overloading the column must be avoided (136). However, reproducible results can be obtained.

Typical elution profiles for the polymers are shown in Figure 6. A number of polymers elute in one peak, although the width of the peak can vary depending upon the polymer. Using such profiles, the elution volumes of several polymers whose molecular weights were known from light scattering photometry were determined and are shown in Table 4. Using this data, a calibration curve was constructed and is shown in Figure 7. The points of the curve are somewhat scattered which is not unexpected given the errors present in light scattering photometry data and the technical difficulties of the Sephacryl gel filtration procedure. However, the predicted general trend is observed and the molecular weights of other polymers can be estimated using this curve.

In the process of determining the molecular weights of the polymers by Sephacryl column chromatography, it was found that several of the polymers produced two peaks in the elution profile. This is illustrated in Figure 8. This result indicates the existence of two different size classes of polymer chains within the same polymer preparation. Furthermore, the number of peaks present in the elution profile and their relative proportions varied among different batches of the same polymer

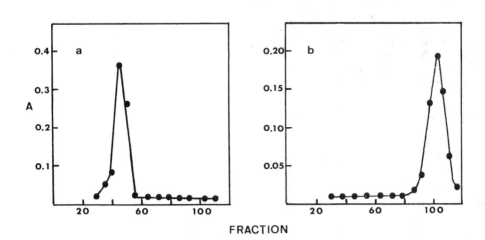

FRACTION

Figure 6. Elution profiles for platinum polymers. One milliliter of a polymer solution in dimethyl sulfoxide was loaded onto a 2.0 x 20 cm Sephacryl S-200 column. Fractions of 0.5 ml were collected and the absorbances determined. Elution profiles for (a) a 0.1% solution of polymer 18 (folic acid) and (b) a 0.05% solution of polymer 21 (tilorone) are shown.

Table 4. Elution Volumes of the Polymers

Polymer Number	Molecular Weight[a]	Ve/Vo
11-batch 1	18,200	2.35
13-batch 1	7,100	2.67
13-batch 2	18,700	2.19
14-batch 1	13,100	2.06
16-batch 1	95,000	1.00
18-batch 1	133,500	1.05
21-batch 1	13,000	2.53

[a]All values determined from light scattering photometry.

depending upon the reaction time of polymer synthesis. This phenomenon is demonstrated in Figure 9 which shows the elution profiles for three different batches of polymer 17. Batch 2 was prepared with a 3.5 hour reaction time, batch 3 was prepared with a 19 hour reaction time, and batch 4 was prepared with a 90 hour reaction time. At 3.5 hours, only one peak with a relatively small molecular weight is seen. At 19 hours, a second peak with a larger molecular weight appears and, at 90 hours, this second peak is greater than the first. Thus the relative proportion of polymer chains in the larger size class increases with increasing reaction time. Other polymers also produced bimodal size distributions. For polymers prepared in multiple batches, these size distributions varied depending upon the reaction time. Examples are given in Table 5; in all cases, a batch prepared with a longer reaction time contains more polymer chains in the larger size class than does a batch prepared with a shorter reaction time.

The existence of two distinct size classes of polymer chains present in a polymer preparation is an interesting observation. That the size of a polymer increases with increasing time of the synthesis reaction is not remarkable. However, it would have been expected that such size increases would be continuous, producing one peak in the elution profile which gradually moved toward larger molecular weights with increasing reaction times. The presence of two separate peaks could be explained by two

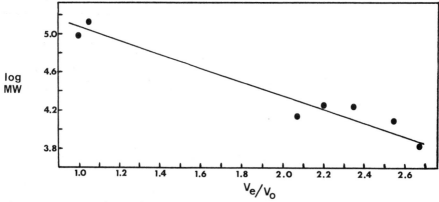

Figure 7. Calibration curve for Sephacryl S-200 gel filtration.

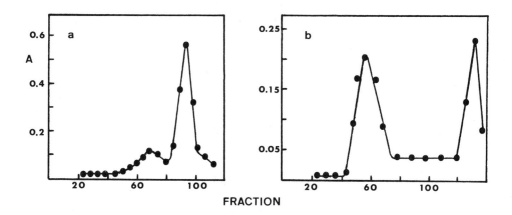

Figure 8. Elution profiles for platinum polymers. One
milliliter of a polymer solution in dimethyl sulfoxide was loaded
onto a 2.0 x 20 cm Sephacryl S-200 column. Fractions of 0.5 ml
were collected and the absorbances determined. Elution profiles
for (a) polymer 8 and (b) polymer 15 are shown.

different modes of polymer synthesis. The first mode of synthesis may occur in solution where the polymer molecule grows to
a certain chain length determined by such factors as solubility
and the diamine component. Once the molecule reaches a certain
size, it precipitates. The second mode of synthesis may then
occur in the solid phase as precipitated polymer chains react
together to form one longer chain. When a polymer is synthesized
with a short reaction time, most of the polymer chains are formed
by the first synthesis mode, giving rise to a peak corresponding
to a small molecular weight. The longer the reaction time, the
more polymer chains can react in the solid phase, and this is
reflected in a shift toward the larger molecular weight peak.
Since polymers of this type can react in aqueous suspension to
form larger chains (143), synthesis in the solid phase is a
reasonable possibility. It will, however, require further study

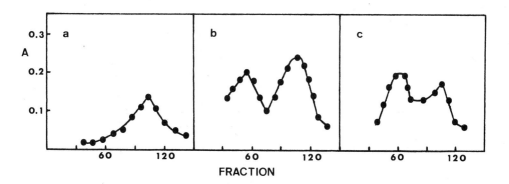

Figure 9. Elution profiles for polymer 17. Three different
batches of polymer 17 were synthesized using (a) a 3.5 hour
reaction time, (b) a 19 hour reaction time or (c) a 90 hour
reaction time. The products were run on Sephacryl S-200 and the
elution profiles are shown.

Table 5. Molecular Weights from Sephacryl Column Chromatography

Polymer Number	Peak	Molecular Weight	Degree of Polymerization
8-batch 1	1	89,100	236
	2	49,000	130
10-batch 1	1	112,000	331
10-batch 2	1	112,000	331
	2	10,600	31
13-batch 3	1	83,200	127
	2	29,500	45
15-batch 1	1	9,600	26
15-batch 2	1	37,200	101
	2	7,100	19
17-batch 2	1	13,000	35
17-batch 3	1	50,000	135
	2	14,000	38
17-batch 4	1	50,000	135
	2	14,000	38

to confirm this theory. Additional questions, such as whether all the platinum polymers can display two size classes under appropriate synthesis conditions and whether there is an upper limit to the size these polymers can attain, remain to be answered. It is apparent, though, that the results of Sephacryl gel filtration support the earlier conclusion that there is no direct correlation between the molecular weight of a polymer and its biological activity.

Degradation of the Polymers

Another factor which could influence a platinum polymer's degree of biological activity is the tendency of the polymer to degrade into smaller molecules. Such smaller molecules might enter the cell more easily and perhaps bind more readily to cellular macromolecules. Polymers which degrade would then be more active than polymers which remain large. It is also possible that the degradation products might be less active than the original polymer chain, particularly if the smaller products are charged, modified by hydrolysis, etc. In this case, polymers which did not degrade would be more active. Ideally, degradation of the platinum polymers should be studied in an aqueous solution of physiological pH and ionic strength. However, the low solubility of the polymers in water combined with the presence of other ions make it technically very difficult to run the degradation reaction under these conditions and then determine the molecular weight of the products by light scattering photometry. For this reason, initial degradation studies were done using dimethyl sulfoxide as a solvent.

Several polymers were dissolved in dimethyl sulfoxide and the molecular weights of the polymers were measured by light scattering photometry over a period of several days. The results are shown in Table 6. All of the polymers tested undergo degradation to a significant extent. Polymers 11 and 14 both

Table 6. Degradation of the Polymers in Dimethyl Sulfoxide

Polymer Number	Initial MW	24 hour MW	48 hour MW	72 hour MW
11-batch 1	18,200	11,300	10,100	10,000
11-batch 3	25,000	12,000	8,800	11,200
13-batch 1	7,100	6,900	7,100	4,500
13-batch 2	18,700	16,300	9,400	11,900
14-batch 1	13,100	3,400	5,500	1,700
16-batch 2	250,000	115,000	85,000	85,000

have little biological activity and both degrade rapidly during the first 24 hours. Polymer 16 is very active and it too degrades quickly. Polymer 13 is also very active, but it appears to degrade more slowly than the others. Thus no consistent trends are seen relating polymer degradation in dimethyl sulfoxide to biological activity for the polymers tested. It will be necessary to study additional polymers in circumstances approaching physiological conditions before definite conclusions can be made regarding the correlation between extent of degradation and biological effects of a polymer.

To explore the mechanism of the observed polymer degradation, a detailed study of the breakdown of polymer 16 (containing methotrexate) was performed. The solvent employed was dimethyl sulfoxide. However, since this solvent is hygroscopic, the actual system is dimethyl sulfoxide/water. The results of the study are shown in Table 7. The behavior of the polymer is consistent with random scission degradation. Condensation and coordination polymers generally undergo depolymerization by this mechanism and follow the relationship

$$1/DP_t = 1/DP_o + kt$$

where DP_t is the degree of polymerization at time t, DP_o is the original chain length, and k is related to the number of bonds broken per unit volume per unit time. Polymer 21 (containing tilorone) gives similar results, also breaking down by random scission. This mechanism of polymer degradation has an important implication regarding the biological activity of the platinum polymers. If the polymers all degrade by random scission and yet degradation is not complete (as indicated in Table 6), then the actual mechanism of breakdown is partial random scission. Partial random scission generates only small amounts of monomeric

Table 7. Degradation of Polymer 16

Time	Degree of Polymerization
0 hours	23
6	7
12	4
24	3
36	2
170	2

Table 8. Degradation of Polymer 29 in Dimethyl Sulfoxide/Water

Time	Degree of Polymerization
0 hrs	25
6	110
12	160
24	175
48	50

units. This suggests that the biological effects of the active polymers are not due to released monomeric units since such monomers would be present only at low levels. The active form of the compounds would therefore be polymeric, or at least oligomeric.

However, not all platinum polymers behave in the same fashion in dimethyl sulfoxide/water. A polymer containing 1,6-hexanediamine (polymer 29-batch 1) actually increases in size in this solvent system, with the maximum chain length seen at 12-24 hours (Table 8). When the experiment is repeated using a nitrogen purge to reduce the water content of the dimethyl sulfoxide, the chain length of the polymer (batch 2) decreases throughout the study but follows the random scission relationship only approximately (Table 9). When molecular (water, 4Å) sieves are used to remove even more of the water, chain length of the polymer (batch 3) decreases with a good linear relationship between $1/DP_t$ and time (Table 10). Thus this polymer undergoes random scission in dimethyl sulfoxide, but the presence of water significantly changes or interferes with the degradation reaction.

Studies with other polymers also indicate that water produces complicated behavior by the polymers. When the polymers (11,13, or 16) are suspended in water and stirred for 24 hours, 90% of the polymer can be recovered from the suspension and the molecular weight of the undissolved material can be determined by light scattering photometry. Results of two runs of this

Table 9. Degradation of Polymer 29 in Dimethyl Sulfoxide (N_2 Flush)

Time	MW	Time	MW
0 hrs	2×10^4 (DP=54)	23.5	3.9×10^3
0.5	1.5×10^4	24	3.8×10^3
1.5	1.3×10^4	25.5	3.5×10^3
2.0	1.1×10^4	26.5	3.5×10^3
2.5	1.0×10^4	31.5	3.4×10^3
3.0	1.0×10^4	33.5	3.2×10^3
4.0	9.0×10^3	48.5	2.8×10^3
4.5	8.5×10^3	49.5	2.6×10^3
5.0	7.8×10^3	53	2.6×10^3
5.5	7.6×10^3	64	2.5×10^3
6.0	7.4×10^3	65	2.3×10^3 (DP=6.5)
14	6.6×10^3		

Table 10. Degradation of Polymer 29 in Dimethyl Sulfoxide
(Molecular Sieves)

Time	Molecular Weight
0 hrs	2.4×10^5 (DP=660)
1	5.5×10^4
6	1.5×10^4
11	9.4×10^3
24	4.2×10^3
48	2.4×10^3
72	1.9×10^3
120	1.4×10^3 (DP=4)

experiment are shown in Table 11. While the size of polymer 13
always decreased, the size of polymer 11 once decreased and once
stayed about the same and the size of polymer 16 once decreased
and once increased. It is possible that these inconsistent
results are due to competition between polymer synthesis and
polymer degradation. An aqueous polymer suspension is basically
what exists during the synthesis reaction. It is therefore a
reasonable possibility that polymer synthesis can continue when a
polymer is suspended in water. Evidence for this reaction comes
from experiments with polymer 29. When solid polymer (DP=5) is
added to water containing dissolved tetrachloroplatinate and
stirred for 24 hours, the chain length increases to a DP=1700.
When the same solid polymer is stirred in water containing
dissolved 1,6-hexanediamine (the diamine component), the size
increases to a DP=32. When the solid polymer by itself is
suspended in water and stirred, the DP increases to 135. In
contrast, solid polymer in the dry state maintains a constant
chain length over a period of months. Thus chain elongation can
take place in aqueous suspension, and it is likely that when the
platinum polymers are suspended in water there is competition
between chain growth and degradation. The conditions that favor
one process over the other are yet to be elucidated, but further
study of this aspect of polymer behavior may eventually help to
explain why some platinum polymers are biologically active while
others are not.

CONCLUSIONS

     The platinum polymers that have been synthesized so far show
a wide range of biological activity. Some compare favorably to
cis-DDP in effectiveness and so are potential chemotherapeutic
agents. The factors that control the extent of a polymer's
biological activity have yet to be determined. Polymer size does
not appear to be a significant factor since active polymers vary
greatly in molecular weight and degree of polymerization, and
different batches of the same polymer with different sizes
exhibit the same biological activity. The tendency of a polymer
to break down may or may not be related to its biological
activity. The degradation of the polymers in dimethyl sulfoxide
does not correlate with biological effectiveness, but this
solvent is not analogous to aqueous physiological conditions.
The degradation of the polymers in water is complicated and the
factors controlling breakdown under these conditions have not
been defined.

Table 11. Behavior of the Polymers in Aqueous Suspension

Run	Polymer Number	Initial MW	24 Hour MW
1	11-batch 1	18,200	5,600
2	11-batch 1	18,200	20,700
1	13-batch 1	7,100	1,500
2	13-batch 1	7,100	3,000
1	13-batch 2	18,700	6,500
2	13-batch 2	18,700	1,400
1	16-batch 1	95,000	202,000
2	16-batch 1	95,000	11,200

These studies have revealed several interesting aspects of the polymers' chemical behavior. They apparently undergo solvolysis at a much slower rate than does the monomeric cis-DDP. The size of a polymer depends upon the synthesis conditions, and it is possible that two distinct mechanisms of polymer synthesis can occur, one in solution and one in the solid phase. In dimethyl sulfoxide, the polymers undergo breakdown by partial random chain scission, while in water, there seem to be competing reactions of chain growth and degradation. Continued study of the platinum polymers is likely to reveal additional information pertinent to polymer chemistry as well as aid in designing better platinum cancer drugs.

REFERENCES

1. B. Rosenberg, L. Van Camp and T. Krigas, Nature, Lond., 205, 698 (1965).
2. B. Rosenberg, L. Van Camp, E. G. Grimley and A. J. Thomson, J. Biol. Chem., 242, 1347 (1967).
3. S. Resolva, Chem.-Biol. Interact., 4, 66 (1971).
4. L. Kutinova, V. Vonka and J. Drobnik, Neoplasma, 19, 453 (1972).
5. J. E. Trosko, "Platinum Coordination Complexes in Cancer Chemotherapy. Recent Results in Cancer Research," 48, (T. A. Connors and J. J. Roberts, Eds.), Springer-Verlag, Berlin, 1974, p. 108.
6. D. J. Beck and R. R. Brubaker, Mutat. Res., 27, 181 (1975).
7. C. Monti-Bragadin, M. Tamaro and E. Banfi, Chem.-Biol. Interact., 11, 469 (1975).
8. W. R. Leopold, E. C. Miller and J. A. Miller, Cancer Res., 39, 913 (1979).
9. B. Rosenberg, L. Van Camp, J. E. Trosko and V. H. Mansour, Nature, Lond., 222, 385 (1969).
10. A. W. Prestayko, J. C. D'Aoust, B. F. Issell and S. T. Crooke, Cancer Treat. Rev., 6, 17 (1979).
11. E. Wiltshaw and B. Carr, "Platinum Coordination Complexes in Cancer Chemotherapy. Recent Results in Cancer Research," 48, (T. A. Conners and J. J. Roberts, Eds.), Springer Verlag, Berlin, 1974, p.178.
12. P. J. Loehrer and L. H. Einhorn, Ann. Int. Med., 100, 704 (1984).
13. J. W. Reishus and D. S. Martin, Jr., J. Am. Chem. Soc., 83, 2457 (1961).

14. W. Schaller, H. Reisner and E. Holler, Biochemistry, 26, 943 (1987).
15. M. C. Lim and R. B. Martin, J. Inorg. Nucl. Chem., 38, 1911 (1976).
16. F. Basolo, H. B. Gray and R. G. Pearson, J. Am. Chem. Soc., 82, 4200 (1960).
17. N. P. Johnson, J. D. Hoeschele and R. O. Rahn, Chem.-Biol. Interact., 30, 151 (1980).
18. P. T. Daley-Yates and D. C. H. McBrien, Chem.-Biol. Interact., 40, 325 (1982).
19. A. J. Thomson., R. J. P. Williams and S. Reslova, Struct. Bonding (Berlin), 11, 1746 (1972).
20. H.C. Harder and B. Rosenberg, Int. J. Cancer, 6, 207 (1970).
21. H. W. van den Berg and J. J. Roberts, Chem.-Biol. Interact., 12, 375 (1976).
22. H. N. A. Fraval and J. J. Roberts, Chem.-Biol. Interact., 23, 99 (1978).
23. H. N. A. Fraval and J. J. Roberts, Chem.-Biol. Interact., 23, 111 (1978).
24. J. M. Pascoe and J. J. Roberts, Biochem. Pharmacol., 23, 1345 (1974).
25. J. M. Pascoe and J.J. Roberts, Biochem. Pharmacol., 23, 1359 (1974).
26. J. J. Roberts and M. F. Pera, Jr., "Platinum, Gold, and Other Metal Chemotherapeutic Agents," (S. J. Lippard, Ed.), American Chemical Society, Washington, D. C., 1983, p.3.
27. P. Horacek and J. Drobnik, Biochem. Biophys. Acta, 254, 341 (1971).
28. S. Mansy, B. Rosenberg and A. J. Thomson, J. Amer. Chem. Soc., 95, 1633 (1973).
29. P. J. Stone, A. D. Kelman and F. M. Sinex, Nature, Lond., 251, 736 (1974).
30. L. L. Munchausen and R. O. Rahn, Biochim. Biophys. Acta, 414, 242 (1975).
31. C. J. L. Lock, J. Bradford, R. Faggiani, R. A. Speranzini, G. Turner and M. Zvagulis, J. Clin. Hemat. Oncol., 7, 63 (1977).
32. A. J. Thomson, "Platinum Coordination Complexes in Cancer Chemotherapy. Recent Results in Cancer Research," 48, (T. A. Connors and J. J. Roberts, Eds.), Springer-Verlag, Berlin, 1974, p. 38.
33. P. K. Ganguli and R. Theophanides, Eur. J. Biochem., 101, 377 (1979).
34. T. J. Kistenmacher, J. D. Orbell and L. G. Marzilli, "Platinum, Gold, and Other Metal Chemotherapeutic Agents," (S. J. Lippard, Ed.), American Chemical Society, Washington, D. C., 1983, p. 191.
35. Z. M. Banjar, L. S. Hnilica, R. C. Briggs, J. Stein and G. Stein, Biochemistry, 23, 1921 (1984).
36. R. B. Ciccarelli, M. J. Solomon, A. Varashavsky and S. J. Lippard, Biochemistry, 24, 7533 (1985).
37. R. Olinski, A. Wedrychowski, W. N. Schmidt, R. C. Briggs and L. S. Hnilica, Cancer Res., 47, 201 (1987).
38. C. R. Morris and G. R. Gale, Chem.-Biol. Interact., 7, 305 (1973).
39. B. Odenheimer and W. Wolf, Inorg. Chim. Acta, 66, L41 (1982).
40. P. C. Dedon and R. F. Borch, Biochem. Pharmacol., 36, 1955 (1987).
41. A. C. M. Plooy, M. van Dijk and P. H. M. Lohman, Cancer Res., 44, 2043 (1984).

42. J. J. Roberts and F. Friedlos, Biochim. Biophys. Acta, 655, 146 (1981).
43. K. V. Shooter, R. Howse, R. K. Merrifield and A. B. Robbins, Chem.-Biol. Interact., 5, 289 (1972).
44. M. C. Strandberg, E. Bresnick and A. Eastman, Chem.-Biol. Interact., 39, 169 (1982).
45. L. A. Zwelling, T. Anderson and K. W. Kohn, Cancer Res., 39, 365 (1979).
46. L. A. Zwelling, M. O. Bradley, N. A. Sharkey, T. Anderson and K. W. Kohn, Mutat. Res., 67, 271 (1979).
47. L. A. Zwelling, S. Michaels, H. Schwartz, P. O. Dobson and K. W. Kohn, Cancer Res., 41, 640 (1981).
48. A. Eastman, Biochemistry, 22, 3927 (1983).
49. R. O. Rahn, J. Inorg. Biochem., 21, 311 (1984).
50. F. J. Dijt, A. M. J. Fichtinger-Schepman, F. Berends and J. Reedijk, Cancer Res., 48, 6058 (1988).
51. W. I. Sundquist, S. J. Lippard and B. D. Stollar, Biochemistry, 25, 1520 (1986).
52. A. M. J. Fichtinger-Schepman, J. L. vander Veer, J. H. J. den Hartog, P. H. M. Lohman, and J. Reedijk, Biochemistry, 24, 707 (1985).
53. M. C. Poirier, E. Reed, R. F. Ozols and S. H. Yuspa, UCLA Symp. Mol. Cell. Biol., New Ser., 40, 303 (1986).
54. A. L. Pinto and S. J. Lippard, Biochim. Biophys. Acta, 780, 167 (1985).
55. G. Villani, U. Huebscher and J. L. Butour, Nucleic Acids Res., 16, 4407 (1988).
56. J. D. Gralla, S. Sasse-Dwight and L. G. Poljak, Cancer Res., 47, 5092 (1987).
57. E. C. Friedberg, "DNA Repair," W. H. Freeman, New York, 1985.
58. H. N. A. Fravel and J. J. Roberts, Cancer Res., 39, 1793 (1979).
59. M. F. Pera, C. J. Rawlings and J. J. Roberts, Chem.-Biol. Interact., 37, 245 (1981).
60. V. Kleinwachter, O. Vrana, V. Brabec and N. P. Johnson, Stud. Biophys., 123, 85 (1988).
61. O. Vrana, V. Brabec and V. Kleinwachter, Anti-Cancer Drug Des., 1, 95 (1986).
62. J. A. Howle and G. R. Gale, Biochem. Pharmacol., 19, 2757 (1970).
63. C. M. Sorenson and A. Eastman, Cancer Res., 48, 4484 (1988).
64. C. M. Sorenson and A. Eastman, Cancer Res., 48, 6703 (1988).
65. J. S. Hoffmann, N. P. Johnson and G. Villani, J. Biol. Chem., 264, 15130 (1989).
66. S. E. Sherman and S. J. Lippard, Chemical Reviews, 87, 1153 (1987).
67. T. A. Connors, M. Jones, W. C. J. Ross, P. D. Braddock, A. R. Khokhar and M. L. Tobe, Chem.-Biol. Interact., 5, 415 (1972).
68. P. D. Braddock, T. A. Connors, M. Jones, A. R. Khokhar, D. H. Melzack and M. L. Tobe, Chem.-Biol. Interact., 11, 145 (1975).
69. T. Tashiro, Nippon Kagaku Kaishi, 4, 684 (1988).
70. J. J. Roberts and M. P. Pera, Jr., "Molecular Aspects of Anti-Cancer Drug Action," (S. Neidle and M. J. Waring, Eds.), MacMillan, London, 1983, p. 183.
71. M. J. Cleare and J. D. Hoeschele, Platinum Metals Rev., 17, 2 (1973).

72. J. P. Macquet, J. L. Butour and N. P. Johnson, "Platinum, Gold, and Other Metal Chemotherapeutic Agents," (S. J. Lippard, Ed.), American Chemical Society, Washington, D.C., 1983, p. 75.

73. N. P. Johnson, J. D. Hoeschele, R. O. Rahn, J. P. O'Neill and A. W. Hsie, Cancer Res., 40, 1463 (1980).

74. A. L. Pinto and S. J. Lippard, Proc. Natl. Acad. Sci. U.S.A., 82, 4616 (1985).

75. S. J. Lippard, Pure Appl. Chem., 59, 731 (1987).

76. N. P. Johnson, P. Lapetoule, H. Razaka and G. Villani, "Biochemical Mechanisms of Platinum Antitumour Drugs," (D. C. H. McBrien and T. F. Slater, Eds.), IRL Press, Oxford, 1986, p.1.

77. A. J. Thomson, I. A. G. Roos and R. D. Graham, J. Clin. Hemat. Oncol., 7, 242 (1977).

78. M. Zak, J. Drobnik and Z. Rezny, Cancer Res., 32, 565 (1972).

79. J. A. Gottlieb and B. Drewinko, Cancer Chemother. Rep., 59, 621 (1975).

80. D. D. Von Hoff, R. Schilsky, C. M. Reichert, R. L. Reddick, M. Rozencweig, R. C. Young and F. M. Muggia, Cancer Treat. Rep., 69, 1527 (1979).

81. J. Ward, D. Young, K. Fauvie, M. Wolpert, R. Davis and A. Guarino, Cancer Treat. Rep., 60, 1675 (1976).

82. N. V. Aguilar-Markulis, S. Becklsy and R. Priore, J. Surg. Oncol., 16, 123 (1981).

83. G. H. Barker, J. Orledge and E. Wilshaw, Br. J. Obstet. Gynaecal., 88, 690 (1981).

84. D. J. Hall, R. Diaso and D. R. Gopleurud, Am. J. Obstet. Gynecol., 141, 309 (1981).

85. C. Litterst, T. Gram, R. Dedrick, A. Leroy, and A. Guarino, Cancer Res., 36, 2340 (1976).

86. Z. H. Siddik, D. R. Newell, F. E. Boxall and K. R. Harrap, Biochem. Pharmacol., 36, 1925 (1987).

87. C. Cvitkovic, J. Spaulding and V. Bethune, Cancer, 39, 1357 (1977).

88. N. Imura, A. Naganuma, M. Satoh and Y. Koyama, Experientia, Suppl., 52, 655 (1987).

89. G. S. Baldew, C. J. A. Van den Hamer, G. Los, N. P. E. Vermeulen, J. J. M. De Goeij and J. G. McVie, Cancer Res., 49, 3020 (1989).

90. F. Zumino, G. Pratesi, A. Micheloni, E. Cavalletti, F. Sala and O. Tofanetti, Chem.-Biol. Interact., 70, 89 (1989).

91. J. M. Onoda, K. K. Nelson, J. D. Taylor and K. V. Honn, Cancer Lett., 40, 39 (1988).

92. R. J. Fram, N. Robichaud, S. D. Bishov and J. M. Wilson, Cancer Res., 47, 3360 (1987).

93. T. Uchida, K. Okamoto, K. Nishikawa and K. Takahashi, Gan to Kagaku Ryoho, 13, 75 (1986).

94. K. J. Scanlon, Y. Lu, M. Kashani-Sabet, J. X. Ma and E. Newman, Adv. Exp. Med. Biol., 244, 127 (1988).

95. P. A. Steerenburg, G. Storm, G. De Groot, A. Claessen, J. J. Bergers, M. A. M. Franken, Q. G. C. M. Van Hoesel, K. L. Wubs and W. H. De Jong, Cancer Chemother. Pharmacol., 21, 299 (1988).

96. R. Reszka, I. Fichtner, E. Nissen, D. Arndt and A. M. Ladhoff, J. Microencapsulation, 4, 201 (1987).

97. J. Freise and P. Magerstedt, "Liposome Drug Carriers Symp.," (K. H. Schmidt, Ed.), Thieme, Stuttgart, Fed. Rep. Ger., 1986, p. 182.

98. B. Hecquet, G. Depadt, C. Fournier and J. Meynadier, Anticancer Res., 6, 659 (1986).

99. Y. Okamoto, A. Konno, K. Togawa, T. Kato, Y. Tamakawa and T. Amano, Br. J. Cancer, 53, 369 (1986).

100. S. Yolles, R. M. Roat, M. F. Satori and C. L. Washburne, "Biological Activities of Polymers," (C. E. Carraher and C. G. Gebelein, Eds.), American Chemical Society, Washington, D. C., 1982, p. 233.

101. B. Schechter, R. Pauzer, M. Wilchek and R. Arnon, Cancer Biochem. Biophys., 8, 289 (1986).

102. B. Schechter, M. Wilchek and R. Arnon, Int. J. Cancer, 39, 409 (1987).

103. A. Eastman and V. M. Richon," Biochemical Mechanisms of Platinum Antitumor Drugs," (D. C. H. McBrien and T. F. Slater, Eds.), IRL Press, Oxford, 1986, p. 91.

104. W. R. Ward, Cancer Res., 47, 6549 (1987).

105. Y. Kikuchi, I. Iwano and M. Miyauchi, Gan to Kagaku Ryoho, 15, 2895 (1988).

106. A. Eastman and N. Schulte, Biochemistry, 27, 4730 (1988).

107. H. Masuda, R. F. Ozols, G. M. Lai, A. Fojo, M. Rothenberg and T. C. Hamilton, Cancer Res., 48, 5713 (1988).

108. S. Sekiya, T. Oosaki, S. Andoh, N. Suzuki, M. Akaboshi and H. Takamizawa, Eur. J. Cancer Clin. Oncol., 25, 429 (1989).

109. S. L. Kelley, A. Basu, B. A. Teicher, M. P. Hacker, D. H. Hamer and J. S. Lazo, Science, 241, 1813 (1988).

110. L. Endresen and H. E. Rugstad, Experientia, Suppl., 52, 595 (1987).

111. P. A. Andrews, M. P. Murphy and S. B. Howell, Cancer Chemother. Pharmacol., 19, 149 (1987).

112. A. D. Lewis, J. D. Hayes and C. R. Wolf, Carcinogenesis, 9, 1283 (1988).

113. A. H. Calvert, "Biochemical Mechanisms of Platinum Antitumor Drugs," (D. C. H. McBrien and T. F. Slater, Eds.), IRL Press, Oxford, 1986, p. 307.

114. K. R. Harrap, M. Jones, C. R. Wilkinson, H. M. Clink, S. Sparrow, B. C. V. Mitchley, S. Clarke and A. Veasey, "Cisplatin: Current Status and New Developments," (A. W. Prestayko, S.T. Crooke and S. K. Carter, Eds.), Academic Press, New York, 1980, p. 193.

115. A. Rahman, J. K. Roh, M. K. Wolpert-DeFilippes, A. Goldin, J. M. Venditti and P. V. Wooley, Cancer Res., 48, 1745 (1988).

116. C. E. Carraher, J. Schroeder, D. J. Giron and W. J. Scott, J. Macromol. Sci.-Chem., A15, 625 (1981).

117. C. E. Carraher, D. J. Giron, I. Lopez, D. R. Cerutis and M. J. Scott, Org. Coatings Plast. Chem., 44, 120 (1981).

118. J. Davidson, D. Faber and R. Fischer, Cancer Chemother. Rep., 59, 287 (1975).

119. C. E. Carraher, W. J. Scott and D. J. Giron, "Bioactive Polymeric Systems," (C. G. Gebelein and C. E. Carraher, Eds.), Plenum Press. New York, 1985, p. 587.

120. B. J. Corden, Inorg. Chim. Acta, 137, 125 (1987).

121. J. J. Roberts, R. J. Knox, F. Friedlos and D. A. Lydall, "Biochemical Mechanisms of Platinum Antitumor Drugs," (D. C. H. McBrien and T. F. Slater, Eds.), IRL Press, Oxford, 1986, p. 29.

122. P. T. Daley-Yates, "Biochemical Mechanisms of Platinum Antitumor Drugs," (D. C. H. McBrien and T. F. Slater, Eds.), IRL Press, Oxford, 1986, p. 121.

123. A. Allcock, Science, 193, 1214 (1976).

124. A. Allcock, "Organometallic Polymers," (C. E. Carraher, J. Sheats and C. Pittman, Eds.), Academic Press, New York, 1978, Chap. 28.

125. C. E. Carraher, T. Manek, D. J. Giron, D. R. Cerutis and M. Trombley, Polym. Prepr., 23, 77 (1982).

126. D. W. Siegmann, C. E. Carraher and A. Friend, J. Polym. Mater., 4, 19 (1987).

127. D. W. Siegmann and C. E. Carraher, J. Polym. Mater., 4, 29 (1987).

128. D. W. Siegmann, D. Brenner and C. E. Carraher, Polym. Mater. Sci. Engineer., 59, 535 (1988).

129. C. E. Carraher, W. J. Scott, I. Lopez, D. R. Cerutis and R. Manek, "Biological Activity of Polymers," (C. E. Carraher and C. G. Gebelein, Eds.), American Chemical Society, Washington, D.C., 1982, p. 221.

130. F. A. Cotton and G. Wilkenson, "Advanced Inorganic Chemistry," Interscience, New York, 1978.

131. J. E. Huheey, "Inorganic Chemistry," 3rd ed., Harper and Row, New York, 1983.

132. H. Into, J. Fugita and K. Sato, Bull. Chem. Soc. Japan, 40, 1584 (1967).

133. A. A. Grinberg, M. Serator and M. L. Gellfman, Russian J. Inorg. Chem., 13, 1695 (1968).

134. P. J. Debye, J. Phys. Pol. Chem., 51, 1 (1947).

135. R. Seymour and C. E. Carraher, "Polymer Chemistry: An Introduction," Dekker, New York, 1982.

136. D. W. Siegmann, C. E. Carraher and D. Brenner, "Progress in Biomedical Polymers," (C. G. Gebelein, Ed.), Plenum, in press.

137. H. Bourne and E. Rozengurt, Proc. Natl. Acad. Sci., U.S.A., 73, 4555 (1976).

138. B. Roth, E. Bliss and C. R. Beddell, "Molecular Aspects of Anti-cancer Drug Action," (S. Neidle and M. J. Waring, Eds.), MacMillan, London, 1983, p. 363.

139. W. A. Denny, B. C. Baguley, B. F. Cain and M. J. Waring, "Molecular Aspects of Anti-cancer Drug Action," (S. Neidle and M. J. Waring, Eds.), MacMillan, London, 1983, p. 1.

140. D. Giron, M. Espy, C. Carraher and I. Lopez, "Polymeric Materials in Medication," (C. G. Gebelein and C. E. Carraher, Eds.), Plenum, New York, 1985, p. 165.

141. M. Tonew, E. Tonew, W. Gutsche, K. Wohlrabe, A. Stelzner, H. P. Schroeer and B. Heyn, Zentralbl. Bakteriol. Mikrobiol. Hyg., Ser. A, 257, 108 (1984).

142. W. I. Sundquist, K. J. Ahmed, L. S. Hollis and S. J. Lippard, Inorg. Chem., 26, 1524 (1987).

143. D. W. Siegmann, D. Brenner, C. E. Carraher and R. E. Strother, Polym. Mater. Sci. Engineer., 61, 214 (1989).

METAL-CONTAINING POLYETHYLENE WAXES

Dan Munteanu

Chemical Research Institute, Center for Plastics
Str.Gării 25, R-1900 Timişoara, Romania

INTRODUCTION

Polyethylene (PE) waxes, i.e. low molecular weight polyethylenes, are useful materials in various applications. PE
waxes available on the market are usually polymers of 1000
to 10000 molecular weight obtained by telomerization, i.e.
the polymerization of ethylene under conditions able to assure
only the formation of low-molecular-weight polymers. Some of
these waxes are obtained by the thermal degradation of the
usual-molecular-weight polyethylenes. No information seems
to be available about the use of waxes resulting as by-products
in high-density polyethylene (HDPE) plants.

Metal-containing or ion-containing polymers often have
complicated structures developed for sophisticated tailor-
made applications.[1-3] However, even simple structures of
metal-containing polymers may have interesting and useful
properties. It was the purpose of this work to develop metal-
containing PE waxes by a two-step modification of low-molecular-weight polyethylenes resulting as by-products in a HDPE
plant. The PE wax was first oxidized in order to attach free
carboxyl groups on the polymer backbone. Metal-containing PE
waxes were then obtained from the oxidized PE waxes and
metallic hydroxides.

OXIDIZED POLYETHYLENE WAXES

In the HDPE plant ethylene is polymerized in hexane,
at 70-90°C and 4-10 bars in the presence of titanium-based
$MgCl_2$-supported catalysts and alkyl aluminium co-catalysts.
Together with the slurry of HDPE powder in hexane a small
amount of PE wax is formed. Due to its low molecular weight

this by-product is soluble in hexane so that it is separated from the HDPE powder by centrifuging the slurry at $65^\circ C$. The PE wax is finally separated as a polymer melt in the hexane recovery section of the plant. The low molecular weight polyethylene, i.e. $\overline{M}_n=1000$ (vapour pressure osmometry, $80^\circ C$, toluene) and $\overline{M}_v=3000$ (viscometry, $130^\circ C$, decalin), resulting as by-product in the HDPE plant was used without further purification as raw material for the synthesis of modified PE waxes.[4]

Synthesis of oxidized polyethylene waxes

The oxidation of the PE wax was carried out by bubbling air into the polymer melt maintained under stirring at temperatures from $130^\circ C$ to $170^\circ C$.[5] Due to the low molecular weight of the PE wax and its low melt viscosity (10-30 cP at $130^\circ C$) the oxidation was performed in vessels fitted with a usual impeller-type stirrer. From a lab-scale glass vessel of two liters capacity it was scaled up to a stainless-steel reactor of one cubic meter capacity (Figure 1). The main requirement for the oxidation vessel was the need to assure a very intensive mixing of the air bubbles in the wax melt. The small diameter of the holes in the bubbling ring, i.e. 1 mm, combined with a usual stirring fulfilled this condition. For the same reaction conditions a faster oxidation and more reproducible results have been obtained when the reactor was used.

The oxidative splitting of paraffins, i.e. C_{18}-C_{30} alkanes of 250 to 450 molecular weight, for the synthesis of fatty acids[6] may be regarded as a model mechanism for the oxidation of low-molecular-weight polyethylenes.

(1) $PH \longrightarrow P\cdot$

(2) $P\cdot + O_2 \longrightarrow POO\cdot$

(3) $POO\cdot + PH \longrightarrow POOH + P\cdot$

(4) $R_1-CH_2-\underset{\underset{\underset{\underset{H}{|}}{O}}{\overset{|}{\underset{|}{O}}}{CH}-CH_2-R_2 \longrightarrow R_1-CH_2-\overset{\overset{O}{\|}}{C}-CH_2-R_2 + H_2O$

main reaction

$\longrightarrow R_1-CH_2-\underset{OH}{\overset{|}{CH}}-CH_2-R_2 + 1/2\ O_2$

side reaction

(5) $R_1-CH_2-\underset{\overset{\|}{O}}{C}-CH_2-R_2 + O_2 \longrightarrow R_1-\underset{\overset{|}{OOH}}{CH}-\underset{\overset{\|}{O}}{C}-CH_2-R_2$

(6) $R_1-\underset{\overset{|}{OOH}}{CH}-\underset{\overset{\|}{O}}{C}-CH_2-R_2 \longrightarrow R_1-CH{=}O + R_2-CH_2COOH$

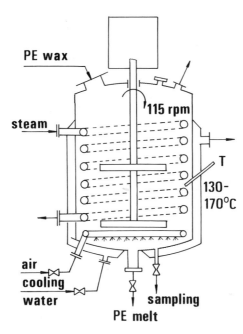

Figure 1. The oxidation reactor

(7) $R_1-CH{=}O + O_2 \longrightarrow R_1-C{\overset{O}{\diagdown}}_{OOH} \longrightarrow R_1-COOH$

(8) $R-C{\overset{O}{\diagdown}}_{OOH} + R_1-\underset{\overset{\|}{O}}{C}-CH_2-R_2 \longrightarrow R-COOH + R_1-\underset{\overset{\|}{O}}{C}{-}O-CH_2-R_2$

(9) $R-COOH + R'-CH_2OH \longrightarrow R-COO-CH_2-R' + H_2O$

(10) $R_1-\underset{\overset{|}{OOH}}{CH}-\underset{\overset{\|}{O}}{C}-R_2 \longrightarrow R_1-\underset{\overset{\|}{O}}{C}-\underset{\overset{\|}{O}}{C}-R_2 + H_2O$

(11) $R_1-\underset{\underset{O}{\|}}{C}-\underset{\underset{O}{\|}}{C}-R_2 + R_3-C\overset{\nearrow O}{\searrow_{OOH}} \longrightarrow R_1-\underset{\underset{O}{\|}}{C}-O-\underset{\underset{O}{\|}}{C}-R_2 + R_3COOH$

(12) $R_1-\underset{\underset{O}{\|}}{C}-O-\underset{\underset{O}{\|}}{C}-R_2 + 2 R_4OH \longrightarrow R_1-COOR_4 + R_2-COOR_4 + H_2O$

The first reaction steps of the free radical mechanism involved the formation of macroradicals (1), peroximacroradicals (2) and hydroperoxides (3). Then, during different subsequent reactions various oxygen-containing compounds are formed: ketones (4a); alcohols (4b); α-ketonehydroperoxides (5); aldehydes (6); acids (6)-(8); peracids (7); esters (8), (9), (12); diketones (10) and anhydrides (11). However, the final oxidation product contained practically only free carboxyl and ester groups. The content of these groups in the oxidized PE wax was measured by acidity and saponification indices, respectively.

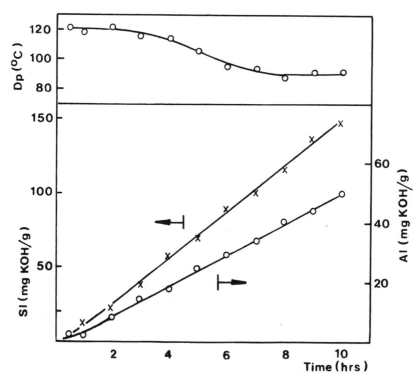

Fig. 2. Influence of the oxidation time on the acidity index (AI), saponification index (SI) and drop point (Dp) of the PE wax (melt temperature 160°C; air flow rate 0.1 Nm³/Kg.PE wax)

From the technical point of view the oxidation of the PE wax was followed by measuring from time to time the acidity and saponification indices (AI and SI, respectively). Following a short inhibition period both acidity and saponification indices linearly increased with the oxidation time (Fig. 2). The slope of each curve depended on the reaction conditions, i.e. temperature, air flow rate, catalysts. Oxidized PE waxes with acidity indices up to 60 mg.KOH/g and saponification indices up to 200 mg.KOH/g have been thus obtained.

When antioxidants such as amines or sterically hindered phenols were added in the PE wax an increased induction period and a lower oxidation rate were noticed. On the contrary, in the presence of free-radical generators, especially organic peroxides such as dibenzoyl peroxide and dicumyl peroxide the oxidation proceeded faster. These results are the consequence of the free-radical mechanism of the PE wax oxidation.

During the oxidation of the PE wax both degradation and crosslinking of the polymer chains may occur. However, only a slight decrease of the molecular weight due to the splitting of polymer chains was noticed. The drop point of the PE wax slightly decreased with the oxidation time until it reached a constant value (Fig. 2).

Characterization of the oxidized polyethylene waxes

The oxidized polyethylene (OXPE) waxes have been characterized by infrared (IR) spectroscopy and nuclear magnetic resonance (^{1}H-NMR) spectrometry. IR spectroscopy is able to show the formation of various oxygen-containing groups during the oxidation of the PE wax: 3555 cm^{-1} hydroperoxide (R-O-O-H); 3380 cm^{-1} hydroxyl (R-OH); 1785 cm^{-1} peracid (R-CO-O-O-H); 1743 cm^{-1} ester (R-COOR'); 1730 cm^{-1} aldehyde (R-CH=O); 1720 cm^{-1} ketone (R-CO-R'); 1712 cm^{-1} acid (R-COOH); 1673 cm^{-1} perester (R-CO-O-O-R'). However, the IR spectrum of the final oxidation product showed only the strong absorption of the carbonyl group (1720 cm^{-1}) with two shoulders for the carboxyl and ester groups (Fig. 3). The ^{1}H-NMR spectrum of the OXPE waxes showed also the presence of carbonyl groups, i.e. the resonance line at 2.4 ppm for the -C$\underline{H}_2$-CO- protons (Fig. 4).

The intensities of both IR absorbtion band and ^{1}H-NMR resonance line followed the oxidation degree of the wax (Fig.5). Thus, the carbonyl absorbtion band was used as the analytical band to estimate the oxidation degree. The ratio between the extinction of the carbonyl absorbtion band at 1720 cm^{-1} and the extinction of a reference band for the PE chain (1460 cm^{-1}) was calculated for wax samples of different oxidation degrees (Fig. 6). Plotting the ratio E_{1720}/E_{1460} versus oxidation degree expressed as saponification index a linear relationship was found (Fig. 7). Indeed, both the carbonyl absorption and the saponification index estimate the total carbonyl content from the carboxyl and ester groups of the OXPE waxes.

Similar results have been obtained when the oxidation degree was estimated from the ^{1}H-NMR spectra. The number of

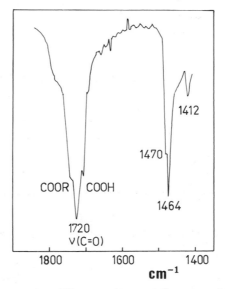

Fig. 3. The IR spectrum of an oxidized
PE wax (wax melt deposited on
KBr disc and cooled)

$-CH_2-CO-$ groups per 1000 $-CH_2-$ groups was calculated from the integrals of the corresponding resonance lines, i.e. $\delta=2.40$ ppm (b) and $\delta=1.27$ ppm (c), respectively (Fig. 4). The linear relationship CO=0.388SI was foundwhen the number of $-CH_2-CO-$ groups/1000 $-CH_2-$ groups (CO) was plotted against the saponification index (SI) for wax samples of different oxidation degree (SI up to 200 mg.KOH/g).

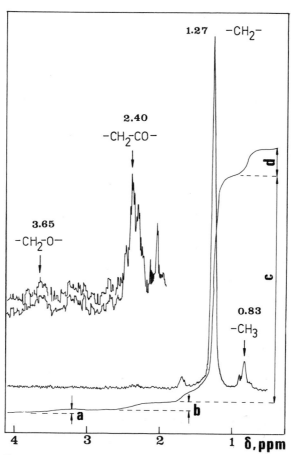

Fig. 4. The ^{1}H-NMR spectrum of an oxidized PE wax ($C_6D_5NO_2$ solution at 110°C, hexamethyldisiloxane as standard)

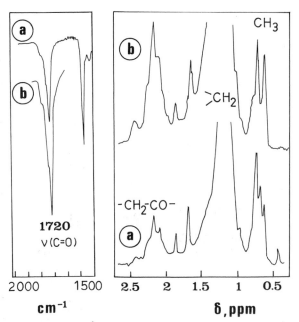

Fig. 5. The IR and ^{1}H-NMR spectra of two oxidized PE
samples with different oxidation degrees:
(a) AI=30 mg.KOH/g, SI=55 mg.KOH/g;
(b) AI=55 mg.KOH/g, SI=110 mg.KOH/g.

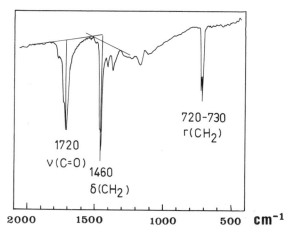

Fig. 6. Estimation of the oxidation degree from IR spectra.

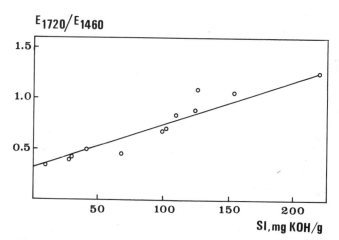

Fig. 7. Correlation between the carbonyl absorption and
the saponification index of oxidized PE waxes.

METAL-CONTAINING POLYETHYLENE WAXES

Metallic soaps, i.e. salts of fatty acids such as myristates, palmitates, stearates (Me: Ca, Cd, Ba, Mg, Zn, Al) and especially the partially saponified montan waxes may be regarded as model compounds for the metal-containing PE waxes.

Synthesis of metal-containing polyethylene waxes

Metallic soaps are usually obtained by the wet (precipitation) method, i.e. a two-step process. First the fatty acid is converted into its sodium salt by a reaction performed in hot water or water/alcohol mixture. A solution containing the salt of the desired metal, i.e. sulphates, chlorides, nitrates, acetates, is then slowly added with agitation. The insoluble metallic soap immediately precipitates from solution and is recovered by filtration. It is obvious that from both technical and economical points of view this route may not be used to obtain metal-containing PE waxes. Therefore, it was attempted to develop a more convenient synthesis method, i.e. the direct reaction of the OXPE waxes with metallic compounds.[7]

The synthesis of calcium-containing PE waxes was studied. The reaction was performed by adding a calcium compound into the OXPE wax and maintaining the melt under stirring. The free carboxyl groups were able to react with certain calcium compounds and the neutralization reaction was followed by measuring the decreasing of the acidity index. In order to favour the elimination of water resulting from the reaction air was bubbled continuously into the wax melt. A cooling roller was fed with the molten wax from the reactor. Thus, the saponified wax was obtained in the form of flakes of uniform thickness (about 0.6 mm) and approximately even shape and size.

Table 1. Synthesis of Calcium-Containing Polyethylene Waxes

Oxidized Waxes AI (mg.KOH/g)	SI	Ca Compound Type p/100p OXPE wax		Temp. ($^{\circ}$C)	Time (hrs)	Saponified Waxes AI (mg.KOH/g)	SI
33.9	115	CaCO$_3$	1.0	135	1.0	30.7	105.4
33.9	115	CaO	1.0	135	1.0	27.6	110.9
33.9	115	Ca(OH)$_2$	1.0	135	1.0	9.5	92.1
48.4	113	Ca(OH)$_2$	3.5	130	1.0	11.9	78.4
48.4	113	Ca(OH)$_2$	3.5	152	1.0	14.8	77.2
48.4	113	Ca(OH)$_2$	3.5	175	1.0	15.5	79.2
38.9	108	Ca(OH)$_2$	1.0	135	0.5	14.2	84.3
38.9	108	Ca(OH)$_2$	1.0	135	2.0	14.1	86.1
38.9	108	Ca(OH)$_2$	1.0	135	5.0	13.5	86.3

Practically, no reaction occured with calcium carbonate or calcium oxide. On the contrary, when calcium hydroxide was added in the wax melt a significant decrease of the initial acidity index of the OXPE wax was noticed (Table 1). The IR spectrum clearly showed the formation of the carboxylate

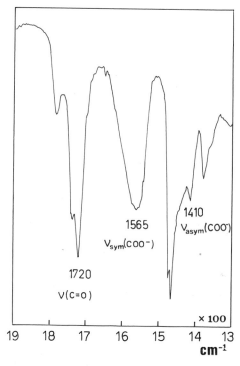

Fig. 8. The IR spectrum of a calcium-containing PE wax (AI=9 mg.KOH/g, SI=92 mg.KOH/g, 0.5% Ca)

bonds, i.e. the two absorption bands at 1565 cm^{-1} and 1410 cm^{-1} due to the symmetrical and asymmetrical stretching vibrations of the COO$^-$ group, respectively. The temperature of the wax melt seemed to have practically no influence on the reaction yield. The neutralization reaction was very fast so that the increasing of the reaction time over 30 minutes was unnecessary (Table 1).

Starting from OXPE waxes of different acidity indices and adding increasing proportions of calcium hydroxide a linear relationship between the values of the acidity index and the proportion of the calcium hydroxide was found (Fig.9). The resulting straight lines have the same slope, i.e. a decreasing of the acidity index of about 11.3 mg.KOH/g per one

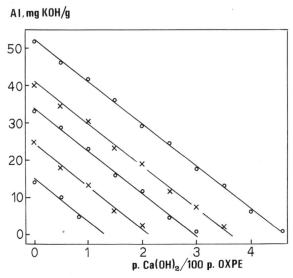

Fig. 9. The dependence of acidity index on the proportion of calcium hydroxide added into OXPE waxes of different oxidation degrees.

part $Ca(OH)_2$ added to 100 parts OXPE wax. This value corresponded to the theoretical one, i.e. the value resulting from the stoichiometric reaction between two -COOH groups and a $Ca(OH)_2$ molecule. Moreover, for all the Ca-containing PE waxes the content of the ester groups, expressed as esterification index EI=SI-AI, was about the same as for the OXPE waxes (see for example Table 1). Therefore, it may be concluded that the formation of the carboxylate groups occured only by the neutralization of COOH groups and not by the saponification of the ester groups.

Potential applications of calcium-containing polyethylene waxes

A wide range of calcium-containing PE waxes with different acidity index values and calcium contents have been obtained by the proper choice of the OXPE wax type, i.e. the oxidation degree, and the proportion of the added $Ca(OH)_2$. For some of them useful properties and, consequently, applications were found.

Oxidized PE waxes partially saponified with $Ca(OH)_2$ were found useful as lubricants for plastics, release agents, solvent-based polishes, matting and antisedimentation agents in organic coatings. Thus, a partially saponified OXPE wax (AI=13.9 mg.KOH/g, SI=78 mg.KOH/g, 1.4% Ca content) was tested as a lubricant for poly(vinyl chloride) (PVC) in comparison with a partially saponified montan wax of similar characteristics (AI=12.5 mg.KOH/g, SI=100 mg.KOH/g, 1.6% Ca content). A mixture obtained by dry-blending of 100 parts rigid suspension-grade PVC (K value 58), 2 parts thermal stabilizer (dibasic lead stearate) and 0.2-1.5 lubricant (partially saponified montan wax or OXPE wax) was charged into the mixing head of a Brabender Plasticorder torque-rheometer. The rheometer chart of torque versus time (Figure 10) showed first an increase to a maximum torque due to the beginning of the fusion process. As the PVC particles fused and formed a plastic melt the torque increased again to a peak, called the fusion peak. The time required to reach this peak is called the fusion time.

Table 2. Brabender Evaluation of Partially Saponified Waxes as PVC Lubricants

Wax Conc. (x)	Temp. (^{o}C)	Max.Torque PE (Nm)	MO (Nm)	Min.Torque PE (Nm)	MO (Nm)	Fusion time PE (min)	MO (min)
0.2	128	30.6	31.4	25.2	25.0	0.66	0.70
0.4	134	24.7	27.6	12.7	18.3	1.57	1.00
0.6	131	25.2	28.5	9.6	20.6	2.40	1.10
0.8	132	20.4	22.1	7.5	11.3	4.73	2.16
1.0	135	18.6	22.9	6.9	13.4	5.60	2.13
1.5	134	15.7	19.1	4.9	9.3	9.20	3.20

(x) parts/100 parts PVC; PE: partially saponified OXPE wax; MO: partially saponified montan wax.

In spite of their similar values of acidity and saponification indices and of the calcium content the two partially saponified waxes have a different behaviour as PVC lubricants

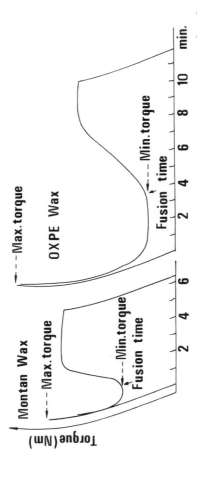

Fig. 10. Brabender rheometer chart of two partially saponified waxes (0.8 p.wax/100 p.PVC).

(Figure 10, Table 2). For the same wax concentration lower values of minimal torque and longer fusion times were obtained with the PE wax. This behaviour showed that the saponified OXPE wax is a more "external" lubricant than the saponified montan wax. Depending on structure and composition lubricants reduce the adhesion of polymer melts to hot metal surfaces and friction at hot machine parts ("external" lubricating effect) or they improve the flowability of polymer melts ("internal" lubricating effect). External lubricants are usually non-polar compounds such as PE waxes, and internal lubricants are polar compounds such as esters, therefore compatible with the PVC chains. Montan waxes are based on montanic acids which predominantly consist of straight chain and saturated carboxylic acids with a chain length of $C_{28}-C_{32}$, i.e. a molecular weight of about 400-500. The longer non-polar chains of the saponified OXPE waxes are responsible for the more "external" lubricating effect.

Completely saponified waxes, i.e. OXPE waxes with all the free carboxylic groups converted to calcium carboxylate groups, may be used as antacid agents (acid scavengers) in polyolefins such as HDPE and PP obtained with chlorine-containing catalytic sytems. In these polyolefins the calcium-containing PE waxes are able to act as acid scavengers for the hydrochloric acid resulting from the thermal decomposition of the catalyst traces during the polymer processing. Starting from OXPE waxes of different acidity index values three samples of completely saponified PE waxes with different calcium contents have been synthesized and tested as acid scavengers for HDPE in comparison with calcium stearate, i.e. the usual antacid agent (Table 3).

An additive-free injection-molding-grade HDPE powder was dry-blended with 0.15% antioxidant (Irganox 1010) and various proportions of different acid scavengers. Following extrusion at 150°C through a single screw extruder some rheological and mechanical properties of the polymer were determined. The maximal admissible chlorine level in the HDPE obtained with the specified catalytic system is 150 ppm. To neutralize the corresponding hydrochloric acid the polymer has to contain 85 ppm calcium, i.e. 1280 ppm calcium stearate. Therefore, in usual HDPE grades 0.15% calcium stearate is added. Some HDPE grades, e.g. the raffia grade, contain higher amounts of calcium stearate, i.e. up to 0.4%, because this antacid agent acts also as a lubricant. Because the calcium content of the saponified waxes was 2-5 times lower than in calcium stearate higher amounts of waxes were added in HDPE in order to reach the required level of the calcium content, i.e. min. 85 ppm. However, the addition of up two percent wax has practically no influence on the basic rheological and mechanical properties of HDPE (Table 3). The completely saponified Ca-containing waxes may be therefore used as acid scavengers in HDPE and other polyolefins.

Table 3. Testing of Calcium-Containing Polyethylene Waxes (CaPE) as Acid Scavenger in HDPE

Type	Acid Scavenger Ca Content (%)	Conc. in HDPE (%)	Ca Content in HDPE (ppm)	$MFI_{2.16}$ (g/10min)	$MFI_{21.6}$ (g/10min)	$\dfrac{MFI_{21.6}}{MFI_{2.16}}$	Tensile strength (Kg/cm^2)	Elongation at break (%)	IZOD impact strength $(Kg\ cm/cm^2)$
–	–	–	–	2.64	56.8	21.5	307	1096	7.11
CaSt	6.60	0.15	100	2.74	58.6	21.4	293	972	7.41
CaSt	6.60	0.25	165	2.70	57.6	21.3	283	1012	7.28
CaSt	6.60	0.40	265	2.54	54.4	21.4	314	984	7.57
CaPE	2.00	0.15	30	2.58	54.2	21.0	278	930	7.29
CaPE	2.00	0.25	50	2.24	54.0	24.1	328	1108	7.02
CaPE	2.00	0.40	80	2.43	53.5	22.0	306	1172	6.89
CaPE	1.45	0.50	73	2.46	55.0	22.4	319	1068	7.89
CaPE	1.45	1.00	145	2.62	59.0	22.5	278	980	7.56
CaPE	1.45	2.00	290	2.65	59.0	22.3	277	980	7.02
CaPE	3.85	0.50	193	2.75	52.7	19.2	287	1052	7.88
CaPE	3.85	1.00	385	2.60	49.4	19.0	332	1092	7.23
CaPE	3.85	2.00	770	2.87	56.3	19.6	275	980	7.01

CONCLUSIONS

Calcium-containing PE waxes are metal-containing polymers of simple structure with useful properties and applications. They are obtained by a simple method involving a two-step modification of a cheap raw material, i.e. the low-molecular-weight by-product of a HDPE plant. The free carboxyl groups attached on the polymer backbone by the controlled oxidation of the PE wax are neutralized with calcium hydroxide. Other metal-containing PE waxes (Ba, Cd, Mg, Zn, Al) may be synthesized by the same approach. Their synthesis, characterization and testing are under progress and will be reported later.

ACKNOWLEDGMENT

The author is greatly indebted to Prof. John E. Sheats for his continuous interest and kindness, for his support and encouragement.

REFERENCES

1. A. Eisenberg and M. King, "Ion-Containing Polymers", Academic Press, New York (1977).
2. A. Eisenberg (ed.), "Ions in Polymers", American Chemical Society, New York (1980).
3. J. E. Sheats, C. E. Carraher Jr., and C. U. Pittman Jr. (eds.), "Metal-Containing Polymeric Systems", Plenum Press, New York (1985).
4. D. Munteanu, Rom.Pat., 94196 (1987).
5. D. Munteanu, Rom.Pat., 91787 (1986).
6. T. Dumas and W. Bualni, "Oxidation of Petrochemicals: Chemistry and Technology", Applied Science Publishers, London (1984).
7. D. Munteanu, Rom.Pat., 97522 (1989).

ORGANOTIN ANTIFOULING POLYMERS : SYNTHESIS AND PROPERTIES OF NEW COPOLYMERS WITH PENDENT ORGANOTIN MOITIES

R.R. Joshi, J.R. Dharia and S.K. Gupta

Centre for Environmental Science and Engineering
Indian Institute of Technology
Powai, Bombay-400 076, INDIA

ABSTRACT

A new monomer, tributyltin α-chloroacrylate (TCA) synthesised in our laboratory is copolymerized with ethyl methacrylate, butyl methacrylate and cyclohexyl methacrylate in solution at 55°C using azobisisobutyronitrile initiator. Copolymer compositions are determined by tin analysis. Monomer reactivity ratios are calculated by the Kelen-Tüdös method. The copolymers are characterized by infrared spectroscopy, PMR and ^{13}C NMR spectroscopy, thermal analysis, x-ray diffraction, solubility and viscosity. Biocidal effect of the copolymers is studied with two microorganisms viz., Pseudomonas aeruginosa and Sarcina lutea and with brine shrimp Artemia salina nauplii II.

INTRODUCTION

Controlled release of biological agents is a concept that has gained increased attention in the past few decades. The basic concept of this technology is the prolonged delivery of the effective agent to a particular site. It has gained importance in various fields like pharmaceuticals for controlled drug release, in agricultural industries for controlled release of fertilizers and pesticides etc. This field of technology began with the advent of antifouling paints/coatings for ship hulls. With gradual improvement over the past few decades, it has led to the discovery of long term release of antifouling agents from the matrix by diffusion - dissolution phenomenon.

A large number of organometallic compounds of copper, arsenic and lead have been tested for antifouling activity. Copper compounds have good antifouling activity, but because of uncontrolled release and inability to form polymeric compounds, their use as antifoulants has been hampered. Though organometallic compounds of arsenic and lead are found to be good antifoulants, they are hazardous to man as well as to aquatic flora and fauna and also cause environmental pollution. Their use in the formulation of antifouling coatings, therefore, has been banned. With the discovery in 1953 of organotin compounds including tributyltin compounds as antifoulants, a large number of derivatives have been synthesised and tested for their antifouling activity. These compounds form the best choice for anti-

Inorganic and Metal-Containing Polymeric Materials
Edited by J. Sheats *et al.*, Plenum Press, New York, 1990

fouling formulation because they show biocidal activity towards a
wide range of marine fouling organisms. They are easily biodegradable
and form nontoxic tin oxide (1).

The actual break-through in the field of antifouling coatings
was achieved by Montermoso et.al. (2) in 1958 who first described
the synthesis of acrylic polymers containing pendent hydrolyzable
organotin moieties. They reported the presently known monomers
viz., tributyltin acrylate (TBTA) and tributyltin methacrylate (TBTMA)
(3). The basic reaction involved in the synthesis of these compounds
is to bind the TBT moiety to the carboxylic end of the compound
so that it can be easily hydrolysed releasing the toxicant and preventing or at least repelling the approaching/attached fouling organisms.
Homopolymers of these monomers are unsuitable as film formers.
Besides this they have too high concentration of toxicant in the matrix,
resulting in pollution and disturbance to the surrounding micro and
macro flora and fauna. This drawback is overcome by careful co
and terpolymerization which allows incorporation of the toxin, in
desired amounts, control of its distribution within the polymer chain,
and better control of its release (4-7). Many desirable properties
can be achieved through fundamental studies on the copolymerization
parameters under specified reaction conditions. Literature about
the reactivity ratio values and detailed analysis of the copolymerization reactions of organotin monomers is limited. Various analyses
like ^{13}C NMR studies, thermal analysis, viscosity, GPC analysis,
and biotoxicity studies of the various organotin polymers have not
been studied in detail. Based on the above analyses it is possible
to design a polymer system and predict the life period of the coating
and its efficiency as antifoulant.

EXPERIMENTAL

Tri-n-butyltin α-chloroacrylate (TCA) was prepared according
to the method of Dharia (7) by the reaction of tri-n-butyltin oxide
with α-chloroacrylic acid. The details of the synthesis are given
below.

The synthesis of TCA involves three main steps viz.,

(i)   Chlorination of methyl acrylate to methyl 2-3 dichloropro-
        pionate (MDP)

(ii)   Dehydrohalogenation of MDP to α-chloroacrylic acid (CAA)

(iii)   Esterification of CAA to TCA with TBTO

The reaction scheme is outlined below

$$CH_2 = \overset{\overset{\displaystyle H}{|}}{C}\text{-COOCH}_3 \xrightarrow[0^\circ C]{Cl_2} ClCH_2 - CHClCOOCH_3 \xrightarrow[\text{(ii) } H_2SO_4 \text{ or HCl}]{\text{(i) } Ba(OH)_2 .8\ H_2O}$$

$$\rightarrow CH_2 = CClCOOH \xrightarrow[\text{in Benzene}]{TBTO} CH_2 = CCl\text{-COOSnBu}_3$$

I. Synthesis of Methyl 2, 3 Dichloropropionate (MDP)

500 ml of methyl acrylate (Fluka A-G) and 200 ml of distilled methanol was placed in a reaction vessel. The solution was bubbled with chlorine gas for more than 10 h while the temperature of the reaction was maintained below $30^{\circ}C$. After the completion of reaction, methanol was distilled out and the product MDP, a colourless, ester smelling liquid was distilled under vacuum. The yield of product was more than 80%.

II. Synthesis of α -Chloroacrylic acid (CAA)

To the solution of barium hydroxide (275 g) in distilled water (500 ml), MDP (83 ml) was added over a period of 30 min with constant stirring. The reaction was carried for more than 3 h, while the temperature was maintained below $30^{\circ}C$. After the completion of the reaction, the reaction mixture was acidified with Conc. HCl till the pH of the solution was reduced to 2.0. The compound α -chloroacrylic acid was then extracted with several portions of a solution containing 1 g of hydroquinone for every 200 ml of diethyl-ether. The ether extract was then dried over anhydrous magnesium sulfate overnight. The product was then isolated by evaporating the ether under reduced pressure. The crude product was recrystalli-zed from hexane at low temperature to get 60 g of α -chloroacrylic acid (MP 64-65$^{\circ}C$).

III. Synthesis of Tributyltin α -Chloroacrylate (TCA)

To a three necked flask containing 450 ml of dry benzene, 49.41 g of CAA, 119.00 ml bis-tributyltin oxide was added dropwise with stirring over a period of 10 min while cooling. The water formed during the reaction was azeotropically distilled under reduced pressure and fresh dry benzene was added. This procedure was repeated until water from the reaction ceased to exist. The crude monomer obtained after the removal of all the benzene was crystallized from dry hexane at low temperature. The monomer was recrystallized twice in the form of white, needle shaped crystals having a melting point of 52$^{\circ}C$. The purity of the monomer was confirmed by gravi-metric analysis of tin as tin oxide and also by PMR studies.

IV. Copolymerization of TCA with alkyl methacrylates

TCA (M_1) was copolymerized with alkyl methacrylates (M_2). TCA was copolymerized with ethyl methacrylate (EMA), butyl methacry-late (BMA) and cyclohexyl methacrylate (CMA). EMA, BMA and CMA (Fluka A.G.) were purified by the standard procedure (8).

All the solvents used during polymerization were BDH make and were purified as described in Vogel (9).

Different mole ratios of TCA, alkyl methacrylate, AIBN (0.1% w/v) and constant volume of dry tetrahydrofuran (THF) were charged into polymerization tubes. This was followed by repeated degassing and freeze thawing cycles. Finally the tubes were sealed and then placed in a thermostatically controlled water bath adjusted at 55 $\pm$ 0.1$^{\circ}C$. The polymerization reaction was restricted to less than 10% as the Kelen-Tüdös differential form of the copolymer equation was to be used for calculating the reactivity ratio (10). The r_1 and r_2 values calculated by this method do not take account the internal points on the straight line and so even if some data points deviate slightly from the theoretical value, the reactivity ratio values would remain invariant.

After less than 10% conversion the reaction tubes were cooled and the vacuum was released. The viscous solution was then precipitated in excess of dry hexane. The polymers were further purified by repeated dissolution and precipitation from THF into n-hexane so as to remove the unreacted comonomers and initiator species. Finally the resulting copolymers were dried in vacuum at 35°C to a constant weight and then overall conversion was calculated from the weight of the copolymers. The tin content of the copolymers was determined by the Gilman and Rosenberg method (11). The data of the % conversion, mole composition in feed and in copolymer, tin % and the reaction conditions adopted during the course of polymerization are given in Tables 1-2.

The copolymerization scheme is given below :

$$CH_2 = C\text{-}ClCOOSnBu_3 + CH_2 = CCH_3 - COOR \xrightarrow[55^\circ C]{THF}$$

$$\sim\sim\sim(CH_2\text{--}\underset{\underset{\displaystyle Sn(Bu)_3}{\overset{\displaystyle |}{\underset{|}{\overset{|}{C=O}}}}}{\overset{\overset{\displaystyle Cl}{|}}{C}})_n \text{--} (CH_2\text{--}\underset{\underset{\displaystyle R}{\overset{\displaystyle |}{\underset{|}{\overset{|}{C=O}}}}}{\overset{\overset{\displaystyle CH_3}{|}}{C}})_m\sim\sim\sim$$

R = Et, Bu, Cy

The copolymers with different mole compositions synthesised were TCA-ethyl methacrylate ($TCEM_1$-$TCEM_4$) TCA - butyl methacrylate ($TCBM_1$ - $TCBM_4$) and TCA-cyclohexyl methacrylate ($TCCM_1$-$TCCM_4$). Subscript 1 to 4 indicate the increasing mole composition of TCA (M_1) in the copolymer. Besides reactivity ratio determination, the copolymers were characterized by IR, PMR, ^{13}C NMR, solubility, viscosity, thermal analysis and x-ray spectroscopy. Biotoxicity of the copolymers was studied with two bacteria and a brine shrimp. These newly synthesised copolymers were compared with earlier reported copolymers of TBTAA and TBTMA with particular reference to their toxic nature.

RESULTS AND DISCUSSION

The infrared spectra of MDP, CAA, TCA and copolymers were recorded by using a Perkin-Elmer Model 621 grating infrared spectrophotometer. MDP showed the absence of an unsaturated band at 1600 cm^{-1} and the formation of a C-Cl band at 740 cm^{-1}. The broad band at 3200 cm^{-1}, 1602 cm^{-1} and 740 cm^{-1} indicates the formation of α-chloroacrylic acid monomer. The band at 1640 cm^{-1} and 1580 cm^{-1} corresponding to COOSn group indicates complete esterification of CAA with TBTO. Absence of CAA_1 impurity is reflected by the absence of a broad band at 3200 cm^{-1}. The infrared spectra of MDP, CAA and TCA are shown in Fig. 1. All the copolymers showed more or less the same spectral peaks such as 1740 cm^{-1} for carbonyl stretching, 1625 cm^{-1} due to tin ester bending and 680 cm^{-1} for the C-Cl band. In many of the copolymers the two carbonyls were well separated. It was therefore possible to determine copolymer composition by paper cutting and weighing. Table 1 shows some of mole composition values for various copolymer systems.

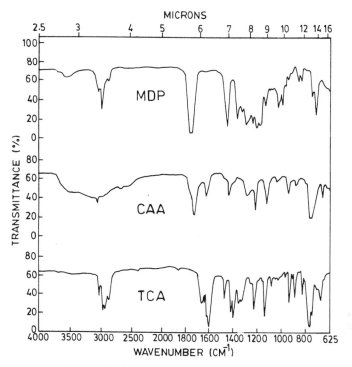

FIG.1: IR Spectrum of MDP, CAA, TCA.

The copolymer composition was calculated from tin analysis. The mole fraction of TCA (M_1) was calculated by estimating the percentage of tin as tin oxide in the copolymer. From this value, the mole fraction of (M_2) was calculated. The copolymerization parameters for TCEM, TCBM and TCCM were determined from the feed composition and copolymer composition relationship. Table 2 shows the % conversion, mole fraction in feed and in copolymer determined by gravimetric analysis. Table 3 gives the monomer reactivity ratios for the three systems calculated by Kelen-Tüdös (10) method. Fig.2 shows the Kelen-Tüdös plot for the three systems. The r_1 r_2 values (Table 3) indicate that the copolymer would be expected to contain a significant number of alternating monomer sequences along the polymer chain. The copolymerization reaction shows azeotropic composition at f_1 = 0.55 for TCEM, f_1 = 0.61 for TCBM and f_1 = 0.93 for TCCM respectively. The molar composition curves shown

Table 1   Comparison of Copolymer Composition by Tin Analysis, IR and PMR Spectroscopy.

Polymer Code	Mole Fraction of TCA (M_1) in copolymer by		
	Tin analysis	IR Spectroscopy	PMR Spectroscopy
$TCEM_1$	0.1344	0.1470	0.1470
$TCEM_2$	0.2604	0.1848	0.1828
$TCEM_{2.5}$	0.3169	0.2918	0.2010
$TCEM_3$	0.4010	0.4811	0.4515
$TCBM_1$	0.0633	0.1010	0.0601
$TCBM_2$	0.2901	0.3010	0.2162
$TCBM_{2.5}$	0.3660	–	0.3701
$TCBM_3$	0.4501	0.3925	0.4110
$TCBM_4$	0.6337	0.6151	0.6107
$TCCM_1$	0.1900	0.1858	0.1638
$TCCM_2$	0.3573	–	0.3818
$TCCM_{2.5}$	0.4308	0.3991	0.4112
$TCCM_3$	0.4513	0.4816	0.4314
$TCCM_4$	0.7200	0.6382	0.7819

in Fig. 3 intersect approximately at the corresponding azeotropic compositions, yielding homogeneous copolymer regardless of % conversion. The inverse of r_1 which gives the relative reactivity of the comonomer shows that EMA > BMA > CMA. From this it can be concluded that the TCA radical has greater affinity for the initiator radical. As the length of the side chain and the bulkiness in the comonomer increases, r_2 decreases. Table 1 shows that the mole ratios calculated by IR and PMR spectroscopy were in close agreement with the tin analysis.

All the copolymers were analyzed by ^{13}C NMR. Assignment of all carbons was done by comparing the NMR of corresponding homopolymers. In all the spctra it was observed that as the TCA content in the copolymer increases, the splitting pattern becomes more complicated. It was possible to fit the Bernoullian statistics model to the growing polymer chain. More details regarding sequence distribution and average sequence length in these copolymers has been explained elsewhere (12).

Solubility of fresh samples was tested in solvents of varying polarity such as acetone, dimethyl sulfoxide, toulene, chloroform and tetrahydrofuran. It was observed that fresh samples were soluble more readily and quickly in polar solvents. Solubility decreases with the aging of the polymer sample. This may be due to the crosslinking in the copolymer which is a typical property of α -Chloroacrylates (13).

It is well known that chloro compounds are light sensitive. Because of the exposure to light the Cl-atom must have been cleaved and a five membered lactone and/or dilactone ring formed which prevented the penetration of solvent molecules inside the polymer matrix. Formation of lactone rings was confirmed by taking IR spectrum of a sunlight-exposed and of a thermally treated sample. The same experience has been reported by Dharia (7).

Table 2 Copolymerization of Tributyltin α -Chloroacrylate (M_1) with Alkyl Methacrylates (M_2)

Polymer Code	% Conversion	Mole fraction of TCA in feed	% of tin in copolymer	Molefraction of TCA in copolymer
TCA(M_1) - EMA (M_2)				
$TCEM_1$	5.78	0.0730	10.50	0.1344
$TCEM_2$	5.64	0.1735	16.50	0.2604
$TCEM_{2.5}$	6.63	0.2395	18.51	0.3169
$TCEM_3$	7.16	0.3208	20.98	0.4010
TCA (M_1) - BMA (M_2)				
$TCBM_1$	7.08	0.0914	11.16	0.0633
$TCBM_2$	6.16	0.2115	15.97	0.2901
$TCBM_{2.5}$	10.04	0.3155	19.10	0.3860
$TCBM_3$	6.09	0.3765	20.86	0.4501
$TCBM_4$	5.08	0.6167	24.08	0.6337
TCA (M_1) - CMA (M_2)				
$TCCM_1$	3.70	0.1002	10.68	0.1900
$TCCM_2$	8.01	0.2294	17.02	0.3573
$TCCM_{2.5}$	3.20	0.3097	19.23	0.4308
$TCCM_3$	5.66	•0.4018	19.80	0.4513
$TCCM_4$	5.35	0.6400	25.77	0.7200

Polymerization conditions : Initiator AIBN 0.01% w/v

Temperature - 55 $\pm$ 0.1°C

Table 3 Reactivity Ratios of Tributyltin α -Chloroacrylate (M_1) with Alkyl Methacrylates (M_2) by Kelen-Tüdös Method.

M_1	M_2	r_1	r_2	$1/r_1$
TCA	EMA	0.55	0.45	1.81
TCA	BMA	0.65	0.44	1.53
TCA	CMA	0.96	0.41	1.04

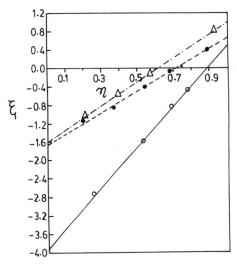

FIG.2 : Kelen Tüdös Plot for Copolymer—
ization of TCEM (—•—) ,
TCBM (--•--) and TCCM (--△--)

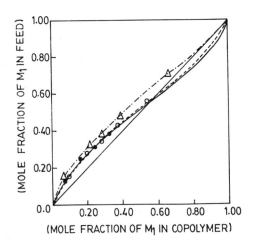

FIG.3 : Mole feed curves of TCEM (--•--),
TCBM (--•--) and TCCM (--△--)

The viscosity data of the copolymers are shown in Table 4. The intrinsic viscosity data were found to be in the range of 0.23 to 0.91 dl/gm in toulene at $26^{\circ}C$. This reveals that as the TCA content in the copolymer increases the viscosity decreases. This is due to the greater reactivity of TCA radical at the growing chain end facilitating the propagation rates as compared to termination. This is proved by the reactivity ratio values shown in Table 3. The order of viscosity observed was TCEM > TEBM > TCCM.

Table 4 Viscosity Data for Tributyltin α -Chloroacrylate (M_1) with Alkyl Methacrylate (M_2) Copolymer System.

Polymer Code	Mole fraction of TCA in copolymer	dl/gm
TCEM$_1$	0.1344	0.413
TCEM$_2$	0.2604	0.281
TCEM$_3$	0.4010	0.230
TCBM$_1$	0.1751	0.635
TCBM$_2$	0.2901	0.560
TCBM$_3$	0.4501	0.360
TCCM$_1$	0.1900	0.915
TCCM$_{2.5}$	0.4308	0.430
TCCM$_4$	0.7200	0.361

Note : Solvent flow rate (to) = 362 sec., Temp. = $26^{\circ}C$

Thermal properties of the copolymers were assessed on V2.21. Dupont thermal analyser. The thermograms of the copolymers are shown in Fig.4. The results are summarized in Table 5. The relative thermal stability of the copolymers has been compared using Integral procedure decomposition temperature values. The data gives the order of thermal stability of TCEM< TCBM < TCCM.

Table 5 Thermal Properties of Copolymers of Tributyltin α -Chloroacrylate with Alkyl Methacrylate.

Polymer Code	IDT $0^{\circ}C$	50% DT$^{\circ}C$	IDPT $^{\circ}C$	FDT $^{\circ}C$
TCEM$_1$	189.1	381.1	329.4	403.4
TCEM$_2$	180.7	334.3	301.9	389.4
TCEM$_{2.5}$	174.9	335.7	284.6	383.0
TCEM$_3$	175.1	330.9	281.1	381.2
TCBM$_1$	174.0	345.2	328.4	395.5
TCBM$_2$	163.4	338.5	326.2	390.5
TCBM$_3$	143.6	317.7	292.7	385.4
TCBM$_4$	136.5	267.5	285.0	371.7
TCCM$_{2.5}$	120.1	250.4	430.0	302.5
TCCM$_3$	150.3	295.4	450.4	331.3
TCCM$_4$	130.1	272.6	448.6	318.1

NOTE : IDT : Initial decomposition temperature,
       IDPT : Integral procedure decomposition temperature
       FDT : Final decomposition temperature

In general it is seen that the thermal stability decreases with the increase in TCA content in the copolymer. This may be due to the presence of the chlorine atom at the α-position which, on heating, easily breaks off, resulting in formation of a free radical site, which during the later stage of reaction forms a 5 membered lactone ring (13). The formation of such ringed structures is reported to decrease the thermal stability of the polymer. DTA thermograms of all the copolymers with different molar compositions were more or less the same.

Biotoxicity of Copolymers

The copolymers were tested for toxicity against two different species of bacteria viz., P. aeruginosa and S. lutea. A bioassay of these compounds was also performed on brine shrimp A. salina, whose life cycle resembles that of barnacles, a severe fouling organism.

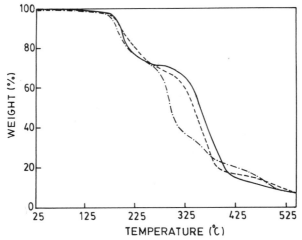

FIG. 4: TGA of Copolymers for TCEM$_2$(——), TCBM$_2$(-----) and TCCM$_{2.5}$(—·—·—)

A modified culture plate method was used to assess the toxicity of copolymers towards two microorganisms. The media and test organisms were (a) Citramide agar and aspargine broth, P. aeruginosa and (b) nutrient agar and nutrient broth with 0.5% NaCl , S. lutea. The agar medium and broth were prepared, sterilized, and cooled to 45°C. The test organism was then added to the cooled medium in concentrations of approximately 2 ml of 24 h shake-flask culture of organism for every 10 ml of medium. The medium was then poured into sterlized petriplates in 10 ml aliquots. The plates were transferred to the refrigerator for solidification. A cylindrical hole of 3 mm was bored in the centre of the petriplate and 5 mg of the compound having particle size of 500 µm was placed into the well. After this, the petriplates were incubated at 37°C for 96 h and observed for inhibition zone.

The toxicity test of copolymers towards A. Salina was performed. Artemia nauplii II was chosen to estimate 48 h LC_{50} values in artificially prepared seawater recommended by Kester (14).

Before determination of LC_{50} value for each compound, a screening test was conducted to estimate the range in which the compound would be toxic to nauplii larva. The following procedure was adopted.

50 mg of shrimp cysts were placed in a hatching tray in 500 ml of aerated artificial seawater of salinity 34%. The tray was covered with a cardboard having a small window. A light source was kept near the window. Most of the cysts hatched within 48 h.

After shrimp hatching was complete, 25 healthy larvae were transferred into a beaker containing 80 ml artificial seawater. A known quantity of compound was dissolved in tetrahydrofuran and used to coat a 7.5 x 2.5 cm^2 glass slide. The volume of the beaker was made upto 100 ml with the addition of artificial seawater. Duplicates were run for each concentration. At the end of 48 h, the number of dead shrimp were noted to calculate the percentage mortality. The above procedure was repeated for each compound to obtain the 48 h LC_{50} value.

Results of the toxicity of copolymers to two microorganisms and A. salina larvae are summarized in Tables 6 and 7 respectively.

Table 6   Inhibition Zones for Different Copolymers

Polymer Code	Mean Inhibition zone after 48 h in cm	
	Sarcina lutea	Pseudomonas aeruginosa
$TCEM_1$	1.30	–
$TCEM_4$	3.06	0.86
$TCBM_1$	2.35	0.55
$TCBM_4$	2.30	0.70
$TCCM_1$	1.20	0.60
$TCCM_4$	1.26	1.20

NOTE : Standard deviation for two microorganisms was within $\pm$ 0.047.

Table 7   Bioassy   results of copolymers with A. Salina

Polymer Code	Mole fraction of TCA	r*	log Y = mx + c	48 h LC_{50} value μg/100 ml
$TCEM_1$	0.12	0.99	1.2×10^{-3} x + 1.37	274
$TCEM_4$	0.8	0.99	9.0×10^{-4} x + 1.63	77
$TCBM_1$	0.17	0.99	7.0×10^{-4} x + 1.56	199
$TCBM_4$	0.59	0.99	6.0×10^{-4} x + 1.63	82
$TCCM_1$	0.19	0.99	1.3×10^{-3} x + 1.23	361
$TCCM_4$	0.92	0.99	1.2×10^{-3} x + 1.48	183

* Correlation Coefficient

The results show that all the copolymers except $TCEM_1$ were toxic towards both the microorganisms. $TCEM_1$ did not show any toxicity towards P. aeruginosa. This may be due to biodegradation of the test compound as this organism has the capability to degrade a wide variety of organic compounds (15). The order of toxicity observed for S. lutea was TCEM > TCBM > TCCM and for P. aeruginosa it was TCCM > TCEM > TCBM. The LC_{50} values for A. salina showed that all the copolymers were toxic towards the larvae. The order of toxicity was TCBM > TCEM > TCCM. The data also show that the toxicity of these copolymers towards all the organisms increased with the increase in TCA content in the copolymer.

Thus from copolymerization studies it can be concluded that TCA monomer is more reactive than TBTAA and TBTMA. Copolymer studies have also shown that TCA and alkyl methacrylates have almost the same reactivity. This can be seen from the reactivity ratio values. The copolymers formed were of the alternating type. Above all, copolymerization reaction shows that substitution of Cl at the α -position increases the reactivity of the monomer. Because of the tendency of chloroacrylates to undergo crosslinking the copolymers are found to have a decreased thermal stability as compared to TBTMA copolymers. But biotoxicity studies reveals that the copolymer system is less toxic than those systems previously reported. This suggests that these compounds undergo crosslinking, decreasing the rate of hydrolysis of the TBT moiety. So, a controlled release of biocide is achieved within the copolymer systems, which is absent in the case of other systems.

REFERENCES

1. A.T. Phillip, Prog. Org. Coat., 2, 159 (1973).
2. J.C. Montermoso, Organotin Acrylic Polymers, U.S. Patent., 3, 016369 (1973).
3. J.C. Montermoso, T.M. Andrews and L.P. Marinelli, J. Poly. Sci., 32, 523 (1958).
4. J.A. Montemarano, E.J. Dyckman, J. Paint. Tech., 47, 59 (1975).
5. R.V. Subramanian and R.S. Williams, J. Appli. Poly. Sci., 26 1681 (1981).
6. N.A. Ghanem, N.N. Messiha, N.E. Ikladious and A.F. Shaaban, Eur. Poly. J., 15, 823 (1979).
7. J.R. Dharia, Ph.D. Thesis, Indian Institute of Technology, Bombay, India (1989).
8. J.A. Riddick and U.B. Bunge, Organic Solvents, Wiley, New York (1970).
9. A.I. Vogel, Practical Organic Chemistry, ELBS, London, pp. 264-272 (1971).
10. T. Kelen and F.Tüdŏs. J. Macromol. Sci. Chem., A(9), 1 (1975).
11. H. Gilman and D. Rosenberg, J. Am. Chem. Soc., 75, 3592, (1953).
12. R.R. Joshi and S.K. Gupta, Microstructure and Sequence Length Determination of Antifouling Polymers (unpublished results).
13. C.P. Pathak, Ph.D. Thesis, Indian Institute of Technology, Bombay, India (1985).
14. D. Kester, Limnol. Oceanogr., 12, 176 (1967).
15. R.Y. Stainer, E.A. Adelberg, and J.L. Ingraham, Microbial World, Prentice-Hall, Englewood Cliffs, New Jersey (1976).

SYNTHESIS, CHARACTERIZATION AND MICRO-STRUCTURE DETERMINATION

OF ORGANOTIN POLYMERS

J.R. Dharia, S.K. Gupta and R.R. Joshi

Centre for Environmental Science and Engineering
Indian Institute of Technology
Powai, Bombay-400 076, INDIA

ABSTRACT

A new organotin monomer tributyltin α-chloroacrylate (TCA) was synthesized in our laboratory. Detailed studies on homopolymerization and copolymerization were undertaken. Copolymerization was carried out with styrene (ST), methyl methacrylate (MMA) and acrylonitrile (AN). Both homopolymer and copolymers were characterized by IR, $1H$ and C-13 NMR and tin analysis. Reactivity ratios were determined using Kelen-Tüdös method. Reactivity ratios were $r_1 = 0.500$ and $r_2 = 0.170$ for TCA-ST, $r_1 = 1.089$ and $r_2 = 0.261$ for TCA-MMA and $r_1 = 1.880$ and $r_2 = 0.243$ for TCA-AN respectively. Micro-structures of homopolymer and copolymers were studied using C-13 NMR spectroscopy. Data obtained were compared with those of tributyltin methacrylate (TBTMA) and its corresponding copolymers. The results indicate that TCA is more reactive than TBTMA.

INTRODUCTION

Organotin polymers find their application as antifouling coatings in the marine environment. Extensive studies have been pursued with regard to the copolymerization behaviour of tributyltin (meth)acrylate (1-7). Despite many efforts, fouling still remains a problem in the marine environment. Development of a new polymer system is necessary in order to have a more effective antifouling coating. In our laboratory we have synthesized a new organotin monomer viz. tributyltin α-chloroacrylate (TCA) in order to design coatings with superior antifouling properties (8). We propose to synthesize different copolymers based on this new monomer. As a first step, we intend to synthesize and characterize the resulting polymers. TCA was polymerized in THF using AIBN as radical initiator and also copolymerized with styrene, methyl methacrylate and acrylonitrile in THF at $50^{\circ}C$ using AIBN as radical initiator. Monomer reactivity ratios were determined using Kelen-Tüdös equation. In this paper, we present the results on the copolymerization behaviour and micro-structural features of these interesting polymers. Microstructure of polyTCA and polyTBTMA and TCA-ST copolymers having different TCA content were studied using 125 MHz C-13 NMR spectroscopy with unhydrolyzed polymers.

Inorganic and Metal-Containing Polymeric Materials
Edited by J. Sheats *et al.*, Plenum Press, New York, 1990

EXPERIMENTAL

Materials

All the solvents used were purified by standard procedures and were distilled prior to use (9). Styrene (SISCO), methyl methacrylate (BDH) and acrylonitrile (Aldrich) were dried over CaH_2 and distilled under reduced pressure prior to use. Azobisisobutyronitrile (AIBN) was recrystallized from methanol and its stock solution was prepared in chloroform and stored in refrigerator.

Preparation of tributyltin methacrylate (TBTMA)

Tributyltin methacrylate was synthesized according to the method described by Montermoso et al (10). Crude monomer obtained was purified by recrystallization from hexane at low temperature (M.P.14°C).

Synthesis of tributyltin α-chloroacrylate (TCA)

TCA was synthesized by treatment of tributyltin oxide with α-chloroacrylic acid (CAA) (8). Crude monomer was recrystallized twice from hexane (m.p. 52-53°C). The monomer was characterized by IR, 1H-NMR and tin analysis (observed tin %29.90, calculated tin % 30.03).

Homopolymerization

2g monomer, AIBN (0.1% w/w) and dry THF 2 ml (50% w/v were charged into a polymerization tube followed by degassing (three freeze-thaw-pump cycles). The polymerization was conducted at 50°C for 4h. The polymer was precipitated by adding the reactants into excess of aqueous methanol (9:1). A tough rubbery polymer obtained was dried in vacuo at 40°C for 48 h.

Copolymerization

AIBN solution in chloroform (0.1% w/v) was taken in polymerization tube. The chloroform was completely removed under vacuum, and predetermined amounts of monomers and THF (1:1 v/v) were charged into polymerization tubes. The reaction mixtures were made oxygen free by repeated freeze-thaw-pump cycles from liquid nitrogen temperature. Polymerization reactions were carried out at 50°C. After desired degree of conversion (<10%), polymerization was terminated and polymers were precipitated in a suitable non-solvent system. The monomer TCA (M_1) was copolymerized with styrene (M_2), methyl methacrylate (M_2) and acrylonitrile (M_2) and designated at TCST, TCMA and TCAN respectively. The subscript 1 to 4 indicate the increasing mole fraction of TCA in the copolymer. Detailed conditions of polymerization, and non-solvent system used are summarized in Table 1.

Characterization

Infrared spectra were recorded with a Perkin-Elmer 685-IR spectrophotometer. Thin film of polymer in $CHCl_3$ or THF was cast on NaCl pellet which was used for this purpose. 1H-NMR were recorded using Varian FX-100 (100 MHz) FT NMR or Bruker AM 500 MHz FTNMR spectrometer using TMS as internal standard. Noise decoupled C-13 NMR spectra were recorded using Bruker AM 500 MH$_z$ FTNMR spectrophotometer in $CDCl_3$ or in DMSO-d$_6$ (10-15% solution) as solvent. All the chemical shifts were assigned with respect to solvent $CDCl_3$ or DMSO-d$_6$.

Table 1 Copolymerization Data of TCA with Styrene, Methyl Methacrylate and Acrylonitrile

Polymer Code	Mole fraction of M_1 in feed	Yield %	% of tin in copolymer	Mole fraction M_1 in copolymer by GA	by PMR
TCST$_1$	0.067	12.0	19.50	0.330	0.260
TCST$_2$	0.161	10.1	20.50	0.360	0.390
TCST$_{2.5}$	0.224	11.0	21.84	0.410	–
TCST$_3$	0.302	10.0	23.30	0.470	–
TCST$_4$	0.536	12.0	25.12	0.570	0.510
TCMA$_1$	0.063	12.0	14.72	0.200	0.180
TCMA$_2$	0.153	9.5	19.46	0.320	0.330
TCMA$_{2.5}$	0.212	11.5	22.61	0.440	0.460
TCMA$_3$	0.288	11.0	24.00	0.500	–
TCMA$_4$	0.418	10.0	24.63	0.530	–
TCAN$_1$	0.041	5.0	16.20	0.130	x
TCAN$_2$	0.100	10.0	22.00	0.261	x
TCAN$_{2.5}$	0.141	6.0	23.90	0.350	x
TCAN$_3$	0.200	10.0	25.50	0.432	x
TCAN$_4$	0.400	9.0	26.40	0.490	x

Temp : 50±0.1°C, Solvent : THF, Initiator : AIBN (0.1% w/v)
Non-solvent : Aq. Methanol (90%) for TCST, Hexane for TCMA and TCAN.
x - not in agreement with tin analysis, GA-Gravimetric Analysis.

RESULTS AND DISCUSSION

Both the organotin monomers TBTMA and TCA were synthesized by treatment of bis (tributyltin oxide) with corresponding acid in benzene at room temperature. α- chloroacrylic acid (11) was synthesized according to the scheme given below.

$$CH_2 = \overset{\overset{\text{H}}{|}}{C} - COOCH_3 \xrightarrow[0°C]{Cl_2} CH_2 - \overset{\overset{\text{Cl}}{|}}{CH} - COOCH_3 \xrightarrow[\text{(ii) } H_2SO_4]{\text{(i) } Ba(OH)_2 \cdot 8H_2O}$$

$$CH_2 = \overset{\overset{\text{Cl}}{|}}{C} - COOH \xrightarrow[\text{in Benzene}]{TBTO} CH_2 = \overset{\overset{\text{Cl}}{|}}{\underset{|}{C}}$$
$$COOSnBu_3$$

IR and NMR spectra of TBTMA and TCA are shown in Fig.1 and 2 respectively. NMR spectrum of TCA showed unsaturated protons at 6.2 ppm and 6.9 ppm. The IR spectrum of TCA showed an absorption

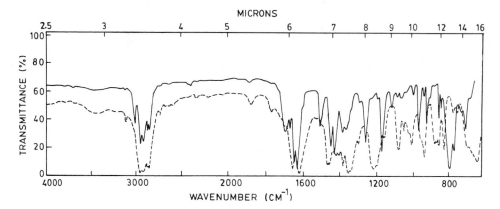

FIG. 1. IR SPECTRUM OF (——) TRIBUTYLTIN ∝−CHLOROACRYLATE
(TCA) AND (----) TRIBUTYLTIN METHACRYLATE (TBTMA)

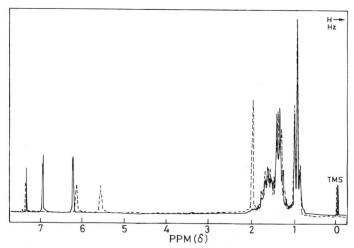

FIG. 2. 100 MHz SPECTRUM OF (----) TRIBUTYLTIN METHACRYLATE
IN CDCl₃ AND (——) TRIBUTYLTIN ∝−CHLOROACRYLATE IN CDCl₃

band at 1610 cm^{-1}, 1630 cm^{-1} and at 1720 cm^{-1} due to double bond, COOSn and carbonyl bond respectively. Tin analysis of TCA and PTCA were in agreement with theoretical values (observed values : 29.90% for TCA and 29.88% for PTCA, calculated: 30.03%) indicating that no loss of tributyltin has occurred during the course of polymerization.

PTBTMA is soluble in various common organic solvents like benzene, toluene, acetone, THF, chloroform etc. On the other hand, PTCA is soluble in polar solvents like acetone, THF and DMSO. This indicates that substitution of chlorine atom at α-position appears to have considerable influence on the solubility of this polymer.

Copolymer compositions were determined by tin analysis (12) and 1H NMR. Except for the TCA-AN copolymers, results obtained from tin analysis were in close agreement with those obtained from 1H NMR. This exception is probably due to the undesired overlapping of backbone methylene protons which affected the preciseness of quantitative estimation by the NMR method.

Using mole composition data, the monomer reactivity ratios were calculated by means of Kelen-Tüdös method (13). The results are summarized in Table 1-2. Copolymer composition curves obtained using r_1 and r_2 data (14) are shown in Fig.3.

Table 2. Reactivity Ratios Data of TCA with Styrene, Methyl methacrylate and Acrylonitrile

Monomer	r_1	r_2
Styrene	0.500	0.170
Methyl Methacrylate	1.089	0.261
Acrylonitrile	1.880	0.243

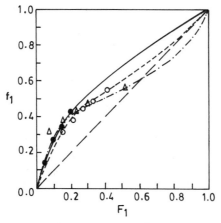

FIG.3. MOLE FEED CURVE OF TCA-ST
(-·▲·-) , TCA-MMA (--o--) AND
TCA-AN (—●—).

From the mole composition curve, it is evident that TCA has higher reactivity than the corresponding vinyl monomers. The higher reactivity of TCA may be attributed to the presence of the chlorine atom at the α-position which activates its polymerizability. TCA-ST copolymer system shows an azeotropic composition of $f_1 = 0.62$ having r_1 and r_2 values less than unity for both the monomers. This indicates a tendency of alternation in its copolymerization process (15). In the case of TCA-MMA and TCA-AN copolymers, r_1 is greater than 1. This implies that these copolymers contain blocks of TCA with intermittent units of MMA or AN. Earlier copolymerization studies of various α-chloroacrylates with (meth) acrylate monomers have revealed that α-chloroacrylates are usually more reactive than the corresponding vinyl and (meth) acrylate monomers (16, 17). In the present case, higher reactivity of TCA is observed over vinyl monomers too. This suggests that bulky tributyltin moiety probably does not affect the reactivity of the α-chloroacrylate part of the monomer.

The relative reactivity of these comonomers towards TCA monomer can be deduced by comparing $1/r_1$ values, which suggest the following order of reactivity of vinyl monomers towards TCA.

$$ST > MMA > AN$$

Reactivity ratio studies of TBTMA with ST, MMA and AN (3-5) have shown that the order of reactivity of these monomers towards TBTMA also follows the same pattern. However, for the TBTMA (M_1) copolymers, r_1 was always found to be less than r_2. This means that TCA is relatively more reactive than TBTMA. The higher reactivity of TCA over TBTMA can be explained on the basis of the ability of the TCA radical to undergo resonance stabilization due to the presence of non bonding electrons of the chlorine atom as shown in the following scheme :

This phenomenon is absent in the case of TBTMA, leading to a less stable growing radical for it.

C-13 NMR of Homopolymers

Most of the work on C-13 NMR of organotin polymers has not received much attention. The work reported on C-13 NMR deals with the hydrolyzed polymers. Dambatta and Ebdon (18) have studied the effect of temperature on microstructure of unhydrolyzed PTBTMA by C-13 NMR spectroscopy. Zeldin and Lin have reported (6) C-13 NMR of TBTMA and TBTMA-ST copolymers in hydrolyzed polymers. Manders et al (19) also have reported the microstructure of TBTMA-MMA in polymers using C-13 NMR spectroscopy. These polymers were first hydrolyzed to the corresponding methacrylic acid derivatives for NMR studies.

125 MHz C-13 NMR of PTBTMA, and PTCA are shown in Fig.4-5 respectively along with the expansion of several carbon resonances.

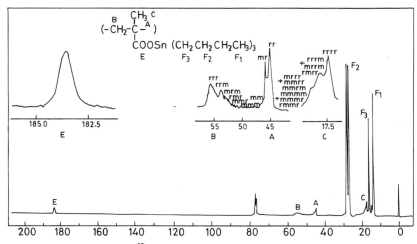

FIG. 4 . 125MHz ^{13}C NMR SPECTRUM OF POLY (TRIBUTYLTIN)
METHACRYLATE

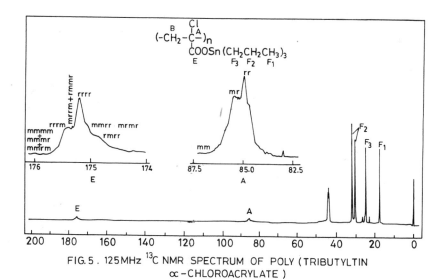

FIG. 5 . 125MHz ^{13}C NMR SPECTRUM OF POLY (TRIBUTYLTIN
α–CHLOROACRYLATE)

From the spectra, it is evident that quaternary carbons of both the homopolymers are most sensitive to tacticity. The carbonyl carbon of PTBTMA appeared as a broad peak only, while the carbonyl carbon of PTCA had shown fine structures. Peak assignments of various carbons of PTBTMA and PTCA are shown in Table 3-4. It can be seen that syndio-tacticity of the polymer is practically remaining constant.

Table 3. Sequence Assignment for Poly(Tributyltin Methacrylate)

Type of sequence	Chemical Shift (ppm)	Peak Intensity (%) (Obs.)	Sequence Intensity (%) (Cal.)[+]
Triad Assignment using Quarternary Carbon			
mm	46.0	6.25	4.41
mr	45.8	31.00	33.18
rr	45.0	62.50	62.50
Triad Assignment using Methyl Carbon			
mmmm			0.19)
mmmr	18.2	5.00	1.46)
rmmr			2.75)
mmrm			1.46)
mmrr	18.0	12.50	5.50)
rmrm			5.50)
rmrr			20.70)
mrrm	17.8	40.0	2.75)
rrrm			20.70
rrrr	17.5	42.5	38.95
Tetrad Assignment using Backbone Methylene Carbon			
mmm	50.0	0.70	1.86
mmr	51.0	8.10	6.96
rmr	52.5	10.80	13.10
mrm	52.0	5.40	3.48
rrm	54.0	21.60	26.21
rrr	55.1	51.50	49.30

(+ calculated using P_r = 0.78, m = meso form, r = racemic form)

C-13 NMR of PTBTMA

The α-methyl carbon showed peaks around 19 ppm which were assigned as partial pentad splitting. Whereas, quaternary carbon and backbone methylene carbons showed triad and tetrad tacticity respectively. P_m and P_r values calculated were in agreement with those previously report-

Table 4. Triad Assignment for Poly(Tributyltin α-chloro Acrylate)

Type of sequence	Chemical Shift (ppm)	Peak Intensity (%) (Obs.)	Sequence Intensity (%) (Cal.)
mm	86.5	6.86	7.77
mr	85.9	38.70	37.70
rr	85.0	54.40	54.40
mmmm+mmmr+ rmmr	176.00	5.50	5.85+(0.47+2.70+ 2.68)
mrrm+rmmr	175.2	8.80	3.75+3.75
mmrr+mrmr	175.0	15.00	7.50+7.50
rmrr	174.3	21.00	21.00
rrrm	175.3	20.00	20.90
rrrr	175.1	29.80	29.50

(+ calculated using P_r = 0.737)

ed by Dambatta and Ebdon (18). This confirms our assignments. In previously reported spectra,only the quaternary carbon has shown splitting but in the present study three carbons are resolved. Thus, at higher field strength better resolution has been obtained.

C-13 NMR of PTCA

Assignments of various carbons of PTCA were made by comparing the C-13 NMR of PTBTMA (19) and polymethyl α-chloroacrylate (PMCA) and other polyalkyl chloroacrylates (20, 21). The carbon atoms sensitive to tacticity are quaternary and carbonyl carbons. Ester carbons resonate in the same region as PTBTMA and PTBTA. The backbone methylene carbon was partially merged up with the solvent DMSO peak hence the tacticity assignment for this carbon was not possible. The carbonyl carbon resonated at around 175 ppm which has shifted to high field as compared to the carbonyl of PTBTMA and PTBTA. On the other hand, quaternary carbon showed triad splitting. These peaks were assigned as rr, mr and mm from high field to low field. This is similar to assignment of the quarternary carbon of PTBTMA having rr triad located at higher field than the mm peak. The carbonyl carbon resonance showed three finer peaks. Thus the effect of α- substitution is also reflected on the chemical shift of the carbonyl carbons. The assignment of various peaks for pentad sequences are shown in Table 3. The P_m value was found to be 0.263 which was slightly higher than that of PTBTMA. Analysis of the configurational sequence statistics of polymers showed that the polymer obeys Bernoullian type chain growth (22,23).

From the NMR of all the three homopolymers, it was also found that all the three homopolymers showed an additional peak besides the C-Sn resonance. PTCA and PTBTA showed a peak on both the side, whereas PTBTMA showed only one peak. These peaks were identified as satellite peaks of 117 & 119-Sn and C-13 nuclei. A coupling constant (J) was found to be in the range of 170 Hz.

MECHANISM OF CHAIN PROPAGATION (22,23)

Analysis of the configurational sequence statistics of polymers usually provides valuable information concerning the propagation mechanism of the polymizeration process. The simplest type of sequence statistics results from building up of the chain by the Bernoullian triad where all the tactic sequences are generated from only one probability P_m (or $P_r = 1 - P_m$). The most important feature of this model is that it is independent of the stereochemistry of the growing end giving rise to a stereochemical chain structure. The configurations of all the polymer synthesized with free radicals were found to follow this propagation model.

The results were tested for conformity with Bernoullian statistics using the following equations

$$P_{m/r} = mr/2(mm) + (mr) \quad \ldots \text{(I)}$$

$$P_{r/m} = (mr)/2(rr) + (mr) \quad \ldots \text{(II)}$$

The chain growth follows Bernoullian model if,

$$P_{m/r} + P_{r/m} = 1 \quad \ldots \text{(III)}$$

This test was applied for both the homopolymers synthesized using raidcal initiator. It was found that both the homopolymers followed Bernoullian statistics. An additional criterion for Bernoullian statistics is that the factor 'Z' which can be calculated using the following equation.

$$Z = \frac{4(mm)\ (rr)}{(mr)^2} \quad \ldots \text{(IV)}$$

'Z' should be either equal to or close to unity. Slight deviation from unity was observed for this quantity. This was probably due to the sensitive nature of the test, particularly its dependence on the value of mr.

C-13 NMR of Copolymers

C-13 NMR of copolymers provides information regarding composition of a copolymer, monomer sequence distribution and end chain propagation mechanism (22-24). In the present studies, TCA-ST copolymers were studied at different mole compositions in order to see the effect of mole composition on the splitting pattern. An alkyl ester group of a copolymer displayed certain influence on the chemical shift of the carbonyl and quaternary carbon resonances. Quarternary carbons in 'ST-ST-ST' triad always resonate at higher field than that of 'TC-TC-TC' triad.

In the present text, Monomer sequences due to TCA are designated as 'TC' whereas, monomer sequences due to styrene are designated as 'ST'.

MICRO-STRUCTURE OF TCA COPOLYMERS

C-13 NMR of TCA-Styrene polymer with different mole ratios viz., $TCST_1$ (33 mole percentage of TCA), $TCST_{2.5}$ (41 mole percentage of TCA) and $TCST_4$ (57 mole percentage of TCA) are shown in Fig.6.

The copolymer spectra recorded at $40^{\circ}C$ showed appreciable splitting. These signals were assigned on the basis of comparison of spectra with that of homopolymers. The assignment of the various peaks was made by comparing the relative intensities of the signals for all the three copolymers.

From the spectra, it is evident that, as TCA content in the copolymer increases, the splitting pattern become more and more complicated along with the broadening of the peak.

Aromatic carbons of styrene appeared as three different peaks viz., peak at 145 ppm due to aromatic-C_1 carbon and peaks at 127 and 129 ppm due to o and m, and p carbons respectively, whereas carbonyl carbons, backbone methylene and quaternary carbons showed resonance in the region 174 ppm, 40 ppm and 78 ppm respectively.

Backbone methylene carbons resonated over 43 to 52 ppm. $TCST_1$ containing 33 mole percentage of TCA clearly exhibited three dyads viz., ST-ST, ST-TC or TC-ST and TC-TC at 43, 44.5 and 47.5 ppm respectively. Now considering the same region of TCST-2.5 containing 41 mole percentage of TCA, it can be seen that peak at 42 ppm diminishes, whereas intensity of peak at 44.5 ppm deceases with concomitant increase in the intensity of a peak at 50.5 ppm. Further increase in TCA content in polymer for $TCST_4$ containing 57 mole percentage of TCA resulted in total disappearance of the peak at 43 ppm with the increase in intensity of peak at 51.5 ppm. This manifests the presence of a TC-TC type dyad at 51.5 ppm. The observed intensities of the signals were in good agreement with those calculated assuming Bernoullian statistics.

The Aromatic C_1 - carbon around 145 ppm should exhibit ST centrered triads in the copolymer. An increase in TCA content in the copolymer causes an increase in peak intensity at 145-144 ppm. Thus, an additional peak is observed indicating presence of TCA centered triads ST-TC-ST in the copolymer.

The carbonyl carbon also showed similar behaviour. An increase in TCA content in the copolymer causes broadening of peak upto 174 ppm. Thus, broadening of peak shows presence of TC-TC-TC triad in the copolymer.

Satellite peaks due to coupling of Sn (117 and 119) and C-13 nuclei were also observed in all three copolymers of TCA. Both styrene and AN copolymers showed satellite peaks on either side of Sn-C resonance whereas, TCA-MMA system did not show satellite peak at low field region due to the possible overlappingof α-methyl peal with Sn-C resonance.

SUMMARY AND CONCLUSIONS

A new organotin monomer tributyltin α-chloroacrylate was synthesized and polymerized using radical initiator at $50^{\circ}C$. TCA was copolymerized with styrene, methyl methacrylate and acrylonitrile. Reactivity ratios data showed that TCA is more reactive than all three monomers

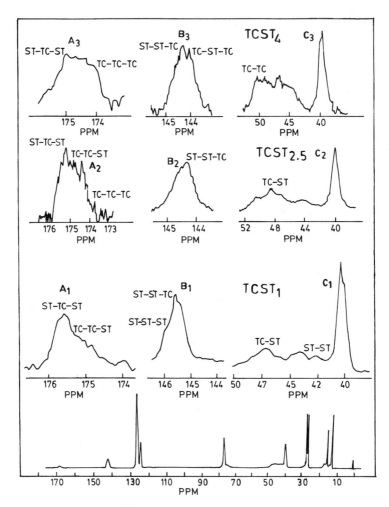

FIG. 6. 125 MHz ^{13}C NMR OF TRIBUTYLTIN α-CHLOROACRYLATE CO-STYRENE COPOLYMERS IN CDCl$_3$ AT DIFFERENT MOLE RATIOS

under study. Reactivity ratios data also suggested that TCA is more reactive than TBTMA. Thus substitution of chlorine at α-position leads to an increase in the reactivity of the monomer.

ACKNOWLEDGEMENTS

The authors would like to thank Dr. G.N. Babu, 3M Centre, Minnesota, and Dr. Pradeep Dhal, NCL, Pune for their valuable suggestions. This work was supported by the Ministry of Environment, Forest and Wildlife, Government of India, New Delhi. 500 MHz NMR spectra were recorded at Tata Institute of Fundamental Research, Bombay.

REFERENCES

1. B.K. Garg, J. Corredor and R.V. Subramanian., J. Macromol. Sci. Chem., A11, 1567 (1977).

2. N.A. Ghanem, N.N. Messiha, N.E. Ikladious and A.F. Shaban., Eur. Polym. J., 15, 823 (1979).

3. N.A. Ghanem, N.N. Messiha, N.E. Ikladious and A.F. Shaban,
    (i) Eur. Polym. J., 16, 339 (1980)
    (ii) Eur. Polym. J., 16, 1047 (1980)

4. P.C. Deb, and A.B. Samui, Die Makromole. Chemie, 80, 137 (1979).

5. P.C. Deb and A.B. Samui, Die Makromole. Chemie, 103, 783 (1982).

6. M. Zeldin and J.J. Lin., J. Polym. Sci., Polym. Chem. Edn., 23, 2330 (1985).

7. J.R. Dharia, C.P. Pathak, S.K. Gupta and G.N. Babu., J. Polym. Sci., Polym Chem. Edn., 26, 595 (1988).

8. J.R. Dharia and S.K. Gupta., (Manuscript submitted to J. Polym. Sci., Polym. Lett., Nov. 1989).

9. D.D. Perrin, W.L.F. Armarego and D.R. Perrin., 'Purification of Laboratory Chemicals' 1st Edn., Pergamon Press, New York p.122 (1966).

10. J.C. Montermoso, T.M. Andrew and L.P. Marinelli., J. Polym. Sci., 32, 423 (1958).

11. C.S. Marvel, J. Dec., H.G. Cook (Jr.) and J.C. Cowan., J. Am. Chem. Soc., 62, 3495 (1940).

12. H. Gilman and D. Rosenburg., J. Am. Chem. Soc., 75, 3592 (1953).

13. T. Kelen and F. Tüdös, J. Macromol, Sci., Chem., A(9), 1 (1975).

14. F.R. Mayo and F.M. Lewis, J. Am. Chem. Soc., 66, 1594 (1944).

15. G. Odian, in "Principle of Polymerization" Academic Press, New York, (1981).

16. C.P. Pathak, M.J. Patni and G.N. Babu, J. Macromol. Sci., Chem., A24, 557 (1987).

17. L.J. Young : "Polymer Handbook", Brandurp, J. and Immergut,E.H. Eds. Interscience, New York, Chap. II, pp 387-404 (1975).

18. B. Dambatta and J.R. Ebdon., British Polym. J. 16, pp. 69-70 (1984).

19. W.F. Manders, and J.M. Bellama et al., J. Polym. Sci., 25, 3469 (1987).

20. C.P. Pathak, M.J. Patni and G.N. Babu., Macromolecules, 19, pp. 1035-1042 (1986).

21. J.M. Bellama and W.F. Manders., ACS Symposium Series 360, Inorganic and Organiometallic Polymers, M. Zeldin, K.J. Wynne, and H.R. Allcock, Eds. NMR Characterization of Compounds and Configuration of TBT Polymers. 483-496 (1988).

22. F.A. Bovey., "High Resolution NMR of Macromolecules" Academic Press, New York (1972).

23. F.A. Bovey., "Chain Structure and Conformation of Macromolecules", Academic Press, New York (1983).

24. J.C. Randell., "Polymer Sequence Determination : C-13 NMR Method", Academic Press, New York (1975).

Bis (tributyltin oxide), 395
Bis(tri-n-butylphosphine)-
    platinum, 5
Bis(ureidosilane), 238-239, 241
Blood, artificial, 8
Boron, see also specific
    derivatives of
  in ceramics, 173, 174, 180-184
  plant growth and, 281-282
  in sol-gel process, 21
Boron-tungsten, 13
Bpy, see Bipyridyl, 2,2'-
    dipyridyl
Brassinolide, 277
Brine shrimp, see Artemia salina
Bromine, 14, 169
p-Bromoethynylbenzene, 153
p-Bromophenol, 134
Butadiene, 12, 50, 51, 62, 71,
    73, 74, 75, 78, 79, 80,
    81, 82, 83, 84, see also
    Polybutadiene; specific
    forms of
Butene, 49, see also specific
    forms of
Butylene, 51, 183, 184, see also
    Polybutylene; specific
    forms of
n-Butyl lithium, 128, 130, 133,
    134, 249
Butyl methacrylate, 381, 383,
    384, 385, 386, 387, 388,
    391, 392

Cadmium, 5, see also specific
    derivatives of
Calcium, see also specific
    derivatives of
  in PE waxes, 372-374, 377, 378,
    379
  plant growth and, 275, 280
  in polyacetylene doping, 87,
    91, 93, 94, 95, 97
Calcium carbonate, 373
Calcium carboxylate, 377
Calcium hydrogen phosphate, 280
Calcium hydroxide, 373, 374, 379
Calcium oxide, 373
Calcium stearate, 377
Callus formation, 274, 287
Cancer, see also specific types
  ferrocene compounds in treat-
    ment of, 139-149
  platinum polyamines in treat-
    ment of, see Platinum
    polyamines
Capillary gas chromatography,
    244, 249
Carbanion chemistry, 233-256

Carbanion chemistry (continued)
  experimental procedures in,
    243-249
  monomer preparation in, 250
  polymer characterization in,
    252-254
  polymer preparation in,
    250-252
  silarylene in, see Silarylene
Carbides, 188
Carbohydrates, 319-324, see also
    specific types
Carbon, 16, see also specific
    derivatives of
  atomic oxygen resistance and,
    225
  in ceramics, 173, 174-180
  palladium catalysis and, 153
  polyimide/metal microcomposites
    and, 109
Carbon dioxide, 259, 263
Carboplatin, 338
Carboranes, 13, 238, 240-241
Carboxylated DVB-crosslinked
    polystyrene (PS-COO), 62,
    63
Carboxylated polyethylene (PE-
    COO), 62, 63
Carboxylate salt, 259
Carboxylic acid, 20, 69
  poly(alkyl/arylphosphazenes)
    and, 259, 263-265
Carboxylmethyl cellulose, 33
Carboxyls, see also specific
    types
  in diene polymerization, 61,
    63, 73
  in PE waxes, 368, 379
Catalysis
  applications in, 11-12
  in diene polymerization, 61-84
    polymerization behavior in,
    72-84
    preparation and character-
    ization of, 62-72
  graft polymerization and, 38
  heterogeneous, 187-195
  MCM-aided, 48-50, 51
  palladium in, 51, 151-158
Cationic initiation, 3
Cellular elongation, 267, 272,
    273, 274
Cellulose, 319, 327
Cellulose phosphate, 33
CEP, see Ethylene-propylene
    copolymer
Ceramics, 173-185, 192
  boron in, 173, 174, 180-184
  polymer backbone and, 20

Cytokinins, 274, 276
Cytosines, 336

2,4-D, *see* 2,4-Dichloro-
    phenoxyacetic acid
DAMN, *see* Diaminomaleonitrile
DCC, *see* N,N'-Dicyclohexyl-
    carbodiimide
DDP, *see* Diamminedichloro-
    platinum
DEVPAc, *see* Dibutyl ether of
    vinylphosphoric acid
Dextran, 319, 327
Dialkyldichlorosilane, 20
Dialkyldihalogermane, 20
Diallyl sulfide, 35
Diamines, 6, *see also* specific
        types
    aliphatic, 173, 182
    BCTD and, 296
    platinum polyamine synthesis
        and, 338, 339, 340-341,
        342, 343, 345, 347,
        348, 351, 355
4,4'-Diamino-diphenylsulfone,
    341
6,9-Diamino-2-ethoxyacridine,
    341
2,4-Diamino-6-hydroxypyrimi-
    dine, 341
Diaminomaleonitrile (DAMN), 101
4,6-Diamino-2-mercaptopyrimi-
    dine, 341
2,6-Diamino-3-nitrosopyridine,
    341
2,5-Diaminopyridine, 341
Diaminosilane, 237
Diamminedichloroplatinum (DDP),
        see cis and *trans* forms
        of
Dibenzophosphole-DIOP-platinum-
    tin catalysts system, 12
Dibenzoyl peroxide, 366
Dibutyl ether of vinylphosphoric
    acid (DEVPAc), 35
Dibutyltin dichloride, 333
Dicarbanions, 233, 234, 243, 249,
    250, 252, 253, 256, *see
    also* specific types
Dicarboxylic acid, 296, 327
Dichlorodialkylsilane, 250, 253
Dichlorodiarylsilane, 250
Dichlorodimethylsilane, 15, 251
Dichlorodiorganosilane, 233
Dichlorodiphenylsilane, 251
Dichlorogermanium, 18
2,4-Dichlorophenoxyacetic acid
    (2,4-D), 271

4,5-Dicyanoimidazole, 102
N,N'-Dicyclohexylcarbodiimide
    (DCC), 146, 148
Dicyclopentadienyldichloro-
    titanium, 6
Dicyclopentadienyltin, 300
Dicyclopentadienyltitanium
    methacrylate, 44
Dienes, *see also* specific types
    catalysts in polymerization of,
        61-84, *see also* under
        Catalysis
    graft polymerization and, 37,
        64, 65, 69, 72
    MCM-aided catalysis and, 50
eta4-Dienetricarbonyliron, 17
1,3-Diethylbenzene, 252
Diethylenetriamine, 147
Differential scanning
        calorimetry
    of ceramics, 183
    of palladium catalyzed
        polymers, 156
    of poly(alkyl/arylphospha-
        zenes), 263, 264-265
    of silicon polymers, 239, 244,
        252, 253, 254
Diffusional restrictions, 32, 33
2,5-Dihydroxy-*p*-benzoquinone, 5
3,4-Dihydroxy-3-cyclobutene-1,2-
    dione, 298, 300
Diisopropyl amide (LDA), 128,
    131-132
Diketones, 161, 162, 164, 169,
    366
N,N-Dimethylacetamide, 112
Dimethylbiphenyl, 251
4,4'-Dimethylbiphenyl, 233, 246,
    247, 249, 250
2,3-Dimethyl-1,3-butadiene, 233,
    256
3,3-Dimethylbutene, 130
Dimethylenebiphenyl, 254
N,N-Dimethylformamide (DMF), 146,
    147, 148
Dimethylsilanediol, 20
Dimethylsiloxane, 19, 235, 236,
    239
Dimethyl sulfoxide (DMSO)
    in diene polymerization, 62,
        73, 74
    platinum polyamine synthesis
        and, 340, 345, 346,
        351, 352-353, 354, 356
    TCA polymerization and, 386
Diorganodichlorosilane, 234
Dioxygen, 139, 140
Diphenyldihydroxysilane, 241
p,p'-Diphenylether, 238, 239

Glutathione, 338
Glycine, 162
Gold, *see also* specific
       derivatives of
  atomic oxygen resistance and,
     225
  in polyimide/metal micro-
       composites, 113, 117,
       119-121
Graft polymerization, 38-39
  biological activity and, 51
  dienes in, 37, 64, 65, 69, 72
  MCM role in, 45-46, 48
  metal-containing complexes in,
     33-38
  of poly(alkyl/arylphospha-
     zenes), 263
Grapes, 275
Grasses, 269, 274
Gravitropism, 276
Guanines, 336, 337
Gymnosperms, 285

Hafnium, 1, *see also* specific
       derivatives of
HDPE, *see* High-density
     polyethylene
Hemicelluloses, 328, *see also*
     specific types
Heptylate, 62, 71
Heteroaromatic polyamides,
      101-107, *see also*
     specific polyamides
Heterobimetallics, 129, 131
Heterogeneous catalysis, 187-195
Hexachlorocyclotriphospezene, 13
Hexamethylcyclotrisiloxene, 263
Hexamethyldisilazane, 194
Hexamethyldisiloxane, 369
Hexamethylene diamines, 183, 184
Hexanediamines, 341, 354, 355
1,6-Hexanediol, 319
Hexanes, 21, 364, 384
Hexaphenyldisilazane, 194
Hexyltrichlorosilane, 19
Hibiscus, 267, 275-276, 282, 283,
     285, 288
High-density polyethylene (HDPE),
     363-364, 377, 378, 379
High resolution electron impact
     studies, 282
Histidine, 341
Holly, 274
Holmium, 63, 71, *see also*
     specific derivatives of
Homoaromatic polyamides, 101, *see*
    also specific polyamides
Homopolycondensation, 237, 252
Homopolymerization, 1, 3
  dienes in, 64

Homopolymerization (continued)
  of MCM, 43, 44, 46, 48
  palladium catalysts in, 153
  polysilane and, 16, 18
  of silicon, 239
  of TCA, 393, 394, 398-400, 401
HSU, *see* N-Hydroxysuccinimide
Hydrazines, 6
Hydrochloric acid, 237, 377
Hydrogen cyanide, 101
Hydrogen peroxide, 139, 140, 162,
     166
Hydroperoxides, 366, 367
Hydroquinone, 21, 319
Hydroxide, 336
Hydroxylamine reductase, 279
Hydroxyls, 139, 140, 367
2,5-Hydroxyquinone, 3
N-Hydroxysuccinimide (HSU), 146,
     147, 148

IAA, *see* Indole-3-acetic acid
IBA, *see* Indole-3-butyric acid
Imidazole, 8
Imidazoline, 145
Indole-3-acetic acid (IAA), 268,
     269-271, 272, 273, 274,
     276, 277, 279, 280
Indole-3-butyric acid (IBA), 267,
     270, 273, 275, 282, 284,
     285, 287, 288
Indole-3-propionic acid (IPA),
     270, 282
Inertial-confinement nuclear
     fusion, 6
Infrared studies
  of BCTD/amylose reactions, 321,
     322
  of BCTD/squaric acid product,
     295
    far region, 310-315
    mid spectra, 300-310
  of ceramics, 173, 182, 183
  of diene polymerization, 61
  of ferrocene compounds, 145
  of PE waxes, 366, 368, 370,
     371
  of PGRs, 282, 284, 285
  of polyacetylene, 94-95
  of poly(alkyl/arylphospha-
     zenes), 264
  of silicon polymers, 191, 239,
     253
  of TCA polymers, 384-385, 386,
     393, 394, 395-397
  of tin polymers, 381
  of triphenylantimony-modified
     xylan, 330
Iodine, 5, 87, 94
Ionizing radiation, 34

Transition monomers, 127-137
Transition polymers, 5
Transmission electron microscopy, 113
Triacontanol, 277
Trialkoxyaluminum, 20
Trialkylaluminum, 20
Tri-alkyl methyl methacrylate, 43
Triaryltin methyl methacrylate, 43
Tributylphosphine, 154, 155, 157, 158
Tributyltin acrylate (TBTA), 382, 384, 392
Tributyltin chloride, 333
Tributyltin alpha-chloroacrylate (TCA), 381-392, 393-405
  biotoxicity of, 390-392
  chain propagation in, 402
  micro-structure of, 403
  synthesis of, 382, 383, 394
Tributyltin methacrylate (TBTMA), 382, 384, 392, 393, 395, 397, 398, 398-401, 405
2,4,5-Trichlorophenoxyacetic acid (2,4,5-T), 271
Trichloro-trimethyl-borazine (TCTMB), 180, 183
Triethylaluminum, 88
Triethylamine (TEA), 330, 331, 332
Triethylorthoformate (TEOF), 12
Triethyltin, 51
Triethyltin methacrylate, 45
Tri-isocyanato borazine, 184
Tri-isocyanatotrimethyl borazine, 173
Tri-isocyanato-trimethyl-borazine (TITMB), 180, 182, 183
Triphenylantimony, 327-333
Triphenyl phosphite, 102, 106
Triphenyltin chloride, 333
Tristriphenylphosphinehydrido-carbonylrhodium, 12
Tryptophan, 280, 341, 347
Tungstacyclobutadiene, 129
Tungsten carbide, 188
Tungsten-rhodium complex, 129

Ultraviolet lithography, 16
Ultraviolet radiation, 7
  atomic oxygen formation and, 225
  graft polymerization and, 34
  lithographic resists and, 168, 169
  metal coordination polymers and, 162, 164

Ultraviolet radiation (continued)
  poly(alkyl/arylphosphazenes) and, 259
  polysilane absorption of, 16, 197-198
  poly(silyne) and, 19
Uranium, 298, see also specific derivatives of
Uranyl, 161, 162

Vanadium, 45, 48, 298, see also specific derivatives of
Vanadyl, 161
Vinyl, 39, 243, 263, see also specific derivatives of
  applications of, 6
  graft polymerization and, 38, 39
  MCM conversion and, 44
  in sol-gel process, 23
Vinyl acetate, 35
Vinylcyclohexane, 12
eta5-Vinylcyclopentadienyl, 1, 3
Vinylimidazole, 43
Vinylmagnesium bromide, 131
1-Vinyl-3-methyl manganese complex, 7
Vinylmethylsiloxane, 239
Vinylpyridine, 35, 37, 43, 44
N-Vinylpyrrolidone, 1
Vinyltetramethylcyclopentadiene, 131
Vitamins, 277

Wheat, 319

X-ray photoelectron spectroscopy
  of polyimide/metal microcomposites, 113, 116, 117, 118
  of silicon catalysis, 191, 192, 194
  of TCA polymers, 381, 384
Xylan, 319, see also specific forms of
  triphenylantimony modification of, 327-333
m-Xylene compounds, 233, 239-240, 244, 249, 250, 251, 253, see also specific compounds

Yeast, 272
Ytterbium, see also specific derivatives of
  in diene polymerization, 63, 67, 69, 70, 71
  in polyacetylene doping, 87, 92, 93, 97